FUNCTIONAL DYNAMICS
OF THE CELL

FUNCTIONAL DYNAMICS OF THE CELL

EDWARD BRESNICK

AND

ARNOLD SCHWARTZ

Baylor University
School of Medicine
Department of Pharmacology
Houston, Texas

ACADEMIC PRESS New York and London

ACADEMIC PRESS, INC.
111 Fifth Avenue, New York, New York 10003

United Kingdom Edition published by
ACADEMIC PRESS, INC. (LONDON) LTD.
Berkeley Square House, London W. 1

LIBRARY OF CONGRESS CATALOG CARD NUMBER: 68-14640

Second Printing, 1969

PRINTED IN THE UNITED STATES OF AMERICA

DEDICATION

We respectfully dedicate this text to our wives, Etta and Ina, whose constant forbearance and assistance are gratefully acknowledged.

PREFACE

Biochemistry, as every dynamic area of knowledge, has passed through many phases. In the beginning, the general concern of investigators was the visual observation of the whole state of the organism under a variety of external stimuli. A number of significant advances, especially in the vitamin and nutrition fields, culminated from such studies. However, these investigations were necessarily limited in their scope since the observations reflected the end result of many interacting forces. The development of the second phase of biochemical pursuit involved the deeper dissection of processes occurring within tissues and organs by the use of tissue slices or homogenates. From these studies emerged an understanding of the catabolic and anabolic pathways, the interrelationships between carbohydrate, protein, and fat metabolism, or what is presently known as "intermediary metabolism."

The third era of scientific enlightenment was principally due to the efforts of the enzymologist who delved into the cell extracting unknown enzymes whose existence had been predicted. This resulted in the purification and characterization of enzymes and, subsequently, in the elucidation of the biochemical mechanisms of enzyme action. In several instances, this approach was instrumental in the definition of metabolic pathways. The field of chemotherapy naturally evolved from such an approach.

We are presently engaged in the fourth phase of biochemical development—the relationship of cell structure to functional activity. It is not sufficient, today, to extract and characterize enzymes or to ascertain the effects of compounds upon isolated enzyme systems. We must re-examine and reconstruct enzymatic activity within the cellular microcosm in a medium approximating the internal milieu. Thus, a new approach has developed from the predication that enzymes, although unique and discrete entities, are integrated within structurally organized systems.

In view of the importance of the latter era, we have compiled this text based in part upon a series of lectures developed for graduate courses. The response of the students has indicated that a text of this type would be a valuable addition to the curriculum of an advanced undergraduate or introductory graduate program. Toward this end, a preliminary course in biochemistry and some knowledge of mammalian physiology would prove helpful prerequisites.

The major objective of the text is to emphasize the importance of viewing, in a correlative way, the structural and biochemical features of cellular components. We also wish to place a limitation on the scope of the objectives we hope to achieve. The mammalian cell will be the only unit to be considered in breadth. Although we realize the importance to cell biology of plant and other cells—and to these biologists an apology would be in order—a space restriction must force us to limit our consideration.

It should be noted that only selected references are employed, i.e., not all text material or data are documented with specific references. The general and specific reference sections contain, it is hoped, enough bibliographic information to afford the reader a sufficient insight to the depth desired.

In an area as rapidly moving as cell biology, it is impossible to include all of the most recent biochemical characteristics of each of the cellular constituents. These omissions, due either to ignorance or lack of space, may indeed prove to be of signal importance in our ultimate understanding of cellular function. While we apologize to those whose data do not appear, we wish to emphasize our primary goal, i.e., to inculcate a sense of cohesiveness in terms of structure and biochemical function or activity.

In the preparation of this text, a pursuit that has taken several years, we have had the help of a number of individuals in our own college as well as in other institutions. Without their aid and encouragement, the formulation of this work would have proved too arduous a task. We owe a particular debt of gratitude to morphologists associated with Baylor University College of Medicine for supplying us with the necessary micrographs to illustrate the various features of the cell. These individuals include: Mr. G. Adams, Drs. E. R. Rabin, J. J. Ghidoni, Z. Blailock, and R. M. O'Neal of the Department of Pathology; Mr. S. MacGregor, Drs. M. Benyesh-Melnick, H. D. Mayor, and K. O. Smith of the Department of Virology; Dr. L. F. Montes of the Department of Dermatology; Drs. C. Stevens and R. A. Liebelt of the Department of Anatomy; Drs. K. Shankar, T. Unuma, and K. Smetana of the Department of Pharmacology.

We also wish to acknowledge Drs. A. Sinha and H. Rosenberg of Texas Children's Hospital; Dr. F. Gyorkey of the Veterans Hospital, Houston; Dr. F. Rapp, of the Department of Virology; Drs. G. Clayton and L. Librik, of the Department of Pediatrics; Dr. W. Plaut of the University of Wisconsin Medical School; and Dr. T. C. Hsu of the M. D. Anderson Hospital and Tumor Institute.

Appreciation is expressed to the following members of the Department of Pharmacology for their invaluable suggestions and critique of sections of the manuscript: Drs. H. Busch, K. Mauritzen (visiting Associate Professor from the University of Melbourne, Australia), H. Stanton, K. Shankar, G. Ungar, and G. E. Lindenmayer.

Special thanks are to be accorded to Mrs. Joyce Callaghan and Mrs. Carol Prinz for skillful typing and editing.

The original studies discussed in this text were supported by grants from the U.S.P.H.S., Welch Foundation, National Science Foundation, American Cancer Society and the Houston and Texas Heart Associations.

Dr. Bresnick is a Lederle Medical Faculty Awardee; Dr. Schwartz is a U.S.P.H.S. Career Development Awardee.

April 1968   EDWARD BRESNICK
                                        ARNOLD SCHWARTZ

CONTENTS

Chapter VIII. **Cell Life Cycle and Cell Division**

Chapter IX. **Special Cells**

Chapter I

ROLE OF TECHNOLOGY IN THE DEVELOPMENT OF CELL BIOCHEMISTRY

I. INTRODUCTION

The evolution of our understanding of the biochemical events that transpire within the mammalian cell is directly related to progress in technology, for both the study of the cell per se and its macromolecular composition. In the primeval stage of development, man depended entirely upon his senses for his knowledge of living processes. His observations were occasionally corrupted by his prejudices and in at least one instance, have been permanently recorded—the theory of spontaneous generation, in which life was accepted as emanating from the nonliving, e.g., maggots from rotting meat, worms from mud. Belief in spontaneous generation was not easily dispelled and even the great philosopher, Aristotle, and the eminent investigator, William Harvey, accepted its tenets without question. The theory was finally interred by the experimentations of the Italian Abbè, L. Spallanzani, in the eighteenth century, and later by Pasteur. Spallanzani succeeded in demonstrating that a sterilized broth would never give rise to microorganisms when the medium was contained in a vessel from which air was excluded. Only when air was present in the vessel would bacteria appear in the broth and rotting occur. Pasteur modified Spallanzani's experiments, removed any further objections, and conclusively established the fundamental fallaciousness of the theory. The promulgation of the hypothesis of spontaneous generation clearly indicated man's desire to understand the origins of his environment, a goal which even today is most anxiously

1

sought. Man desires to reach the moon with the belief that further knowledge of the mysteries of the universe may be awaiting him.

New and exciting vistas were opened with the exploitation of the optical properties of curved transparent surfaces, leading first to the invention of spectacles at the end of the thirteenth century, then to the construction of the compound microscope by two Dutch lens makers, Zachary and Francis Janssen in 1590. Progress in this area continued with the development by Leewenhoek of lenses of magnification sufficient for simple scientific investigations.

Although the microscope was employed by A. Kircher in the study of disease, it was Robert Hooke in 1667 who exploited its utilization in the detection of substructure and who postulated the first concept of the *cell*, namely, a *hollow space*. More than 150 years passed before the cell theory was fully accepted, the result of the publication of a treatise on the microscopic structure of organisms by two biologists, Schwann and Schleiden, in 1839. These investigators concluded that all organisms are composed of cells. Their definition of a cell was based upon the presence of the *nucleus*, a structure that had been just recently recognized by Robert Brown.

The origin of cells proved a source of controversy for several years. Schwann, one of the proponents of the cell theory, believed that cells arose from a liquid, structureless substance, the *cytoblastema*. This concept was gradually replaced by the idea that cells are produced only from the division of preexisting cells—as first formulated by R. Virchow in 1855 (*"omnis cellula e cellula"*). The cell theory now took on a new and more universal meaning; the cell was regarded as a vital intermediate in the perpetuation of life.

As confidence in the newly found technique of microscopy developed, the role of the nucleus as a determinant structure in the cell gradually evolved. Initially, the disappearance of the nucleus during cell division and re-formation in each of the daughter cells led the investigators to doubt the significance of this structure in the transmission of hereditary characteristics.

In 1873, three biologists, Schneider, Bütschli, and Fol, independently described the cellular events that transpired during cell division and which we now recognize as *mitosis*. Interestingly, their important observations were buried in the details of papers describing the morphology of a flatworm, *Mesostoma*, and of a roundworm, *Rhabdifis*. Schneider also made the significant observation of "strands" within the nucleus that passed into the daughter cells during cell division. These "strands" were subsequently rediscovered in 1888 and termed *chromosomes*. From

1878 to 1882, Flemming published extensively on mitosis as an apparent universal feature of all living organisms.

One of the most important contributions to cytology and genetics was offered at a lecture delivered to the Natural History Society of Brunn in 1865, by an Austrian monk, Gregor Mendel. His work dealing with the transmission of hereditary determinants between various strains of peas remained unknown for many years until rediscovered independently in 1901 by the botanists, Correns, Tschermack, and DeVries. Molecular genetics is largely the outgrowth of the laws originally set down by Mendel.

The involvement of the nucleus in the transmission of the hereditary characteristics was proposed first by Haeckel in 1866 and later established by Boveri, Hertwig, Strasburger, Kölliker, and Weissmann. The latter investigators further attributed the physical basis of inheritance to the chromosomes.

Progress also continued in the elucidation of the other cellular structures. Within the cytoplasm, the *cell center* was recognized by Van Beneden and Boveri, the *"chondriome,"* i.e., mitochondrion, by Altmann and Benda, and the *"reticular apparatus,"* by Golgi. Various forms of the endoplasmic reticulum were also recognized within most mammalian cells.

It was soon realized that most mammalian cells were devoid of the cell wall covering present in plants and bacteria but did possess an enveloping membrane. A concept of a cell membrane gradually evolved, spurred by the studies of Overton from 1895 to 1900, evolving into the notion of a discrete lipid-containing entity functioning in the intracellular penetration of substances.

The inquisitiveness of investigators was not completely satiated by the simple description of the morphological characteristics of the cell. Information concerning the composition of cellular structures was required and initially cyto- and histochemistry fulfilled this need. The development of biochemical techniques further amplified our knowledge of the functions of the subcellular organelles. The success of the latter approach owed much to the development of technology for the separation of subcellular structures in relatively pure form. For example, the study of mitochondria led to the understanding of cellular energetics, the central role of adenosine triphosphate (ATP), and the mechanisms for electron transport. The study of nuclear functions was largely responsible for progress in the science of genetics. A concept of the *"inborn errors of metabolism,"* originally proposed by Archibald Garrod, was developed in which the etiology of certain disease states was looked upon as arising

TABLE 1.1

Historical Developments in Biochemistry

Nucleic Acids

1776   Scheele and Bergmann—Isolated first purine, uric acid, from bird excreta
1817   Marcet—Discovered second purine, xanthine
1860's Hoppe-Seyler—Center of biochemical research
1866   Kekulé—Presented ring structure of benzene
1869   Miescher—Isolated "nucleins" from discarded bandages
1874   Piccard—Discovered purine bases in nucleic acids
1879–1891   Kossel—Obtained purines from natural sources. Isolated carbohydrates from nucleic acids
1889   Altmann—Prepared protein-free nucleic acid from yeast. First to use the term *nucleic acid*
1894   Hammarsten—Identified carbohydrate of yeast nucleic acid as pentose
1894   Kossel and Neumann—Prepared DNA from thymus and isolated thymine
1898   E. Fischer—Introduced term *purine*
1900   Ascoli—Isolated uracil from yeast nucleic acid (RNA)
1909   Levene and Jacobs—Showed pentose of yeast nucleic acid was D-ribose
1924   Feulgen—Developed histochemical stain for detection of DNA
1945   Schmidt and Thannhauser—Separated and quantitated RNA and DNA by means of alkaline hydrolysis
1945   Schneider—Hot trichloroacetic acid extraction employed in separation of nucleic acids from protein
1947   Vischer and Chargaff—Paper chromatography used in separation of nucleic acid components. Determined chemical composition of nucleic acids from a number of sources
1953   Watson, Crick, and Wilkins—Double helical structure for DNA proposed
1955   Grunberg-Manago and Ochoa—Polyribonucleotide synthesis by polyribonucleotide phosphorylase
1956   Kornberg—Synthesis of DNA by DNA polymerase
1956   Volkin and Astrachan—Recognition of a DNA-like RNA of short half-life functioning as a messenger RNA
1958   Weiss, Hurwitz—Synthesis of RNA by RNA polymerase
1965   Holley—Structure of alanyl-transfer RNA

Nucleotides

1904   Harden and Young—Found Coenzyme I (nicotinamide-adenine dinucleotide) in alcohol ferments
1927   Embden and Zimmerman—Isolated adenylic acid from natural sources
1929   Lohmann—Isolated adenosine triphosphate (ATP) and subsequently adenosine diphosphate (ADP)
1935   Warburg and Christian—Isolated Coenzyme II (NADP)
1936   Schlenk and von Euler—Proposed structure for NAD
1938   Warburg and Christian—Isolated flavin adenine dinucleotide
1941   Lipmann—Showed the hydrolysis of ATP to be driving force for biochemical processes. Subsequently, structure and function of Coenzyme A (CoA)
1951   Lynen—Acyl CoA isolation
1950's LeLoir—Uridine diphosphate sugars demonstrated as intermediates in sugar interconversions

---

TABLE 1.1 — *continued*

1955 Kennedy and Weiss — Cytidine nucleotides active in fat metabolism
1958 Buchanan, Greenberg — Biosynthesis of purines
1958 Reichard — Biosynthesis of pyrimidines
1964 Reichard — Formation of deoxyribonucleotide by reduction of ribonucleotide
1966 Spiegelman — Synthesis *in vitro* of an RNA virus

Biological Energetics and Metabolism

1760's Scheele — Chemical composition of plant and animal tissues. Isolated number of natural substances, e.g., tartaric, lactic, and citric acids
1780 Lavoisier — Respiration utilizes oxygen and is analogous to combustion of charcoal. Spelled the demise of phlogiston theory.
1780 Pasteur — Introduced concept of aerobic and anaerobic organisms and their associated fermentations
1780's Schwann — Yeast could convert sugars to alcohol and carbon dioxide. Fermentation was therefore of biological origin
1800's Chevreul — Fats composed of fatty acids. Isolated glycerol from them
1828 Wohler — Synthesis of urea. Synthesis of organic compounds no longer required instillation of a vital force, i.e., vitalism
1860—90 Fischer — Structures of carbohydrates, amino acids, and fats
1857—65 Bernard — Inhibition of animal respiration by cyanide. Subsequently studied processes of digestion
1886 MacMunn — Discovered cytochromes in tissues (called them histohematins)
1897 Buchner — Fermentation of sugars by cell-free yeast extract
1903 Wieland — Mechanics of cellular oxidation (dehydrogenations)
1904 Knoop — β-Oxidation of fatty acids
1900's Warburg — Mechanics of cellular oxidation. Iron-containing catalyst involved in utilization of oxygen. Respiratory enzymes
1920's Thunberg — Demonstrated dehydrogenases in animal tissues
1925 Keilin — Rediscovery of cytochromes (see MacMunn)
1925—30 Embden and Meyerhof — Elucidation of glycolytic pathway
1932 Warburg and Christian — Isolation of the "yellow enzyme" from yeast
1937 Braunstein — Transamination of amino acids
1937 Kalckar, Ochoa — First described aerobic or oxidative phosphorylation coupled specifically to respiration
1940's G. and C. Cori — Carbohydrate metabolism
1940's Krebs — Postulation of the tricarboxylic acid cycle
1940's Bloch — Biosynthesis of steroids
1951 Lynen and Reichert — Elucidation of structure of "activated" fatty acids (S-acyl CoA)
1956 Kennedy and Weiss — Formation of triglycerides in liver
1958 Huennekens — Role of folic acid in one-carbon metabolism
1960's Calvin — Carbon dioxide assimilation
1950—60's Green, Slater, Racker, Lehninger, Change — Electron transport in mitochondria. Oxidative phosphorylation
1960's Chance — Respiratory control

(continued)

TABLE 1.1—*continued*

Proteins

1837 Berzelius—First suggested name *protein*. Investigated phenomenon of catalysis

1850–1900 Fischer—Deduced the manner in which amino acids are linked into a protein. Responsible for "lock-and-key" mechanism for substrate specificity

1878 Kühne—First proposed *enzyme*

1908 Garrod—Recognition of "inborn errors of metabolism" as genetic disorder

1913 Henri, Michaelis, and Menten—Kinetics of enzyme-substrate interaction

1925 Svedberg—Theories of ultracentrifugation

1926 Sumner—Crystallized first enzyme urease and showed it to be protein

1931 Lineweaver and Burk—Experimental determination of dissociation constant for enzyme-substrate interaction

1937 Tiselius—Introduced electrophoresis, i.e., movement of charged compound under influence of external electric field, as means for separation of proteins

1951 Pauling—Structure of proteins. Postulated existence of α-helix and of pleated sheet structures

1954 Sanger—Elucidation of amino acid sequence of insulin. Introduced the use of dinitrofluorobenzene as an agent for studying amino acid composition of proteins

1954 Zamecnik—Development of *in vitro* system for study of protein synthesis

1955 Benzer—Construction of fine structure map of gene (rII region) containing large number of alleles

1957 Koshland—Development of flexible active site theory for enzyme-substrate interaction, i.e., induced fit

1957 DuVigneaud—Structure of polypeptide hormones. Subsequently, the synthesis of oxytocin

1958 Ingram—Amino acid composition of hemoglobin. Demonstrated that normal adult hemoglobin (HbA) differs from sickle-cell hemoglobin (HbS) by only one amino acid, i.e., valine is substituted for glutamic acid at unique position of β-chain

1959 Hirs, Stein, and Moore; Anfinson—The amino acid composition and structure of ribonuclease elucidated

1959 Watson, Roberts—Study of ribosomes and involvement in protein synthesis

1960–61 Kendrew, Perutz—Use of X-ray diffraction in the study of tertiary structure of proteins

1961 Monod, Jacob, Pardee—Operon concept, allosteric interactions, and mechanisms of regulation

1961 Nirenberg and Matthaei, Ochoa—Involvement of polyribonucleotides in the genetic code

1963 Rich, Noll—Polyribosomes and their role in protein synthesis

1964 Ames, Martin—Polycistronic messenger RNA, coordination and polarity

1965 Monod, Changeux—Allosteric models

---

from a *single point mutation*, with the resultant inability of the organism to carry out a specific metabolic process. One of the major tasks before us is the definition of all disease and indeed of the normal physiological state in terms of molecular equations.

The probings into the functions of the endoplasmic reticulum have led to the promulgation of a central dogma for protein synthesis centering about the interrelationships of deoxyribonucleic acid (DNA), ribonucleic acid (RNA), and protein. Autoradiography has played an enormous role in establishing the chronological sequence of events in this area.

With the discovery of the *lysosome* came some understanding of the manner in which the dead cells and foreign material are prevented from producing effects detrimental to the organism. The study of lysosomal activities has proved a firm ground for a marriage of classical cytochemical with biochemical techniques.

Concomitant with the developments in cytology, important advances were taking place in biochemistry. These advances are listed in Table 1.1 and many of them will be discussed in greater detail in succeeding chapters.

The fundamental importance of technology in our understanding of the cell should be stressed. Among the more modern techniques that have played significant roles in broadening our concepts of the cell may be included microscopy, cytochemistry, autoradiography, and differential centrifugation. The remainder of this chapter will be devoted to a discussion of these methods. The more quantitative techniques, e.g., chromatography, spectrophotometry, polarography, have played an enormous part in the understanding of cellular events, but a consideration of these processes would entail a monumental task. Additional methods that have proved valuable in the study of specific aspects of cellular activity will be presented in subsequent chapters.

II. MICROSCOPY

One of the most significant contributions to cell biology was the invention of the microscope. The importance of this innovation was immediately apparent from the simple experiments of Robert Hooke who was led to examine the structure of cork with the microscope. Although Leeuwenhoek conducted many experiments during the latter half of the seventeenth century with the newly constructed instrument, it was not until more than a century later that the capabilities of his invention were realized.

The ultimate aims of all types of microscopy are twofold: (1) the formation of an enlarged image that is absolutely free from optical defects and (2) the attainment of contrast, a difference in intensity or color between the object and background. Mere magnification cannot be the sole criterion since the magnified image of the object must be *visible* to the investigator. Let us direct our attention to the methods for the fulfillment of these criteria, that is, the achievement of magnification of an object with as little optical defect as possible. The fundamental limitation of all lenses which arises from the nature of the light wave, however, will not allow the production of a perfect image nor the preservation of the complete object.

A maximum *limit of resolution* — the ability of a lens to distinguish minute details — can be calculated to be 0.25 μ for visible light; for monochromatic light of wavelength equal to 400 mμ, the limit of resolution of the light microscope is 0.17 μ. With the unaided eye, a person can resolve details only of the order of 100 μ. With shorter wavelengths, however, the use of the glass lenses is precluded; for example, with ultraviolet radiation of about 300 mμ, a quartz lens or reflecting optical device must be employed. The resolution with the quartz lens is improved by a factor of 2 to 0.1 μ.

A. Fixation

Increasing the resolving power of the instrument is only half of the problem facing the cell biologist; the other and equally weighty consideration is *contrast*. The eye can discern differences only in the intensity of visible light or in wavelength, i.e., color. Unfortunately, except for pigment-containing bodies, most of the cellular organelles are transparent to light of the visible spectrum. An early attempt to overcome this limitation in the observation of cells made use of the affinity of specific structures for various dyes. The use of these agents, however, does not allow for the preservation of the cell in a living state.

Once the tissue has been removed from its natural environment, it is necessary to stop any further enzyme action, especially the catabolic (degradative) activity, by killing the cell. The latter must be accomplished, however, with the preservation of the integrity of the structural aspects of the cellular constituents and with a minimal introduction of artifacts. It is the art of fixation which addresses itself to this task. A variety of suitable fixatives are at the cytologist's disposal, each of which possesses its own advantages; the particular needs of the investigator generally dictate the choice.

Cellular protein is the favorite target for the action of the fixatives. Formaldehyde, mercuric chloride, and dichromate, constituents of Zenker's, Flemming's, and Regoud's solutions, react with protein by establishing bridges between several molecules. The reaction of formaldehyde with proteins is indicated in Fig. 1.1.

Dichromate, in addition to oxidizing proteins and establishing chromium cross-linkages, also reacts with phospholipid material. Mercuric ions, as well as other heavy metals, attack sulfhydryl, carboxyl, and amino groups, producing mercury linkages between these moieties. Acid fixatives, e.g., Carnoy's solution (absolute ethanol and acetic acid) and Bouin's fluid (picric acid, formaldehyde, and acetic acid), are most useful in the study of nuclei and chromosomes. The acid fixatives cannot be employed in the histochemical demonstration of subcellular enzymic activity because of the destructive nature of the low pH upon the protein. However, a variety of aldehyde fixatives are available, which, when properly buffered, allow for the localization of enzymes *in situ*. These include glutaraldehyde, hydroxyadipaldehyde, crotonaldehyde, pyruvicaldehyde, acetaldehyde, as well as formaldehyde.

$$R - NH_2 + HCHO \longrightarrow R - NH - CH_2OH$$

Protein amino group + formaldehyde

$$R - NH - CH_2OH + R_1 - NH_2 \longrightarrow R - NH - CH_2 - NH - R_1 + H_2O$$

Another amino group of protein   methylene bridge

Fig. 1.1. Reactions of proteins with formaldehyde.

B. Phase Microscopy

The difficulty in viewing the transparent cellular structures has been overcome with the introduction of more advanced technology leading to the development of *phase microscopy, interference microscopy, polarization microscopy,* and *electron microscopy.* Although cellular structures are highly transparent to visible light, the structures do produce alterations in the phase of the transmitted radiation, a phenomenon which forms the basic principle of both phase and interference microscopy. The differences in phase, which result from gradients of refractive index and of the thickness existing between the different cellular organelles, can be made more readily discernible. When a ray of light impinges upon a structure possessing a higher refractive index than the surrounding medium, a delay or retardation of the wave occurs, i.e., a *phase change* (see Fig. 1.2). Upon emerging from the structure, the transmitted wave resumes its original velocity although the lag is maintained. The phase change that has resulted may be measured.

In phase contrast microscopy, originally developed by the Dutch physicist, Zernicke, for the testing of optical components, the small differences in phase are amplified until detectable with the eye or recorded on a photographic plate. Zernicke in 1953 received the Nobel prize for this innovation. The phase effect arises from the interference between the direct geometry of the image emanating from the central part of the objective and the retarded diffracted image. As a result of the inter- ference, the phase changes are first amplified, then are translated into changes of intensity. In *bright or negative contrast*, the two sets of waves may be added with the appearance of objects that are brighter than the surroundings; in *dark or positive contrast*, the two sets of rays may be subtracted, so that the object appears darker than the surroundings (see Fig. 1.2).

Phase microscopy is routinely employed in work on living cells as well as upon fixed but unstained preparations and as such, supplies a valuable addition to the armamentarium of the cell biologist. The in- strument has also proved to be of value in the quantitation of the refrac- tive index of the cell cytoplasm and in the study of the osmotic behavior of living cells.

C. Interference Microscopy

Phase contrast microscopy suffers from several disadvantages related to the bright region or halo surrounding the dark object and to a *zone of action phenomenon*. The latter is a limitation in the region over which the phase contrast operates. In phase microscopy, the contrast is most marked at the edge of an object and at the discontinuities but falls off gradually toward the center. These gradual changes are not detected very readily. A more important criticism is related to an inability to de- termine any exact relationship between the intensity and the optical path. With the use of an interference microscope, continuous small changes in refractive index may be detected that are transformable into vivid color changes. The light emitted from a single source is split into two beams, one of which is allowed to pass through the object, while the other is reflected at the surface of a semireflecting mirror and bypasses the object (Fig. 1.3). Each beam travels through a separate optical system and the two images may be recombined by a semireflecting mirror. Since the two beams are independent until recombined, the relative amplitude and phase may be altered by the introduction of filters or trans- parent plates in the path of either beam. In interference microscopy, the ultimate purpose is the addition of a wave that interferes with the

Fig. 1.2. (A) The light path in a phase microscope. (B) The normal retardation by one- quarter wavelength of light diffracted by an object, and its difference in phase from the light passing through the surrounding medium. By phase optics the two waves are super- imposed to reinforce each other in bright contrast phase as shown in D or to subtract from each other as in dark contrast phase shown in C. From the American Optical Co.

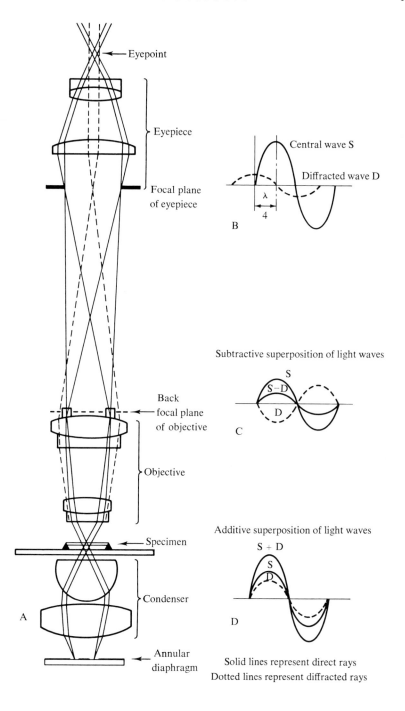

Eyepoint

Eyepiece

Focal plane
of eyepiece

Central wave S

Diffracted wave D

$\dfrac{\lambda}{4}$

B

Back
focal plane
of objective

Objective

Subtractive superposition of light waves

S

S−D

D

C

Specimen

Condenser

A

Additive superposition of light waves

S + D

S

D

D

Annular
diaphragm

Solid lines represent direct rays
Dotted lines represent diffracted rays

transmitted wave (whose phase has been altered by the object) so that the resultant ray will possess a lower amplitude than the incident ray, i.e., the object will appear dark.

Interference microscopy permits the determination of the weight of the dry object since the latter is related to its refractive index. It has also been employed in the estimation of the nucleic acid, lipid, and protein contents of a cell. The technique requires complicated and expensive equipment and hence is not routinely employed in the laboratory.

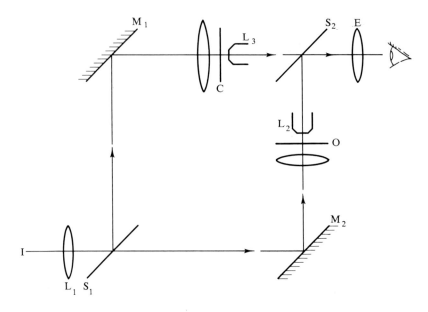

Fig. 1.3. Schematic representation of an "ideal" interference microscope system: S_1 and S_2, semireflecting mirror surfaces; M_1 and M_2, fully reflecting mirror surfaces; L_2 and L_3, microscope lenses: O, object slide; C, comparison or "blank" slide. From P. Barer, *in* "Cytology and Cell Physiology" (G. H. Bourne, ed.), p. 110. Academic Press, New York, 1964.

D. Polarization Microscopy

The oldest technique for the introduction of contrast into an image by employing gradients in refractive indices makes use of polarized light. The material to be viewed may be described as *isotropic*, i.e., polarized light is transmitted through the object with the same velocity independently of the direction of the incident wave, or *anisotropic*, i.e., the velocity of propagation of the polarized beam is dependent upon the direction of incidence. Isotropic substances have a uniform index of refraction in all directions, while anisotropic structures present different refractive indices corresponding to the different velocities of transmission. Accordingly, the latter property is referred to as *birefringence*. Birefringence is measured in the polarization microscope as a retardation of the plane-polarized light relative to the light polarized in another perpendicular plane. Quantitatively, birefringence is expressed as the difference between the two indices of refraction associated with the two rays; the measurement is conducted by a compensator placed into the optical system. The polarizing microscope differs from the light microscope

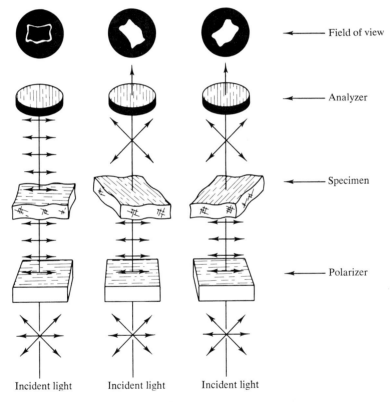

Fig. 1.4. Schematic drawing showing variations in darkness and brightness of an anisotropic object when placed between crossed polarizer and analyzer and rotated ± 45°. From G. B. Wilson and J. H. Morrison, "Cytology." Reinhold, New York, 1961.

by possessing two polarizing devices (see Fig. 1.4), one between the incident light and the specimen, the *polarizer*, the other placed above the objective lens, the *analyzer*. The polarizer transmits only plane-polarized light into the object. Complete extinction of the plane of the incident polarized beam will result if the plane of the latter is perpendicular to the plane of the wave passed through by the analyzer. The intensity of the light passing through the analyzer may vary from 0 to a maximum, depending upon the angle between the planes of the polarization of the polarizer and analyzer. When a birefringent material, e.g., tissue section, is placed on the stage, the plane of polarization will deviate according to the amount of retardation produced by the tissue section; the retardation is measurable by a sensitive compensator in terms of millimicrons or fractions of a wavelength.

Polarization microscopy has proved useful in the measurement of molecular orientation or cell ultrastructure of nonliving material. In particular, investigations with the mitotic spindle of the dividing cell and of the morphology of the muscle cell relied heavily upon this device. The technique as applied to cell ultrastructure has, however, largely been replaced by the electron microscope.

E. Ultraviolet Microscopy

One of the major limitations in the above-cited methods is the utilization of visible light. Since viable cells are transparent to this form of radiation and cannot easily be seen, some attempts have been made to utilize infrared and ultraviolet light, wavelengths at which the cells absorb strongly. The long wavelengths of the infrared spectrum, however, do not lend themselves to cytologic observation because of their low power of resolution. The use of ultraviolet light, however, has received considerable attention beginning with the construction of the first ultraviolet microscope by Zeiss in 1904. The latter resembles an ordinary light microscope in all respects but in the composition of the lenses. Since ordinary glass will not transmit light below 320 mμ, fused quartz must be used as the basis for the ultraviolet lens. Ultraviolet lenses, today, are also constructed of the mineral fluorite, a mixture of calcium and lithium fluorides.

Certain cellular structures, in particular chromosomes, absorb so strongly in the ultraviolet region that excellent contrast is obtainable with either unstained or fixed preparations. Unfortunately, the structures absorbing in this region are also quite sensitive to ultraviolet radiations and in living cells may be irreversibly damaged. The total dosage of ultraviolet light applied to a cellular preparation must therefore be a carefully controlled factor.

A major breakthrough in this field came as a result of the studies of T. Caspersson, who measured the light absorption of each part of the cell as a function of the wavelength. From the absorption curve constructed for each cellular organelle, he noted the resemblance of the curve obtained with the chromosomes to nucleoprotein, e.g., each possessed a maximum absorption at 260 mμ. These observations formed the basis for the technique of microspectrophotometry. The technique is often employed in conjunction with various histochemical procedures (see p. 30), e.g., the Feulgen reaction for DNA localization. With the latter, the amount of nuclear DNA has been determined in a variety of tissues of a number of species.

F. Fluorescence Microscopy

Many substances have the capability of absorbing light energy and reemitting some of the energy as light of a different wavelength, namely, the property of *fluorescence*. Fluorescence may be of two types: *primary* or *autofluorescence*, which is naturally present within a substance, and *secondary*, the consequence of staining the basic structure with fluorescent dyes, referred to as *fluorochromes*. Although many biological materials exhibit autofluorescence, e.g., coenzymes, pyridine nucleotides, and vitamins, most of the research studies have been performed with secondary fluorescence.

The fluorescence spectrum, like an absorption spectrum, or a "fingerprint," is unique for an individual substance and is independent of the mode of excitation. Many of the fluorescent substances exhibit bands of absorption in the near ultraviolet region between 320 and 400 mμ, a convenient range in wavelength since a specimen may be irradiated with an invisible ultraviolet light with the elaboration of a visible fluorescence. These observations form the basis of the technique of *fluorescence microscopy*. Since the fluorescent emission possesses a different wavelength from the absorbed light, observation with high contrast is easily achievable.

The applicability of the fluorescence microscope has been considerably enlarged with the development of Coons technique for the preparation of fluorescein-tagged antibodies. The antibodies normally present in the serum may be reacted with the fluorochrome, fluorescein isocyanate (or isothiocyanate). After staining of the prepared tissues with the fluorescent antibody, the localization of the antigens may be visualized in the fluorescent microscope. The technique makes possible the localization of antibodies at their site of formation within the plasma cells and the establishment of preferential binding of substances such as polycyclic hydrocarbons to particular cellular organelles. The technique has been extensively employed in virology for the localization of viruses or their constituents within subcellular organelles. The localization of specific viral antigens exclusively within the nucleus of the cell may be seen in Fig. 1.5. Little, if any, antigen is apparent in the cytoplasm of these cells.

G. Electron Microscopy

The introduction of electron microscopy to the field of cell biology in 1940 has made possible the extension of our knowledge of cells and the comparison of their biological properties with their ultrastructural features. The instrument allows us to examine the interior of the cell and to examine the subcellular mosaic. The application of autoradiography to ultrastructural analysis has added an even newer dimension. A discussion of the latter is presented on page 46.

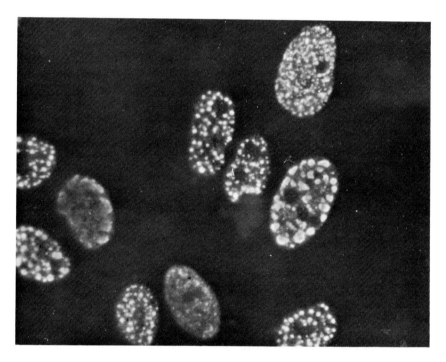

Fig. 1.5. Fluorescence microscopy. Green monkey kidney cells were injected with SV40 virus. At 40 hours postinfection, anti-SV40 monkey kidney serum (reacted with fluorescein isothiocyanate) was applied. The SV40 virus antigen is localized within the nuclei of these cells; no fluorescence is apparent in the cytoplasm ($\times$ 400). Note the absence of any fluorescence in the nucleoli. Courtesy of F. Rapp.

The resolution of the electron microscope is approximately 100 times that of the light microscope. Both light and electrons possess a vibratory character with wavelengths of approximately 550 mμ and 0.005 mμ, respectively (in addition, electrons are electrically charged). The streams of electrons generated by heating a metal filament in an evacuated tube are deflected by an electromagnetic or electrostatic field. The velocity of the electron release may be controlled by regulation of an electron potential applied to the metal filament (see Figs. 1.6 – 1.8). The electrons are focused upon the object by a magnetic coil acting as a condensor; the paths of the electrons may be altered by another magnetic coil acting as an objective lens. The resultant magnified image of the object is even further enlarged by a third coil that functions as an ocular or projection lens. A final magnified image is recorded upon a fluorescent screen or a photographic plate.

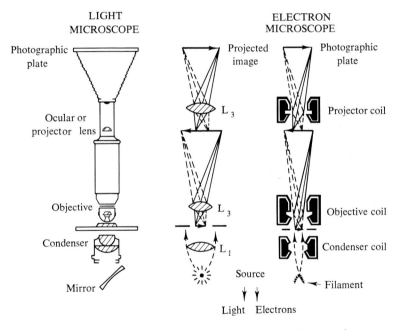

Fig. 1.6. Comparison between the optical microscope and the electron microscope.

In the light microscope, the formation of the image is dependent upon the absorption of light within various parts of the tissue, while in the electron microscope, the scattering of electrons is the responsible factor. The limit of resolution is, in practice, approximately 10 Å although several instruments have recently been marketed with resolutions in the order of 5 Å. Direct magnifications as high as 160,000 times are achievable with the electron microscope, and photographic enlargement will yield an overall magnification approaching 1 million. A comparison of a light and an electron micrograph of a preparation of "lymphoblastoid" cells is presented in Fig. 1.9.

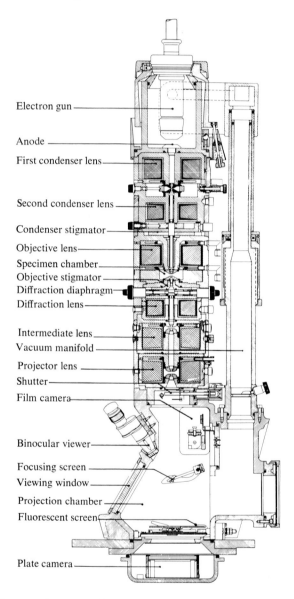

Electron gun

Anode

First condenser lens

Second condenser lens

Condenser stigmator

Objective lens

Specimen chamber

Objective stigmator

Diffraction diaphragm

Diffraction lens

Intermediate lens

Vacuum manifold

Projector lens

Shutter

Film camera

Binocular viewer

Focusing screen

Viewing window

Projection chamber

Fluorescent screen

Plate camera

Fig. 1.7. Electron microscope. Courtesy of Philips Electronic Instrument, Mt. Vernon, New York.

Although the resolution and applicability of the electron microscope make the instrument an important addition to biological technology, several disadvantages may limit its use.

1. The penetrability of electrons is very low, hence, ultrathin tissue

Fig. 1.8. Norelco EM 200 electron microscope. Courtesy of Philips Electronic Instruments, Mt. Vernon, New York.

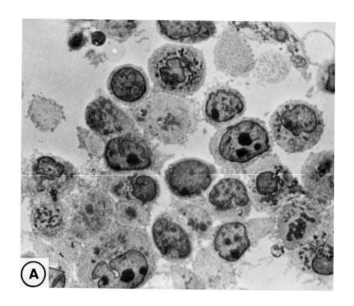

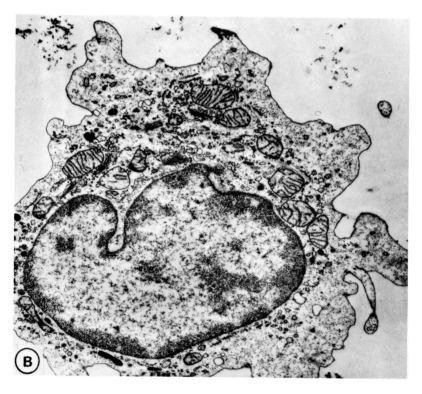

sections must be prepared. A specimen with a thickness in excess of 5000 Å will not transmit electrons and will appear opaque.

2. The tissue preparations must be able to withstand dehydration and evacuation. These procedures destroy the living cell, giving rise to the formerly held belief that electron micrographs can reveal only cellular artifacts.

In the light microscope, the standard procedure for tissue preparation includes fixation, embedding in paraffin, preparation of thin sections by a microtome, and staining of the sections. Although the steps in electron microscopy are similar, the details of the methodology differ markedly. Sections thin enough to allow passage of the electrons *cannot* be obtained with paraffin-embedded tissues. Thinner sections must be cut from hard embedding media such as the acrylic monomers or epoxy resins.

1. FIXATION

Today, the most commonly employed fixative is osmium tetroxide, which interestingly, was the first to be used by electron microscopists. Osmium tetroxide reacts with the double bonds of lipid material, forming unstable osmic esters which decompose and deposit osmic oxides or hydroxides at these sites. The fixative also reacts with proteinaceous substances, producing initially a gel and later after further oxidation, a solubilization of some of the material. Nucleic acids, however, do *not* bind osmium tetroxide.

Formalin has been employed as a fixative in electron microscopy as well as in light microscopy although formalin-fixed tissue is unduly sensitive to electron bombardment and may even decompose while under observation. Acrolein offers the advantage of fine penetrating ability although acrolein fixation must generally be followed by a second treatment with osmium tetroxide. Acrolein is toxic, is a vesicant, and is unstable — properties that are not very desirable in a fixative.

A major objection to the use of chemical fixatives has been voiced, which is related to a certain degree of distortion produced by these agents. As a result, freeze-drying came into use as a means for preserving tissue by purely physical means. In this technique, the tissue is rapidly frozen in liquid nitrogen and dehydrated in a vacuum at low temperatures. Al-

---

Fig. 1.9. Comparison of light and electron microscopy. (A) Light microscopy: A "lymphoblastoid" cell from human leukemic bone marrow grown in tissue culture is presented (× 2400). The cells are stained with toluidine blue. (B) Electron microscopy: A "lymphoblastoid" cell arising from cultured human bone marrow (infectious mononucleosis) was stained with 2% osmium tetroxide (× 12,750). Courtesy of M. Benyesh-Melnick and E. R. Rabin.

though care must be taken to avoid the formation of ice crystals, the advantages of this method are many: (1) no shrinkage of the tissue is observable and the structure is preserved; (2) the fixation is homogeneous; (3) the soluble components of the tissue are not extracted by the fixative and thus the chemical composition is preserved; and (4) cell function is arrested almost immediately by the subfreezing temperatures. A major disadvantage of the method is in the necessity of relatively thick sections to show sufficient contrast, thus resulting in a loss of resolution.

Another physical means for the preservation of cells is freeze-substitution which differs only slightly from freeze-drying. The method offers the advantage of exposing the tissues simultaneously to fixative and staining agents during the infiltration process.

Recently a new method for the preservation and examination of cellular ultrastructure has been developed, the technique of freeze-etching. In brief, freeze-etching involves the preparation and examination of minute replicas of exposed surfaces of the cell which have been ruptured in the frozen state. The exposed surface of a frozen preparation is etched by allowing the water to sublimate until some structures stand out in relief from the hydrated layer. After shadowing of the etched preparation (see p. 23) with carbon and platinum and stabilization of the specimen, the replica is ready for examination. A striking three-dimensional effect is created by this method. Freeze-etching also avoids any production of artifacts by chemical fixation and embedding.

2. EMBEDDING

One of the first embedding materials to be employed in electron microscopy was butyl methacrylate, a substance which when polymerized, possesses fine cutting characteristics. The tissue is infiltrated with the plastic; the latter is catalytically polymerized, permanently embedding the tissue. Thin sections for ultrastructural analysis may easily be obtained with an ultramicrotome.

Under the influence of an electron stream, the methacrylates may partially decompose, a process which enhances the contrast in the specimen although a certain loss in the fine detail is apparent.

Araldite, an epoxy resin, is not subject to the same objection—the ultrastructural detail is kept intact. Araldite, unfortunately, is difficult to cut, and the monomers do not readily penetrate into tissues. With this embedding material, very low specimen contrast is obtained and occasionally the samples must be "stained" to make the cellular organelles visible.

3. STAINING

Staining in electron microscopy refers to the addition of heavy atoms to tissue sections so that the electrons can be scattered from the surface, with the subsequent enhancement in contrast. In addition to its fixative properties, osmium tetroxide is an important stain in electron microscopy. Unfortunately, osmium tetroxide alone *does not* stain nucleic acid structures such as chromosomes and accordingly, other substances have been introduced to complement its fixative properties. Phosphotungstic acid, originally employed in the studies on collagen, is an excellent stain since each molecule yields 24 atoms of tungsten that upon binding to receptor sites can greatly enhance the electron density. Uranyl acetate and alkalinized lead acetate also are useful stains in electron microscopy.

Ultrathin preparations do not effectively disburse electrons without the aid of these stains and several other methods, depending largely upon enhancement of contrast, have been devised to overcome this difficulty. In *shadow-casting*, the specimen is placed in an evacuated chamber and a heavy metal, e.g., chromium or uranium, is evaporated at an angle from a tungsten filament. The deposition of the metal on only one side of the particle surface establishes a shadow on the other side. The shadow aids in contrast (see Fig. 1.10).

The technique of *negative staining* has proved of great utility in the study of viruses. In this technique, the specimen is embedded in a droplet of a heavy metal-containing substance such as phosphotungstate, which then permeates into the empty interstices between the macromolecules, rendering the spaces visible by *negative contrast* (see Fig. 1.11).

A very delicate *replication* of the surface topography may also be employed in enhancing surface contrast. The replica is produced as an ultrathin film transparent to electrons and hence, examinable in the electron microscope — *single-stage* replica. Because of the fragility of single-stage replicas, a two-stage replication procedure has been adapted. A thick primary replication is initially fashioned and is then stripped from the surface of the specimen. The replica is in turn copied, producing a second positive facsimile.

III. HISTOCHEMISTRY AND CYTOCHEMISTRY

The cellular composition and functions have been studied by (1) microscopic cyto- and histochemical techniques, which are concerned with the localization, characterization, and identification of chemical constituents by means of staining reactions. Histochemistry largely considered tissue

Fig. 1.10. Shadow-casting. An electron micrograph of ECHO virus 19 (× 115,000) shadowed with platinum-carbon. For comparison, a latex bead of 88 mμ diameter has been placed in the lower right corner. Courtesy of H. D. Mayor.

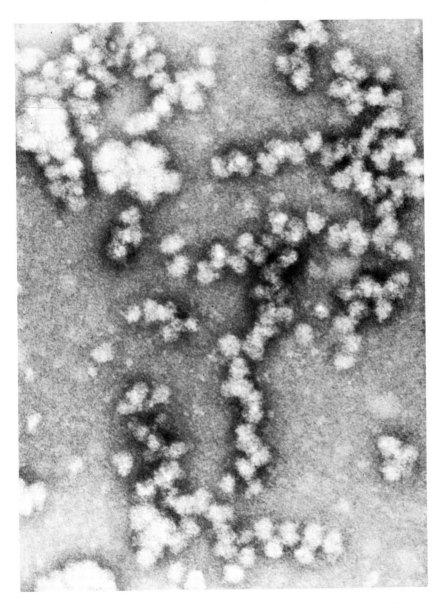

Fig. 1.11. Negative contrast. An isolated polyribosomal preparation from rat liver ($\times$ 200,000) has been negatively stained with uranyl acetate. Courtesy of H. Busch and K. Shankar.

components, while in cytochemistry the cellular constituents were investigated. (2) Homogenization or the disruption of tissues by mechanical means (see p. 42). (3) Microdissection technique, the division of a biological sample by means of a microtome into small units which consist of only one or a few cellular types. The structural barriers of the cell are destroyed by freezing and thawing and good contact between the environment and the cells can be achieved.

A. History

Although isolated reports on the utilization of chemistry in the study of morphological structures in tissue preparations appeared during 1800–1829, histochemistry was largely unknown as a separate science. Histochemistry had its beginnings in the period from 1830–1855 under the leadership of François-Vincent Raspail. The advances at this time were largely in the botanical sciences. With the botanical samples, Raspail was able to demonstrate the presence of starch granules by means of iodine solutions, the presence of protein by a xanthoproteic reaction and devise a test for carbohydrate using the formation of furfural after reaction with HCl. The use of enzymes for tissue digestion was first reported in 1861 by Beale who removed *by gastric juice* excess tissue from nerve fibers which he wished to study. The histochemical demonstration of the presence of enzymes in tissues was made by Klebs in 1868 and Struve in 1872. They showed that pus in the presence of tincture of guaiac gave a blue color, the first demonstration of what was later proved to be peroxidase. The presence of this enzyme in the granules of leukocytes was definitely established by Brandenburg in 1900.

Histochemical techniques were used to show the presence of macromolecules other than protein. In 1868, Heidenhain showed that the endoplasmic reticulum of the secreting gland cells contained material which was precipitable by acetic acid. This substance is now recognized as ribonucleic acid (RNA).

These initial successes in histochemistry spurred a wave of interest in the discovery of new synthetic dyes and in the development of additional staining techniques. In the early days of the twentieth century, the use of the aniline dyes became widespread. The developments in the morphological aspects of the staining reactions grossly overwhelmed the development and understanding of the chemistry of the staining process. In 1896, Daddi was first to use Sudan III for the *in vitro* staining of fat which, after ingestion by animals, could easily be detected in tissues by its red color. It wasn't until several years later that Michaelis proved that the Sudan dyes exhibited their staining capabilities for fat by purely physical means, i.e., the inert dyes dissolved in the fats themselves (lipophilic stains).

The indiscriminate use of the synthetic dyes by individuals not founded in fundamental chemical principles of staining led to the appearance in many cases of precipitation artifacts, structures which did not occur in the cell. Accordingly, by 1910, interest in histochemistry was at a low ebb and investigators were returning to the study of living cells. From 1910 to 1935, important advances arose in cytology.

The period following 1935 was marked by a rebirth of histochemistry. The impetus to this development was the work of Lison, immortalized in the "bible" of histochemistry, published in 1936, "Histochemie Animale." From then, the science advanced by leaps and bounds to a state of sophistication. An insight into some of the histochemical methodology will now be presented.

B. Principles of Staining

In this section, we will consider some of the staining reactions that have proved to be of considerable importance in the development of our knowledge of cellular mechanics. In cytochemistry, the most commonly employed stains are the basic and acidic dyes. The dyes consist of a chemical group which imparts the color, the *chromophore*, and an *auxochromic group*, i.e., one which possesses the property of attaching to some tissue structure. The more common chromophores are usually aromatic compounds, e.g., unsaturated, carboxyl, sulfonic, azo, nitroso, nitro, quinoid, or ethylene groups. The most commonly employed dyes are presented in Table 1.2.

The proteins, nucleic acids, polysaccharides, and phospholipids are mainly responsible for the staining reactions exhibited by the tissue. Proteins, possessing both carboxyl and amino groups, are amphoteric and can dissociate as an acid or a base. The acid groups become ionized at pH's that are above their isoelectric point and the proteins can react with the basic dyes. The basic groups are ionized at pH's below the isoelectric point and accordingly, proteins at these pH's will react with the acidic dyes. The nucleic acids, because of the dissociation of the phosphoric acid groups, possess very low isoelectric points and stain very intensely at low pH's with basic dyes. Indeed, at these pH's the staining reaction may be specific for the nucleic acids, e.g., the use of toluidine blue for staining RNA.

Metachromasia is another property which has proved useful in histochemistry. A number of dyes, when employed as a stain for certain tissues (or cellular components), differentially color the tissue constituents. This property is referred to as metachromasia. These stains obviously do not obey the Beer-Bouguer absorption law, i.e., the absorbance is not directly proportional to the concentration of the dye. An example of a metachromatic dye is toluidine blue, which in aqueous solution will stain nuclei blue while the matrix of cartilage, mucus, and the granules of mast cells will be stained red.

The tissue component believed responsible for evoking a metachromatic response from a suitable dye is referred to as a *chromotrope* and generally is of polyanionic nature. The most suitable intracellular candidates for the latter are heparin, chondroitin sulfate, the nucleic acids, etc. The metachromatic dye, on the other hand, is a cationic substance. On this basis, a more encompassing definition for metachromasia has been offered—a shift of the maximum absorption band to a shorter wavelength when a dilute solution of the dye of one charge type is added to a polyelectrolyte of the opposite charge.

Although the exact mechanism underlying the metachromatic response is unknown, it is believed to involve dye association or aggregation. This hypothesis is predicated upon the elimination or reduction of metachromasia by low dye concentration, high temperature, low pH, high ionic strength, or the addition of alcohol or detergents.

C. Staining of Particular Substances

1. PROTEINS (OTHER THAN ENZYMES)

Several general methods are available for the detection of proteinaceous material in tissue sections. In addition to these reagents, the existence of specific enzymes may be detected by the end products of the reactions which they catalyze. The nonspecific methods will be discussed in this section; the detection of specific enzymes will follow.

The *Millon's test* is one of the oldest methods for the detection of protein and was first applied to histochemistry by Bensley and Gersh in 1933. The reagent contains mercuric nitrate in acid solution and is specific for the tyrosine and tryptophan moieties of proteins, forming their mercurials as red precipitates. The extreme acidity at which the cytochemical demonstration must be conducted has a detrimental effect upon the tissue section and is a serious drawback in the use of this method.

The *sulfhydryl* content of most proteins may also be employed as a basis for the detection of the latter. One of the most popular methods is the utilization of 2,2'-dihydroxy-6, 6-dinaphthyl disulfide, a colorless substance which when allowed to react with sulfhydryl groups is converted into a highly chromatic azo derivative.

The *Sakaguchi* reaction for arginine, a constituent of most proteins, has been employed as a histochemical stain for the detection of proteins. Arginine reacts with α-naphthol and hypochlorite or hypobromite under alkaline conditions to yield a red color. The arginine-rich histones (see Chapter IV) are responsible for the positive test which is observed in the nuclei of many cells.

The presence of tyrosine, tryptophan, or histidine in proteins is demonstrable by their reaction with tetrazotized dianisidine and the subsequent coupling with β-naphthol. The resultant red coloration is stable and very intense. The method may be made more specific with the prior treatment of the tissue, with benzoyl chloride blocking the reaction of tyrosine and tryptophan with the dye.

TABLE 1.2

Commonly Employed Dyes in Histochemistry

Acidic dyes	Basic dyes
Acid fuchsin	Acridine red
Alizarin	Aniline blue
Bismarck brown	The azures
Congo red	Basic fuchsin
Eosin Y	Crystal violet
Erythrosin	Light green
Fluorescein	Malachite green
Janus green B	Methylene blue
Orange G	Methyl green
Phloxin	Pyronine
Picric acid	Safranine
Rose bengal	Thionine
	Hematoxylin
	Carmine

The *periodic acid-Schiff* (PAS) method may under suitable conditions be employed in the detection of special proteins. If the polysaccharides and lipids are extracted prior to the staining reaction, the method becomes relatively specific for mucoproteins. In this technique, the vicinal hydroxyl or amino hydroxyl groups are oxidized to aldehydes by the periodic acid, and the aldehydes are subsequently coupled to leucofuchsin to form a red dye. The PAS technique is most useful in the demonstration of glycogen when combined with the enzymic destruction of the latter and the consequent elimination of the positive reaction. Mucoproteins and hyaluronic acid-containing substances also yield positive PAS reactions. With the utilization of the appropriate catabolic enzymes, the PAS test may be made specific for the particular substance, e.g., hyaluronidase, mucinase.

Several methods are based upon the affinity of protein for certain dyes, such as mercuric bromophenol blue, naphthol yellow S and fast green. The latter has some utility in the localization of histones within the cell.

2. LIPIDS

Lipids are characterized by their stainability with a variety of lipophilic substances, e.g., Sudan dyes, the fluorescent dye, 3,4-benzpyrene, Nile blue, rhodamine B, and phosphine 3R. These substances act by diffusing into the interior of the fat droplets and staining the latter. Sudan black is particularly valuable for its ability to stain, with great contrast, phospholipids and cholesterol.

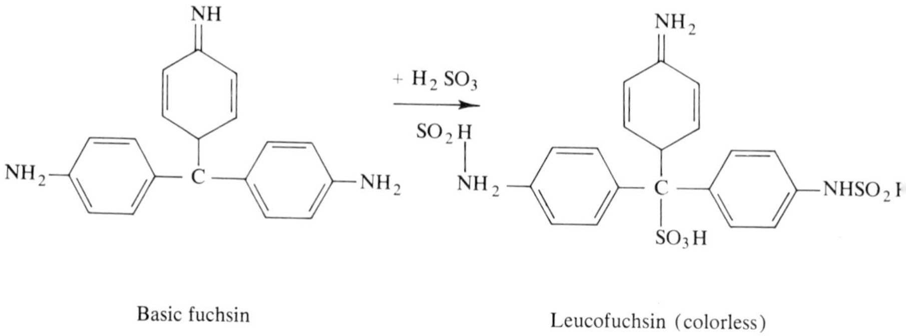

Basic fuchsin   Leucofuchsin (colorless)

Fig. 1.12. Formation of leucofuchsin.

The lipids are rendered visible by their stainability with bichromate or osmium solutions. The unsaturated bonds of the lipids are oxidized by bichromate producing an insoluble substance which becomes discernible after coupling with a fat-soluble stain (see above) or by reaction with a lake-forming dye such as hematoxylin. The unsaturated lipids are oxidized by osmium salts depositing the insoluble osmic oxides and producing an intense osmophilia which is readily viewable.

3. NUCLEIC ACIDS

The Feulgen technique for the demonstration of DNA is one of the most widely employed histochemical stains. Although more than 40 years have passed since the initiation of this technique, no completely satisfactory explanation for its chemical basis has evolved. The method is dependent upon the preferential release of the purines from DNA as a result of an acid hydrolysis and the formation of aldehyde groups on the adjacent deoxyribose sugar. The unmasked aldehyde groups react with decolorized Schiff's base (see below) to yield a colored compound. The decolorized Schiff's base, leucofuchsin (Fig. 1.12), is formed from basic fuchsin by sulfurous acid. The purple compound formed in the presence of the aldehydes is believed to have the structure presented in Fig. 1.13.

Methyl green had been employed for many years as a histochemical tool principally in combination with pyronine before the biochemical basis for the reaction was understood. About 40 years after its introduction into investigation of morphological structure, it was demonstrated that the methyl green stained DNA green and RNA red. A positive methyl green reaction is not only dependent upon the molecular weight or state of polymerization or both, but also depends upon the configuration of the DNA.

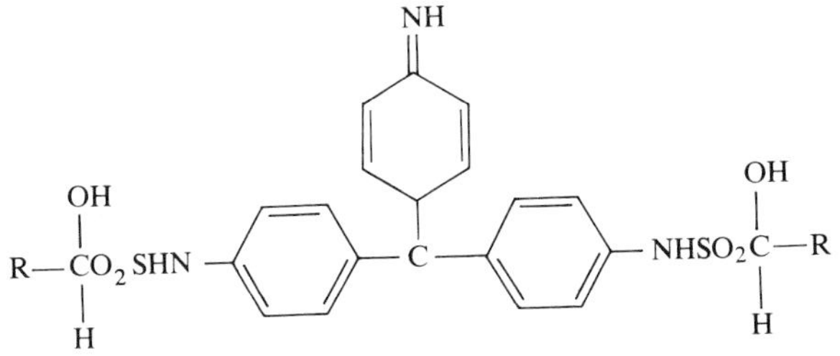

Fig. 1.13. Aldehyde-fuchsin complex.

RNA has affinity not only for pyronine but for azure B and toluidine blue, although a positive reaction is not absolutely specific for this polymer. Acridine orange has been used very effectively as a fluorochrome in the study of viral nucleic acids. The dye imparts a reddish hue to double-stranded RNA and a yellow-green fluorescence to DNA (Fig. 1.14).

4. ENZYMES

Histochemistry has perhaps played its greatest role in the localization and identification of enzymes within the cellular organelles. Until recently, the only enzymes which could in practice be histochemically studied were the oxidative enzymes, dehydrogenases and oxidases. However, in 1939, the technique for demonstrating alkaline phosphatase in tissue sections was reported and the technique has now been extended to encompass a number of other phosphorolytic enzymes.

One of the major advances in enzyme histochemistry was the develop-

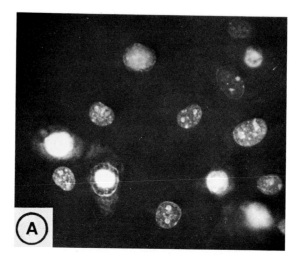

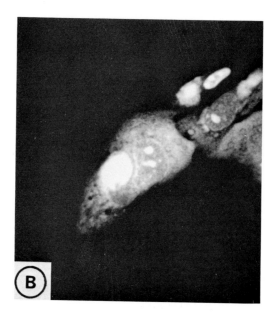

Fig. 1.14. Acridine orange fluorescence. (A) Tissue culture cells were infected with adenovirus, a double-stranded DNA virus, and the preparation was subsequently reacted with acridine orange (× 1000). Note the localization of the fluorescent dye within the nucleus of the cells (yellow-green fluorescence). (B) Tissue culture cells were infected with the cytoplasmic virus, REO, a double-stranded RNA virus. After reaction with acridine orange, the fluorescence typical of RNA appeared within the cytoplasm, i.e., the nucleus appeared yellow-green, the cytoplasm, red. Courtesy of H. D. Mayor.

ment of techniques for fixing tissues without concomitantly inactivating the enzymes (see p. 9). The tissue section is briefly incubated with the appropriate substrate and the product of the reaction must be either visible or easily converted into a compound that is visible under the light microscope.

a. PHOSPHATASES. The interest in the enzyme phosphatase, in 1939, was responsible for the rebirth of enzyme histochemistry. The histochemical technique for the demonstration of alkaline phosphatase was developed by Gomori and employs β-glycerophosphate as a substrate. The tissue sections are incubated in an alkaline medium with a substrate and calcium and magnesium ions. The sections are then treated with cobalt acetate and with ammonium sulfide (Fig. 1.15). The phosphate released as a product of the enzymic reaction is precipitated in the presence of the calcium ions as the calcium salt and the latter is subsequently converted to cobalt phosphate. The cobalt phosphate reacts with the ammonium sulfide, producing cobalt sulfide, which is readily identifiable under the light microscope.

Several other methods for the demonstration of phosphatase activity have been introduced which differ largely in the choice of substrate. An aromatic phosphate ester is allowed to react with the tissue section, releasing during the enzymic reaction the aromatic portion, which may be coupled with a diazonium salt to produce an azo dye, readily visible under the light microscope (Fig. 1.15).

The method for the histochemical determination of *acid phosphatase* was also introduced by Gomori in 1952 and depends upon the trapping of phosphate ions released during enzymic reaction with β-glycerophosphate by lead nitrate. The lead phosphate is then converted to the black precipitate of lead sulfide. A major disadvantage in these methods is the diffusion of the phosphate from the site of enzyme localization to a distal region. Aromatic phosphate esters have also been employed in the determination of acid phosphatase activity (Fig. 1.16).

Several of the other phosphatases that have been studied by histochemical techniques are indicated in Table 1.3.

b. ESTERASES. The esterases which catalyze the hydrolysis of esters to produce the corresponding alcohol and acid can be readily identified by histochemical techniques. Included within this broad category are the *simple esterases*, which hydrolyze the short chain aliphatic esters, the *lipases*, which cleave esters with long carbon chains, and the *cholinesterases*, which attack the esters of choline. The simple esterases and lipases are very conveniently localized with the use of β-naphthyl acetate and the ester of β-naphthyl alcohol and a long carbon chain aliphatic acid. The product common to both reactions is β-naphthyl alcohol, which may be coupled in the presence of a diazonium salt to yield a colored azo component at the site of enzymic activity (see Figs. 1.17 and 1.18). The nonspecific esterases have also been localized with the use of a soluble ester of indoxyl as substrate. The free indoxyl produced during the course of this reaction is rapidly oxidized to the very insoluble blue dye, indigo.

TABLE 1.3

Histochemically Determined Phosphatases

Enzyme	Substrate
Alkaline phosphatase	α- or β-glycerophosphate
	or
Acid phosphatase	Naphthyl phosphate
Adenosinetriphosphatase (ATPase)	Adenosine triphosphate
5'-Nucleotidase	Adenosine 5'-monophosphate
Phosphoamidase	Phosphocreatine
	or
	Naphthyl phosphate diamines
Glucose-6-phosphatase	Glucose 6-phosphate
Thiaminepyrophosphatase	Thiamine pyrophosphate
Pyrophosphatase	Sodium pyrophosphate
	or
	Dinaphthyl pyrophosphate
Ribonuclease	Ribonucleic acid
Deoxyribonuclease	Deoxyribonucleic acid

c. GLYCOSIDASES. The glycosidases catalyze the cleavage of glycosidic linkages. The enzymes comprising this group include β-glucuronidase, β-glucosidase, β-galactosidase, and β-glucosaminidase. The histochemical methods are generally based upon the utilization of a water-insoluble azo dye containing a glucuronide or an 8-hydroxyquinoline glucuronide as substrates. The enzymically released product of the latter is coupled with an iron salt to produce the insoluble ferric hydroxyquinoline which may then be converted into *Prussian blue*. In the former method, the cleavage of the glucuronide of the azo dye results in the appearance of a colored substance which is readily visible under the light microscope. The β-glucosidases may be localized histochemically by the employment of the substrate 3-(5-bromoindolyl)-β-D-glucopyranoside which upon enzymic hydrolysis produces an indigo dye.

d. PROTEASES. The proteolytic enzymes which hydrolyze proteins and amino acids may be either endopeptidases or exopeptidases. Centrally located peptide bonds are hydrolyzed by the endopeptidases while the latter hydrolyze peptide bonds that are adjacent to the terminal α-amino or terminal α-carboxyl group of the proteins. The exopeptidases are best exemplified by the two enzymes aminopeptidase and carboxypeptidase. Aminopeptidase has been localized largely with the use of a synthetic substrate containing a β-naphthylamine as a constituent of a peptide bond (Fig. 1.18). The released chromogenic moiety, β-naphthylamine, may be diazotized and visualized under a light microscope. Chymotrypsin, an

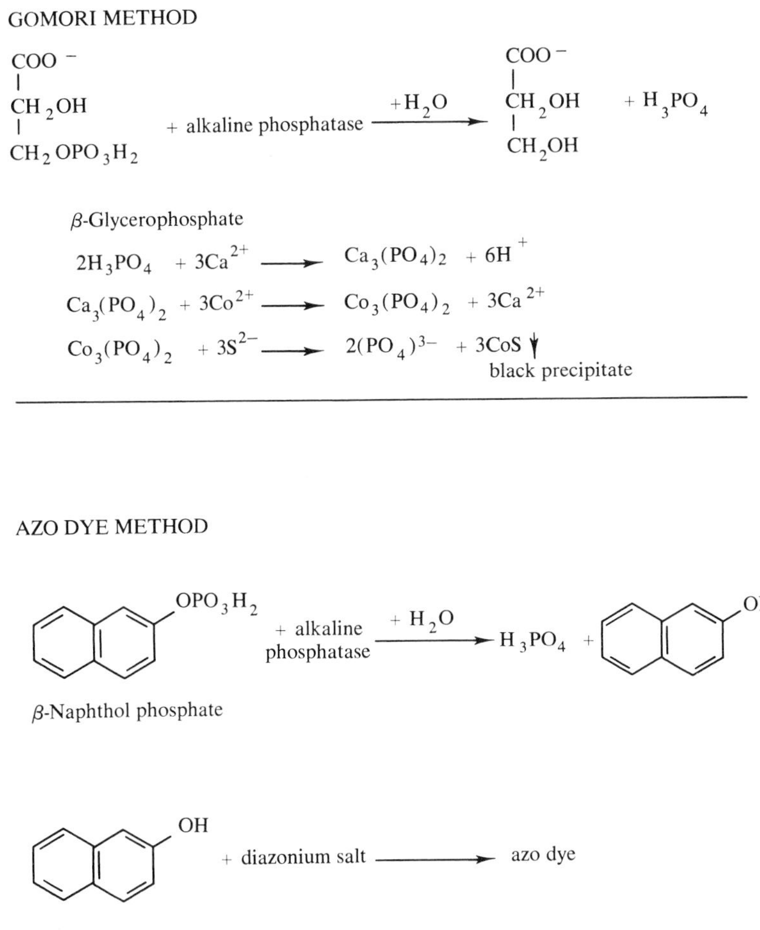

GOMORI METHOD

$$\begin{array}{c}\text{COO}^- \\ | \\ \text{CH}_2\text{OH} \\ | \\ \text{CH}_2\text{OPO}_3\text{H}_2 \end{array} \quad + \text{ alkaline phosphatase} \xrightarrow{\;+\text{H}_2\text{O}\;} \begin{array}{c}\text{COO}^- \\ | \\ \text{CH}_2\text{OH} \\ | \\ \text{CH}_2\text{OH} \end{array} \quad + \text{H}_3\text{PO}_4$$

β-Glycerophosphate

$$2\text{H}_3\text{PO}_4 + 3\text{Ca}^{2+} \longrightarrow \text{Ca}_3(\text{PO}_4)_2 + 6\text{H}^+$$

$$\text{Ca}_3(\text{PO}_4)_2 + 3\text{Co}^{2+} \longrightarrow \text{Co}_3(\text{PO}_4)_2 + 3\text{Ca}^{2+}$$

$$\text{Co}_3(\text{PO}_4)_2 + 3\text{S}^{2-} \longrightarrow 2(\text{PO}_4)^{3-} + 3\text{CoS} \downarrow$$
$$\text{black precipitate}$$

AZO DYE METHOD

β-Naphthol phosphate   + alkaline phosphatase $\xrightarrow{\;+\text{H}_2\text{O}\;}$ H_3PO_4 +

OH   + diazonium salt $\longrightarrow$ azo dye

Fig. 1.15. Alkaline phosphatase demonstration.

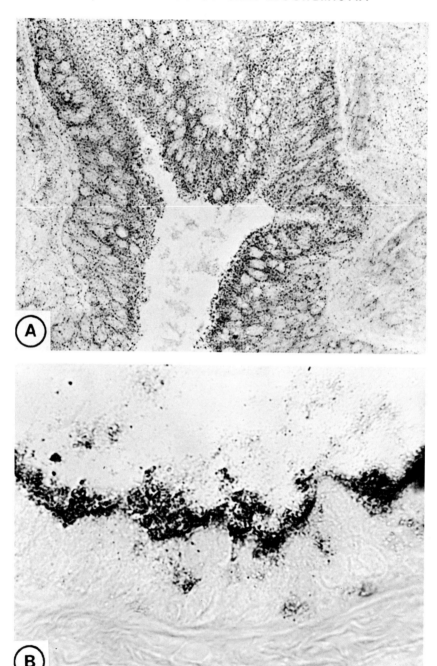

example of an endopeptidase, can readily cleave the chloroacetyl ester of 3-hydroxy-2-naphthoic acid anilide which once again, after diazotization, is rendered visible.

e. OXIDASES. The oxidases which catalyze the transfer of electrons from a donor substance to oxygen comprise the following: cytochrome oxidase, tyrosinase, and polyphenol oxidase, all of which contain copper; peroxidase and catalase which contain iron; and monoamine oxidase.

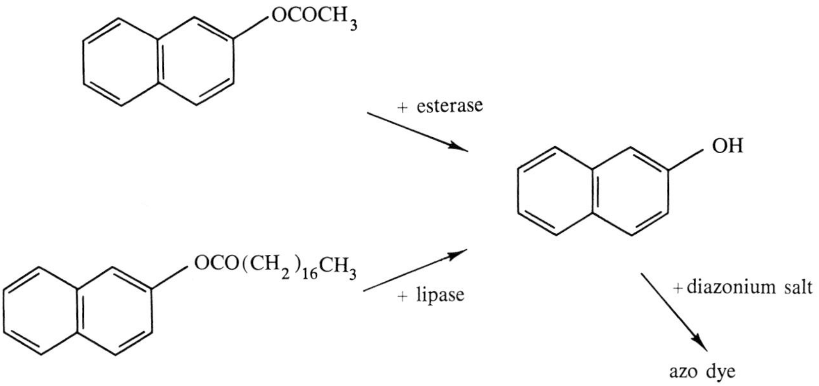

Fig. 1.17. Diagram of reactions leading to demonstration of esterase and lipase activity.

f. DEHYDROGENASES. The transfer of hydrogen atoms or electrons from one molecule to another is a vital function in every living cell. The mechanism by which these substances are shuttled within the cell toward the ultimate acceptor, oxygen, is quite intricate and the details are discussed in Chapter III. The pyridine nucleotide-linked and succinic dehydrogenases are several of the systems which are directly involved in biological oxidations and which are localized within the mitochondria. Several histochemical techniques have been employed in the demonstration of these and other dehydrogenases: the tetrazolium, tellurite, and triazole methods. The tetrazolium technique has received the most widespread application in histochemistry (see Fig. 1.19). A schematic reproduction of the basis by which the tetrazoles accept electrons from the dehydrogenase systems is presented in Fig. 1.20. The technique depends

Fig. 1.16. Histochemical demonstration of acid phosphatase. (A) Cells from a human hyperplastic prostate have been stained for acid phosphatase by the azo dye method. (B) Acid phosphatase has been localized in the human fetal prostate ($6\frac{1}{2}$ months old) by the Gomori method ($\times$ 1250). Courtesy of F. Gjorky.

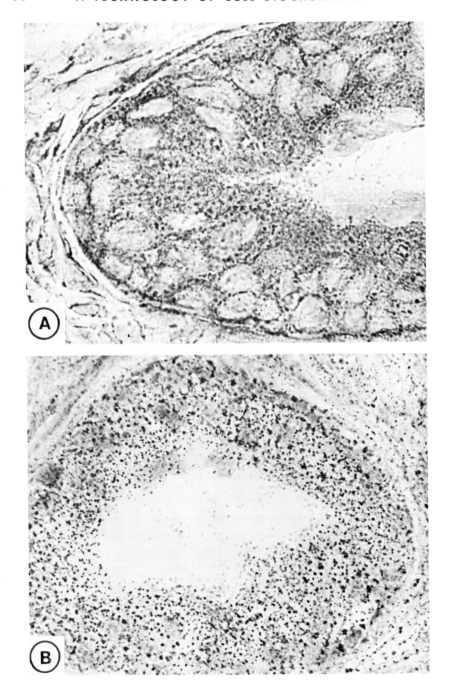

upon the conversion of either a monotetrazole or a ditetrazole into a highly insoluble chromogenic formazan dye by a reducing system.

Cytochemistry and histochemistry have proved to be of immense value in the localization of enzymes and enzyme systems within cellular organelles. The field is dynamic and one that is dependent upon the ingenuity of the investigator. During the next few years, the applicability of the techniques in electron microscopy may be exploited with even more definitive localization.

IV. ISOLATION OF SUBCELLULAR ORGANELLES

Although histochemistry and its variations provided the initial impetus toward the understanding of the cellular events, only with the introduction into cell biology of the ingenious techniques of differential centrifugation could the dynamic and quantitative aspects of cellular organelles be fully exploited. The nineteenth century was marked by isolated reports of the separation of subcellular components. During this period, Kölliker's studies on muscle cell granules were published (1856), Miescher, from Hoppe-Seyler's laboratory, began his work on isolated nuclei (1869), and Engelmann separated free chloroplasts from plant tissues (1881). It was not until 1934, however, when Bensley developed relatively mild methods for the isolation of mitochondria from guinea pig liver that the possibilities of this technology were realized. Bensley's isolation techniques were carefully controlled by microscopic examination correlating the concentrated "mitochondria" with the organelle present *in situ* and thus avoiding any criticism as to the possibility of artifactual structures arising during the course of isolation.

These studies were extended by Claude with the isolation of chromatin, the secretory granules, and the microsomal fraction from tissue preparations by differential centrifugation. With this technique, large quantities of the various cellular components became readily available to the investigator, allowing a closer scrutiny of cell metabolism and the comparison of cellular activities within the different organelles.

Since the studies of Claude and his colleagues, liver has proved to be the favorite starting material for the isolation of the cellular components. The reasons are obvious — the liver comprises approximately 4%

Fig. 1.18 Histochemical demonstration of leucylaminopeptidase and esterase. (A) Leucylaminopeptidase positive reaction is presented ($\times$ 1250). Human fetal prostate ($6\frac{1}{2}$ months old) has been stained by an azo dye method. (B) Cells from normal human prostate have been stained for esterase activity by an azo dye method ($\times$ 750). Courtesy of F. Gjorky.

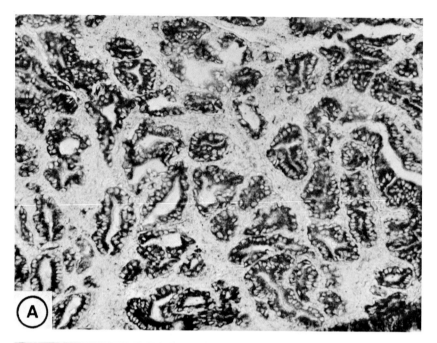

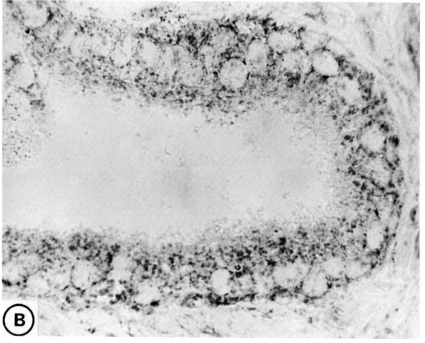

of the wet weight of the mammalian organism, thus representing a considerable mass and furthermore, it is easily dissected free of contaminating tissues. Unfortunately, liver tissue is heterogeneous, being composed of parenchymal cells that constitute 90% of the liver weight and 65% of the total number in the cellular population and the nonparenchymal cells, e.g., the Kupffer and bile duct epithelial cells. The heterogeneity must be continuously borne in mind in any evaluation of effects of environmental or artificially applied stimuli. In addition, the methodology which has been developed for this tissue cannot be applied *in toto* to another tissue.

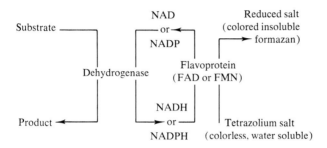

Fig. 1.20. Transfer of electrons to tetrazolium salt. FAD, flavin adenine dinucleotide; FMN, flavin mononucleotide; NAD, nicotinamide-adenine dinucleotide; NADP, nicotinamide-adenine dinucleotide phosphate.

The principal obstacle in the elucidation of the subcellular biochemistry within the mammalian cell is the mammalian cell membrane. Any technique for the isolation of cellular components must, as a prerequisite, incorporate a procedure for breaking this structure. One of the foremost of these procedures is homogenization, which not only accomplishes this task but also succeeds in dispersing the cellular constituents in a selected medium. Upon liberation of the cellular components, a method must be selected by which a separation of these organelles may be achieved. The method of choice for the latter task is centrifugation. Let us consider, in order, these two processes, *homogenization* and *centrifugation*.

Fig. 1.19. Histochemical demonstration of cytochrome oxidase and succinate dehydrogenase. (A) Carcinoma cells from a human prostate have been stained for cytochrome oxidase activity by the *Nadi* method of Burstone (× 1000). (B) A positive succinate dehydrogenase reaction in normal human prostate is indicated (× 1000). Courtesy of F. Gjorky.

A. Homogenization

The ultimate end point of homogenization is the release of the cellular organelles into a medium in a manner which preserves their morphological and functional integrity. Toward this end, the selection of a homogenizing vehicle is crucial and often is dependent upon the particular organelle to be investigated. In *general*, the osmotic pressure of the cell must not be interrupted, and a strict adherence to the intracellular ionic strength and pH must be maintained, although nuclei have been isolated in citric acid solutions of rather low pH's. The selection of a homogenizing medium must also take into consideration the investigations to be performed upon the isolated components, e.g., the medium must not influence any chemical or enzymic analysis.

The "physiological solutions," of potassium chloride or sodium chloride, were first employed although considerable leaching of subcellular constituents and agglutination of the organelles were observed, thus severely curtailing their usefulness. Attention was then directed toward the nonionic media, with the introduction by Hogeboom of sucrose solutions as a homogenizing vehicle. He first employed 0.88 M sucrose and noted the preservation of subcellular form and structure but later advocated the use of the less dense and less viscous 0.25 M sucrose. With the latter, the enzymic distribution could generally be maintained although a loss in some morphological characteristics was apparent.

Homogenizing media have not been restricted to these aqueous solutions but have included several nonaqueous solvents. Behrens, in 1932, introduced their use as an attempt to avoid the loss of the water-soluble constituents from the organelles. The method is based upon grinding lyophilized tissue in a binary mixture of solvents, ether, chloroform, benzene—carbon tetrachloride or cyclohexane—carbon tetrachloride, and has been used to advantage in the isolation of nuclei (see Chapter IV).

Having selected a homogenizing medium, the next problem to be circumvented is the grinding of the tissue with the aim of maximally rupturing the cells with minimum damage to the cellular organelles. Unfortunately, this problem has never been completely solved. The methods have included such physical procedures as freezing and thawing, sudden decompression, and the mechanical means which utilize crushing in ball or colloid mills or in mortars. The most common procedure employs devices which act by shearing the tissue and which are derived from the Potter-Elvejhem apparatus. The latter consists of a glass vessel into which a tightly fitting glass pestle (today, largely replaced by Teflon) is inserted. The pestle may be driven by hand or mechanically (see Fig. 1.21). During this procedure, a considerable amount of heat may be generated, hence, it is most important to keep both the tissue and the medium cold.

B. Centrifugation

The suitably prepared homogenate now consists of a disbursed suspension of cellular components of varying size and density. Accordingly, it is possible to separate these organelles by employing centrifugation techniques. A theoretical consideration of the forces which may be active in the sedimentation of these components will not be discussed in this text.

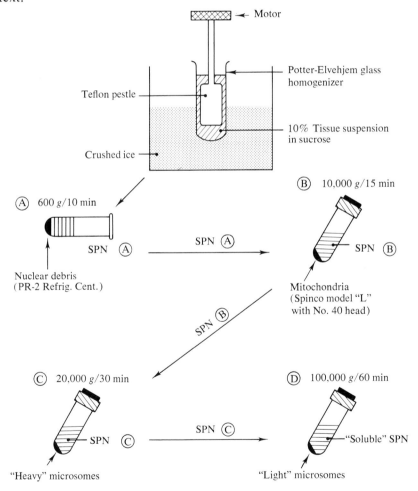

Fig. 1.21. Subcellular fractionation.

A flow sheet for the separation of the cellular organelles from a liver homogenate prepared in 0.25 M sucrose by differential centrifugation is offered in Fig. 1.21. An initial centrifugation at $600g$ (gravity) will sediment a crude nuclear fraction which includes as contaminants unbroken cells and cellular debris. The supernatant fraction (A) contains the remaining cellular components from which the mitochondria may be removed by a second centrifugation at $5000-10,000g$. The mitochondrial pellet may be contaminated with lysosomes and Golgi apparatus. The "heavy" and "light" microsomal fractions may be sedimented by further centrifugation of supernatant (B) at 20,000 and $100,000g$, respectively. The microsomal fraction (see Chapter V) is a homogenization artifact which is a composite of broken bits of "smooth" and "rough" endoplasmic reticulum and free ribosomes.

In classical ultracentrifugation, the migration of the cellular constituents through the homogeneous medium is usually disturbed by convective currents which arise. This annoying phenomenon may be eliminated by a modification of the differential centrifugation technique. A gradient of density which increases with the distance from the axis of rotation is produced within a centrifugation tube. In this technique of *density gradient centrifugation,* the gradient may be formed during the centrifugation run or prior to it by some layering device.

Density gradient centrifugation may be classified into three categories depending upon the manner in which the gradient is used: *stabilized moving-boundary centrifugation, zonal centrifugation,* and *isopycnic gradient centrifugation.* In the first type, a shallow gradient is established in a centrifuge tube, the dilute sample is carefully overlaid and the tube is centrifuged in a swinging-bucket rotor for a period of time. At the completion of the run, fractions are sequentially removed and analyzed to determine the position of the boundary. The technique has proved useful in the determination of the sedimentation coefficient of viruses and in particular, enzymes.

In zonal centrifugation, the suspension of particles is layered over a steep density gradient column prepared from the same solvent and solute as the suspension. The layering device is schematically depicted in Fig. 1.22. Under the influence of a centrifugal force, each substance will sediment at its own rate, forming separate zones in the column. Each material can be drawn off for further analysis. The principal objective of zonal centrifugation is usually the separation of the suspension into its relatively pure components, and the method has achieved prominence in the nucleic acid and mitochondrial fields (see Chapters III and IV).

In isopycnic gradient centrifugation, the gradient within the tube extends over the entire range of densities of the macromolecular constituents and the latter will attain equilibrium positions at regions in the tube corresponding to their own density. Thus, the separation depends exclusively upon the densities of the macromolecules and involves long centrifugation times. Generally, the macromolecules are distributed throughout the solution before beginning the centrifugation run, although the material may be overlaid as well.

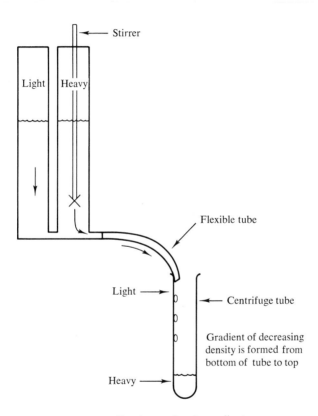

Fig. 1.22. Continuous density gradient.

1. GRADIENT-FORMING MATERIALS

The gradient-forming material ideally should be chemically inert, nontoxic, very soluble in water, of high density and molecular weight, of low viscosity, and should not contain any nitrogen (to avoid any errors in total protein determinations). Sucrose has been the most extensively employed, although concentrated solutions do possess high viscosities. A number of inorganic salts have been utilized, particularly in equilibrium analyses upon DNA. These include the cesium salts, chloride and sulfate;

rubidium chloride; ficoll, a water-soluble, neutral colloid with average molecular weight of 50,000; glycerol; dextrans; and polyvinyl pyrrolidone.

Applications. The specific applications will be treated in the following chapters.

2. ANALYTICAL ULTRACENTRIFUGATION

The analytical ultracentrifuge is employed in the production of high centrifugal forces required for the estimation of the movement or redistribution of sedimenting particles. The instrument has become an invaluable tool in the study of proteins including enzymes, viruses, nucleic acids, and natural and synthetic polymers. With this instrument, for example, the following molecular parameters may be ascertained: sedimentation coefficient; diffusion coefficients; size, shape, and partial specific volume of particles; chemical equilibria; and the compressibility of gels.

The material to be studied is placed in a specially constructed cell which will allow the passage of light rays through the cell's entire length. The cell is placed in a rotor with an appropriate counterbalance and the rotor is accelerated. The speed and temperature are accurately controlled during the centrifugation. As the particles sediment under the centrifugal force, light from an optical source is transmitted through the rotating cell. The movement of the molecular particles is translated into an optical pattern which can be either viewed directly or photographed at selected time intervals. The rate of sedimentation may then be calculated and employed in the estimation of the other parameters.

V. AUTORADIOGRAPHY

Autoradiography had its beginnings in 1896 with the detection by Becquerel of radioactive elements by means of photographic plates. Since that time, the number of radioisotopes which have become commercially available has reached an astronomical figure. Concomitantly, the utilization of autoradiography in the laboratory has expanded. The most useful isotopes have been the β-emitters, a list of which is presented in Table 1.4, along with the half-lives and energy of emission.

Autoradiography may be considered one of the most useful techniques for studying single cells *in situ* and the localization of biochemical processes within the cellular organelles. The technique requires relatively simple equipment although quantitative radiography can prove tedious. The specimen containing the radioactive material is allowed to remain in contact with a covering of a thin layer of a photographic emulsion of a designated type for a defined exposure period. The radioactive specimen, which may consist of either individual cells, sections of tissues, tissue squashes, or imprints of the whole animal, emits radiations which impinge upon the photographic emulsion, activating the silver bromide granules and producing an image that describes the nature of the radioactive source. The activated emulsion is processed photographically and then viewed under the light microscope. In addition to supplying information relative to the location of the radioactive structure, the technique allows for some semiquantitative interpretations of the amount of radioactivity by a comparison of the intensity of the exposed areas.

TABLE 1.4

Useful β-Emitters in Autoradiography

Isotope	Energy (meV)		Half-life	
^{3}H	0.018		12.3	years
^{14}C	0.155		5600	years
^{35}S	0.167		87.1	days
^{32}P	1.71		24.3	days
^{131}I	0.25	(3)[a]	8.04	days
	0.36	(9)		
	0.61	(87)		
	0.82	(1)		
^{45}Ca	0.254		164	days
^{59}Fe	0.27	(46)[a]	46	days
	0.46	(54)		
	1.56	(0.3)		
^{65}Zn	0.32		245	days

[a] Percent of components of radiation.

In practice, the tissue sections may be placed in contact with the photographic emulsion by (1) covering with stripping film, (2) pouring a melted emulsion onto the glass slide over the specimen, (3) immersing the slides into a melted emulsion and (4) floating sections onto emulsion-coated glass slides. The third method has proved to be of great utility in the laboratory.

The sensitive emulsions consist of a gelatin matrix containing radiosensitive silver halide crystals (mostly silver bromide). The silver crystals contain either structural defects or silver sulfide impurities. When a β particle of sufficient energy impinges upon and interacts with a silver grain, a charge displacement occurs within its crystal lattice. The displaced electrons become attracted to the structural defects or to the silver sulfide impurity, creating a locus of negative charge. A movement of silver ions then occurs which results in the deposition of silver metal at these nodes, gradually producing a latent image. The latter may become exaggerated during the development process.

The factors which affect the quantitative measurement in autoradiography have been listed by Perry (1964) as the following: (1) the geometry of the specimen in relation to the photographic emulsion, (2) the density of the specimen, (3) the spectrum of the energy of the β-emissions, (4) the thickness and sensitivity of the emulsion, (5) the adjustment of the grain count so that it is high enough above the background for statistical accuracy but low enough to avoid coincidence of emission.

Electrons radiating from the β-emitters of high energy, e.g., [32]P (see Table 1.4), are scattered in all directions and are not attenuated to any extent by the thin tissue section. The scattering phenomenon can yield photographs in which the grains *do not* lie directly above the β-emitter in the specimen region but are skewed to one side. The spreading of the image is the principal observation in these poorly resolved samples. In the past, this lack of resolution was the factor responsible for the limited use of autoradiography. A major contribution to the solution of this problem has been afforded by the introduction of tritium, ^{3}H, an isotope of hydrogen with a mass of 3, a half-life of 12.3 years and which releases a β particle with an average energy of only 0.018 meV (see Table 1.4). Over 90% of the β particles of tritium are absorbed by 1.2 μ of a medium with the density of 1, while in an average photographic emulsion with a density of 3.5, 99% of the radiations are absorbed in 0.8 μ. Practically all the grains will lie within 1 μ of the tritium source, resulting in excellent resolution. Unfortunately, electron absorption now becomes a factor of significant magnitude. An emission from a point source may not reach the photographic emulsion at all, thus making quantitation more difficult. The thickness and density of the specimen and the proximity to the photographic emulsion markedly influence the observed grain density. One must also ensure that the number of grains per unit area over the specimen does not exceed levels where the probability of coincidence is great. The latter may be controlled by adjusting either the exposure time or concentration of the administered isotope.

An improved resolution may be obtained by decreasing the size of the silver bromide crystals in the photographic emulsion. Today, highly sensitive emulsions are available which make possible excellent resolution with great reproducibility. The resolution is also considerably improved by reducing the thickness of the specimen. In fact, the resolution may be enhanced so much that the preparation can no longer be seen with the light microscope and the electron microscope is required. The major disadvantage of autoradiography is the difficulty in the estimation of intermediate metabolic compounds, agents which may be water soluble and therefore extracted by the wash procedure. One must be aware, when comparisons between different tissues are sought, of the variable preexisting pools of intermediates which may be present in these different tissues.

A. Electron Microscope Radiography

The best resolution which has been obtained at the light microscope level has been approximately 1 μ. With the advent of techniques for the preparation of ultrathin sections of tissues for electron microscopic examination, a further improvement has been afforded; the superior morphology at the electron microscope level of these ultrathin sections is an added bonus.

Autoradiography in electron microscopy was first introduced in 1956 by Liquier-Milward although it was not until 1961 that Van Tubergen demonstrated the superior resolution which was possible with electron microscopic autoradiography of thin sections containing labeled thymidine. Electron microscopic autoradiography has been of considerable utility in the study of protein synthesis in cartilage and in DNA biosynthesis along various loci of the chromosomes. The technique has afforded a better opportunity for the localization and identification of the chronological sequence of events during any particular synthetic period. The specific applications of the technique will be presented in subsequent chapters.

VI. TISSUE CULTURE

Cell and tissue culture techniques have played a vital role in the elucidation of biochemical pathways operating within the cell and their localization within the cellular organelles. Cell culture has also proved invaluable to the virologist in uncovering the sequence of events that takes place subsequent to viral infection. In the following chapters we will have many occasions to note observations which have been made possible by the utilization of this technique.

Cell and tissue culture were natural outgrowths of techniques which were commonplace in embryology. The desire to foster the development of a cell or tissue culture arose from the realization, as stated by Claude Bernard in 1878, of the importance of the external environment upon the control of the cellular activities. The early attempts with this technique in the nineteenth century were plagued with failure and the inability to repeat the work of others. The principal reason for this inability was the choice of the medium in which the cell or tissues were to be suspended. Several outstanding contributions, however, were forthcoming during this time. Von Recklinghausen in 1866 was able to maintain amphibian blood cells alive in sterile containers for 35 days under a variety of conditions. In 1885, W. Roux was first to employ culture techniques on organized tissues, e.g., the neural plate of developing chick embryo.

The actual birth of tissue culture came in 1907 with the experimentations of R. Harrison, who demonstrated the unequivocal continuation of normal function of explanted pieces of the undifferentiated medullary tube region of the frog embryo when cultured in clots of frog lymph. Of even greater significance was the reproducibility of the method. Indeed, when maintained aseptically for several weeks, the formation of nerve fibers was shown to occur by the spinning of neuroblastic protoplasm into filaments at the free end of the preparation.

The major methodology employed in tissue and cell culture is illustrated in Fig. 1.23. One of the simplest methods is the embedding of small bits of tissue in a clot on a cover slip which is then inverted over a depression in a slide and then sealed with paraffin. The ease and convenience make this method ideal for short-term experiments involving large numbers of cultures. Variations have been introduced by the incorporation of various growth-stimulating substances into the medium and the use of frequent transfers. Carrel, who was instrumental in constructing precise techniques during 1914–1931, was able to propagate his strain for almost 40 years by the use of this method.

The double *cover slip method* devised by Maximow in 1925 is used when the tissue is to be kept alive for long periods without transfer. Where the culture growth and morphology are to be followed by an optical method, i.e., microscopy, special slides must be prepared for viewing the tissue, e.g., phase slides, perforated metal, or glass slides.

The *watch-glass technique* developed by Strangeways and Fell in 1926 has been most useful in organ culture. In this method the tissue is grown on a clot in a watch glass which is enveloped by moist cotton-wool in a petri dish. The cotton-wool is required for maintaining the humidity.

Several of the above-cited methods require specialized devices to contain the transplant. Hence, the trend was toward the development of methodology in which the more convenient test tubes could be utilized. Test tubes have been employed in growing large numbers of cultures where continuous monitoring of cellular growths by microscopic examination is not required. The tubes were either rotated at constant speed or kept stationary in a test tube rack. The test tube methods have been most useful in the maintenance of viruses or cells for biochemical study. When optical methods for morphological examination are required, the *flying cover slip* technique may be employed. In this method, the cultures are prepared on narrow cover slips and inserted into roller tubes.

The original *Carrel flask* was designed to allow for the easy maintenance of cultures for long periods without danger of bacterial contamination and to give cultures in which morphological examination could be conducted. Today, these vessels are used only sparingly to start the growth of a particular strain. Modifications of this vessel have been constructed to fulfill a particular need, e.g., the T flask of Earle was designed in 1947 to grow large quantities of cells. These elaborately constructed and expensive vessels have largely been replaced by the simpler Erlenmeyer flask or Roux bottles in which the cells may be grown easily in suspension and in large numbers.

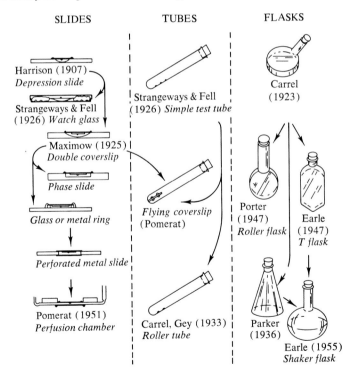

SLIDES   TUBES   FLASKS

Harrison (1907)
Depression slide

Strangeways & Fell
(1926) *Watch glass*

Strangeways & Fell
(1926) *Simple test tube*

Carrel
(1923)

Maximow (1925)
Double coverslip

Phase slide

Glass or metal ring

Flying coverslip
(Pomerat)

Porter
(1947)
Roller flask

Earle
(1947)
T flask

Perforated metal slide

Pomerat (1951)
Perfusion chamber

Carrel, Gey (1933)
Roller tube

Parker
(1936)

Earle (1955)
Shaker flask

Fig. 1.23. Techniques in cell and tissue culture. Vessels used for culturing cells and tissues. From J. Paul, "Cell and Tissue Culture." Williams & Wilkins, Baltimore, Maryland, 1965.

52   I. TECHNOLOGY OF CELL BIOCHEMISTRY

1. MEDIUM

By far one of the most important considerations in cell and tissue culture is the selection of a medium. The medium must provide the physiological state of pH and osmotic pressure and contain the nutrients necessary for the sustenance of the cells, i.e., the medium must provide the chemical substances which cannot be synthesized by the cells rapidly enough to keep pace with their growth.

Ideally, the chemical composition of a medium should be accurately defined, although in practice this has not been completely achieved. Animal cells require a completely *natural* medium or a synthetic medium which is supplemented with a natural product, e.g., serum. The synthetic medium generally consists of a balanced inorganic salt solution containing several essential organic substances. An example of such a medium, Eagle's solution, is presented in Table 1.5.

The natural materials which have been employed in culture to promote growth are: coagula, such as plasma clots; biological fluids such as serum, amniotic fluid, ascitic and pleural fluids or aqueous humor; tissue extracts, such as chick or bovine embryonic extracts. In addition, several supplements to the basal medium may be required, such as lactalbumin hydrolyzate, Difco-bactopeptone or dehydrated yeast extract.

2. CELLS

A wide variety of cells are presently commercially available, which range from primary cultures, i.e., cultures derived from an animal tissue within a short time after the death of the animal, to established cell lines, i.e., cells which have been maintained in tissue or cell culture for many generations. The dictates of the investigator are responsible for the selection of the type of cell.

VII. CONCLUSION

The methodology employed in cell biology has been briefly introduced in this chapter to impress upon the readers the debt of gratitude that is owed to the developers of these techniques and to lay the groundwork for the study of the individual cellular organelles to follow in the subsequent chapters. The discussion is by no means complete and additional methodology will be presented in the individual sections. It is impossible to discuss the methods for quantitative analysis, i.e., spectrophotometry, electrophoresis, X-ray diffraction, optical rotatory dispersion, magnetic spin resonance, etc., because of space limitations.

TABLE 1.5

Eagle's Medium, 1959[a]

	Concentration	
	Milligrams per 1000 ml	Approx. equiv. in millimoles
L-Arginine	105	0·6
L-Cystine	24	0·2
L-Histidine	31	0·2
L-Isoleucine	52	0·4
L-Leucine	52	0·4
L-Lysine	58	0·4
L-Methionine	15	0·1
L-Phenylalanine	32	0·2
L-Threonine	48	0·4
L-Tryptophan	10	0·05
L-Tyrosine	36	0·2
L-Valine	46	0·4
L-Glutamine	292	2·0
Choline	1	
Nicotinic acid	1	
Pantothenic acid	1	
Pyridoxal	1	
Riboflavine	0·1	
Thiamine	1	
i-Inositol	2	
Folic acid	1	
Glucose	2000	
NaCl	8000	
KCl	400	
$CaCl_2$	140	
$MgSO_4 \cdot 7H_2O$	100	
$MgCl_2 \cdot 6H_2O$	100	
$Na_2HPO_4 \cdot 2H_2O$	60	
KH_2PO_4	60	
$NaHCO_3$	350	
Phenol red	20	
Penicillin	0.5	

[a] This version is based on Hanks' BSS instead of on Earle's BSS.

SPECIFIC REFERENCES

THE CELL THEORY

Baker, J. R. (1952). *Quart. J. Microscop. Sci.* **93**, 157-190.
Watson, J. D. (1965). "Molecular Biology of the Gene." Benjamin, New York.

MICROSCOPY

Barer, P. (1964). *In* "Cytology and Cell Physiology" (G. H. Bourne, ed.), pp. 91-158. Academic Press, New York.
De Robertis, E. D. P., Nowinski, W. W., and Saez, F. A. (1965). "Cell Biology." Saunders, Philadelphia, Pennsylvania.
Price, G. R., and Schwartz, S. (1956). *In* "Physical Techniques in Biological Research," (G. Oster and A. W. Pollister, eds.), Vol. 3, pp. 91-148. Academic Press, New York.
Wyckoff, R. W. G. (1959). *In* "The Cell" (J. Brachet and A. E. Mirsky, eds.), pp. 1-20. Academic Press, New York.

CYTO- AND HISTOCHEMISTRY

Burstone, M. S. (1962). "Enzyme Histochemistry and Its Application in the Study of Neoplasms." Academic Press, New York.
Gersh, I. (1959). *In* "The Cell" (J. Brachet and A. E. Mirsky, eds.), Vol. 1, pp. 21-66. Academic Press, New York.
Glick, D. (1959). *In* "The Cell" (J. Brachet and A. E. Mirsky, eds.), Vol. 1, pp. 139-160. Academic Press, New York.
Pearse, A. G. E. (1960). "Histochemistry. Theoretical and Applied." Churchill, London.

ISOLATION OF SUBCELLULAR COMPONENTS

Anderson, N. G. (1955). *Exptl. Cell Res.* **9**, 446-459.
Anderson, N. G. (1956). *In* "Physical Techniques in Biological Research" (G. Oster and A. W. Pollister, eds.), Vol. 3, pp. 300-348. Academic Press, New York.
Brakke, M. K. (1951). *J. Am. Chem. Soc.* **73**, 1847-1848.
de Duve, C., and Berthet, J. (1954). *Intern. Rev. Cytol.* **3**, 225-275.
Moule, Y., and Chauveau, J. (1963). *In* "The Liver" C. Rouiller, ed.), pp. 379-449. Academic Press, New York.
Schneider, W. C., and Kuff, W. (1964). *In* "Cytology and Cell Physiology" (G. H. Bourne, ed.), pp. 19-90. Academic Press, New York.

AUTORADIOGRAPHY

Liquier-Milward, J. (1956). *Nature,* **177,** 619.
Perry, R. P. (1964). *Methods Cell Physiol.* **1,** 305-326.
Taylor, J. H. (1960). *Advan. Biol. Med.* **7,** 107-130.

TISSUE CULTURE

Parker, R. C. (1961). "Methods of Tissue Culture." Harper (Hoeber), New York.
Paul, J. (1965). "Cell and Tissue Culture." Williams & Wilkins, Baltimore, Maryland.

Chapter II

THE CELL MEMBRANE

I. INTRODUCTION

The classic studies of Oparin, Miller, and others have contributed to the generally accepted hypothesis that organic substances were formed on earth by abiogenic methods, before the advent of life. The reactions leading to the synthesis of molecular compounds (amino acids, ATP, NADH, etc.) probably occurred in a reducing environment, and in the oceans that served as a type of "nutrient broth." The emergence of "life," after prolonged evolution from the nutrient broth, must have involved a "simple" separation from the waters, by a type of barrier. Thus, the first primitive, structured element probably was a membrane. The development of such a system might not be particularly exceptional. Goldacre has shown, e.g., that under natural conditions, small, closed bladders of a lipoprotein composition can develop from folds produced by the wind in surface films on certain bodies of water.

After the reductive abiogenic synthesis of some relatively high molecular weight compounds occurred in the primeval seas, coacervate droplets could readily have formed. Bungenberg de Jong has demonstrated that in the presence of lipids, the surface of these droplets can assume a *protein-lipid membrane,* similar to a sandwich.

The capability of interacting with the surrounding medium, i.e., *selective* absorption of substances, was a requisite property of the newly formed membrane. These reactions required specific chemical or biophysical processes, which were probably derived from organic substances reaching the membrane from the external environment. The organic compounds in the broth probably first served as the main sources of energy for the continued operation of these processes. Under the conditions of

55

a reductive atmosphere, redox reactions, i.e., processes associated with the movements of hydrogen and electrons, were of initial significance. The early development of some type of catalyst or catalytic system to effect or accelerate the transfer of energy was also of fundamental importance. Thus, the first biologically active system must have been selectively separated from its environment in such a manner that an equilibrium would not develop, or else further evolution of the original system could not have taken place. A protein-lipid membrane would fulfill these requirements. Consequently, the single *structural* element that is characteristic of all living things is the membrane.

II. HISTORY

Development of techniques of cell staining, such as the introduction of osmium tetroxide and aniline dyes in 1865 by Schultz, was of prime importance in studies of the cell wall and membrane. In 1812, Moldenhauer described a technique for macerating tissues in water and, teasing the tissue into its component parts, examined them under the microscope. This method proved an important tool in the subsequent investigations of Nägeli and von Mohl on the cell wall. Nägeli proposed that a crystallization or precipitation of nitrogenous gum occurred out of the interstitial substance which separated preexisting cells. He referred to the contents of the plant slime as *"Schleimschicht."* Later studies established the identity of the latter as the substance of the plasma membrane.

Nägeli, from intricate studies of the lower plants (mosses, fungi, etc.), correctly described some of the important properties of the plasma membrane that later led Pfeiffer to his classic work on osmotic pressure. Nägeli also recognized that the proteinaceous cell sap, when placed in contact with water, hardened at the contact surface and formed a *protective membrane.* He was firmly convinced that this was true of both plant and animal cells, a point that was contested by most botanists at that time. He noted that the "wound heals itself" when the protoplasm is torn. A new protoplasmic surface, forming quickly around bits of protoplasm, became more dense and viscous than the original protoplasm itself. This Nägeli called *"Plasmamembran"* and found that it was impermeable to several plant pigments. He concluded that the *plasma membrane* endowed the cell with osmotic properties.

Although the monumental efforts of Nägeli clearly established the importance and primary significance of the plasma membrane of the cell, it was almost completely neglected until the microdissection studies of Chambers in the 1920's. Chambers injected small amounts of dilute acid into starfish eggs and observed cytolysis. When the acid was introduced into the medium, however, no obvious effect occurred. His experiments and those of his contemporaries definitively established the existence of a semipermeable membrane surrounding all living protoplasm.

III. THE NATURE OF THE MEMBRANE

At the turn of the nineteenth century and into the first decade of the 1900's, most investigators thought of the cell membrane as an ultrathin layer of lipid material. Since the membrane could not be observed under the light microscope, the width of the membrane could be no greater than between 1200 Å and 1700 Å nor less than 5 Å, the approximate dimensions of a fatty acid molecule. Gough (1924) in a classic paper, analyzed the lipids of erythrocytes, calculated the thickness to be between 20 – 30 Å, and reasoned that the membrane consists of a single layer. Gorter and Grendel in 1925, continuing these studies, extracted the lipids from erythrocytes with acetone, evaporated the residue and then spread it on a Langmuir trough as a unimolecular film (Fig. 2.1). The individual mole-

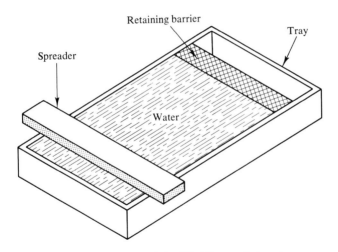

Retaining barrier

Tray

Spreader

Water

Fig. 2.1. Diagram of simplified Langmuir trough.

cules were packed in a coherent film, with the polar, i.e., water-soluble ends of the molecules facing or "attached" to the water and the nonpolar, or hydrocarbon chains pointing outward (Fig. 2.2). Gorter and Grendel measured the dimensions of the erythrocyte and found that the ratio of the *area of lipid* to the *area of cell* was approximately 2, and concluded that the cell membrane consists of a *bimolecular leaflet,* or two-molecular array of lipid (Fig. 2.3).

The structural requirements necessary to invest a lipoidal membrane with *permeability* characteristics, i.e., the ability for this "cellular coating" to permit certain *molecules* to pass into or from the environment, prompted investigators in the early 1930's to search for nonlipid material as an integral part of the membrane. Harvey and Shapiro, among the first to obtain evidence suggesting the presence of a nonlipoidal component, observed that the surface tension at the interface between a single oil drop and the protoplasm of a mackerel egg was much lower than any nonliving oil-water system. Therefore, either the egg oil contained some very active surface substance which was not present in the body oil, or else some sub-

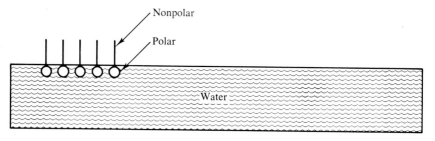

Fig. 2.2. Diagram of unimolecular layer of erythrocyte lipid.

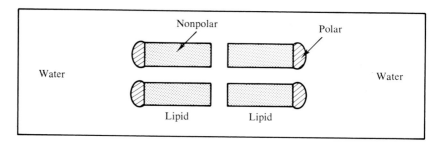

Fig. 2.3. Diagram of early cell membrane concept.

stance in the protoplasm had the capability of greatly reducing the tension at the surface of the oil. To resolve this question, Danielli and Harvey (1934) undertook a detailed investigation of the nature of the substance in the protoplasm or in the egg oil. They mixed the egg material with an oily compound (brom-benzine) and found that the surface tension dropped precipitously (Fig. 2.4). Vigorous shaking "denatured" the egg compound and the surface tension returned (Fig. 2.5). The unknown material was therefore a protein and these investigators concluded that the cell membrane was not purely lipoidal in nature but was probably a lipo-

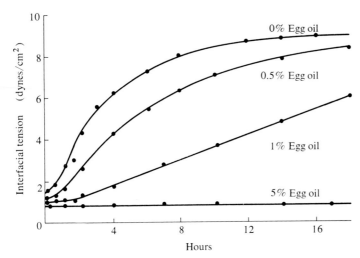

Fig. 2.4 The change of surface tension with time modified by different concentrations of egg oil in presence of brom-benzine. Taken from J. F. Danielli and E. N. Harvey, *J. Cellular Comp. Physiol.* **5**, 483 (1934).

protein, with a specific molecular arrangement (Fig. 2.6). Since it was known that proteins exist in solution in *spherical units*, Danielli and Harvey suggested that the globular proteins (Fig. 2.6) were hydrated and were in juxtaposition to the polar ends of the hydrocarbon molecules, thus separating the oily phase from the aqueous phase of the protoplasm.

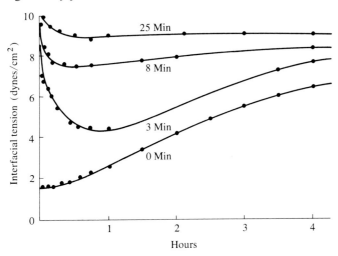

Fig. 2.5 Effect on surface tension of shaking brom-benzine with egg for various periods (minutes). Taken from J. F. Danielli and E. N. Harvey, *J. Cellular Comp. Physiol.* **5**, 483 (1934).

The major factor which determines whether or not a protein material will denature at an oil-water interface is the magnitude of the interfacial tension. When the surface tension reaches a high value, the protein units are pulled out from their spherical structure into a sheetlike structure, thus effecting "denaturation." Using this information, Danielli and Davson (1934) described the basic architecture of a membrane shown in Fig. 2.7. Note that both the exterior and the interior of the cell are delimited by a protein-lipid combination structure. According to this "paucimolecular" scheme, a lipoid film occurring on the surface of the cell would have a layer of protein molecules at least one molecule thick absorbed upon it from the interior of the cell. The protein aspects of the film would be capable of some type of selective permeability toward molecules of different sizes. Anions would pass freely on the acid side of the isoelectric

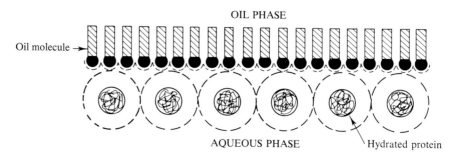

OIL PHASE

Oil molecule →

AQUEOUS PHASE   Hydrated protein

Fig. 2.6. Schematic representation of brom-benzine-aqueous egg contents. From J. F. Danielli and E. N. Harvey, *J. Cellular Comp. Physiol.* **5**, 483 (1934).

point of the protein film while cations would diffuse only on the basic side of the protein.

> Clearly, then, a protein-lipoid film of the type postulated could provide a pore-type structure which would distinguish between molecules of different size (and ions of different charge). [Danielli and Davson (1934).]

This hypothetical picture of the cell membrane remained only a suggestion until the elegant studies of the electron microscopists during the 1950's. Of particular importance are those of Robertson, who in 1958 described the unit-membrane hypothesis which confirmed the *paucimolecular theory* of Davson and Danielli, at least for myelin (see below) and perhaps for the erythrocyte (Robertson, 1959). Thus, the cell membrane, according to Robertson, is a trilamellar structure approximately 75 Å units thick, consisting of two dense lines (osmiophilic, i.e., stains with osmium), 20−25 Å units thick and a light central zone (osmiophobic), approximately 20−25 Å units thick. Until a few years ago (1964), this structure was thought to be a characteristic of all plasma membranes of different types and of all subcellular structures which are delimited by membranes. These include mitochondria, endoplasmic or sarcoplasmic reticulum, lysosomes, Golgi bodies, the nucleus, and possibly other unknown organelles. In fact, it had been postulated by Robertson that an interconnection exists among some or all of the

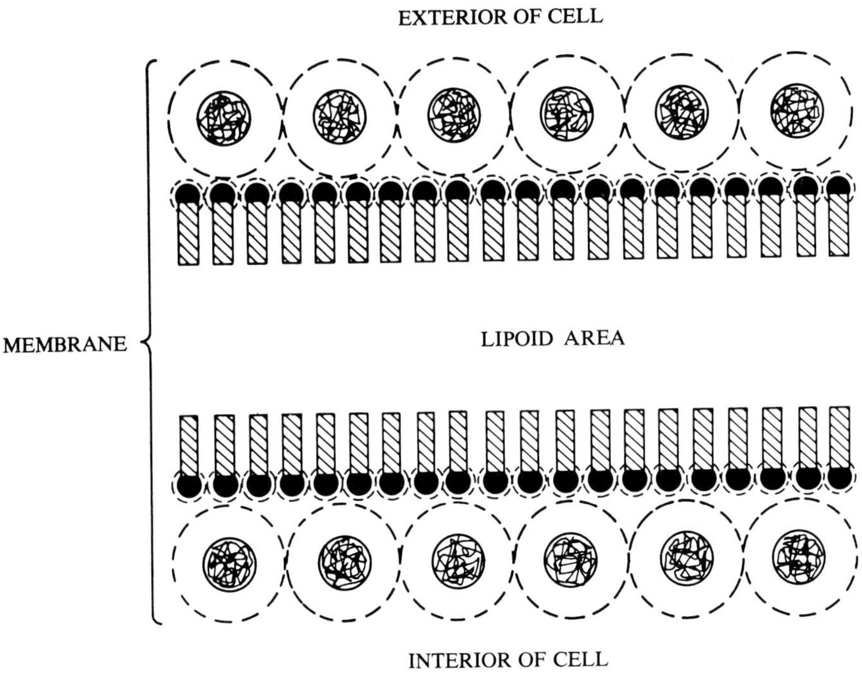

EXTERIOR OF CELL

MEMBRANE

LIPOID AREA

INTERIOR OF CELL

Fig. 2.7. The "pauci-molecular" cell membrane structure. Taken from J. F. Danielli and H. Davson, *J. Cellular Comp. Physiol.* **5**, 495 (1934).

cellular and subcellular membranes as illustrated in Fig. 2.8A. Robertson (1959) originally regarded the cell as a complex aggregate of membranous evaginations and folds, possibly originating or extending from the perinuclear membrane. Thus, in the secretory cell, e.g., secretion might involve the passage of molecules through the cell membrane and thence along specialized membrane-lined conduits to all portions of the interior of the cell. If this is true, pinocytosis and phagocytosis (see pp. 85-87) may not involve the passage or transfer of large aggregates directly out of or into the cytoplasm by means of a "pinching off" or "cytopempsis" of the mem-

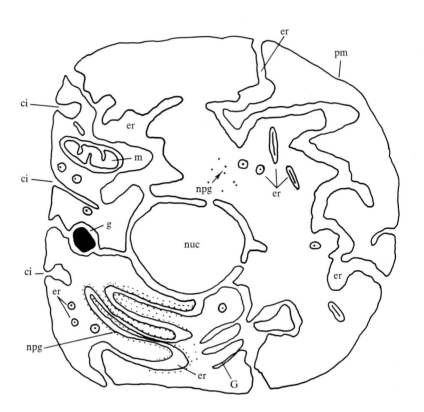

Fig. 2.8A. Diagram of a hypothetical cell (Robertson's original view) illustrating relationships of the cell membrane to various cell organelles. The cell membrane while not shown, exists as a pair of dense lines separated by a light interzone (see p. 60). The invaginations of the cell surface known as *caveolae intracellularis* (ci) are indicated in several areas. Some of these extend for a considerable distance into the cell and they may connect with the endoplasmic reticulum (er). The nuclear membrane is composed of flattened sacs of the endoplasmic reticulum, and by means of the nuclear pores nucleoplasm (nuc) may be in continuity with cytoplasm. The Golgi apparatus (G) is here shown as a modified component of the endoplasmic reticulum. Secretion granules (g) are shown as dense aggregates contained within membranes of the endoplasmic reticulum. Nucleoprotein granules (npg) are shown scattered through the cytoplasm and in some regions attached to the cytoplasmic surfaces of membranes of the endoplasmic reticulum. In some regions the endoplasmic reticulum is shown as tubules, either in longitudinal section or cross-section. It is not clear on present evidence how many of these round membranes are transected tubules and how many, if any, represent isolated vesicles. One mitochondrion (m) is shown with its cristae formed by invagination of its inner membrane. Taken from J. D. Robertson, *Biochem. Soc. Symp.* (*Cambridge, Engl.*) **16**, 3 (1959).

brane. The molecules instead may enter at specific loci in the plasma membrane and aggregate into "granules" which then are transferred via the endoplasmic (or sarcoplasmic in the case of muscle) reticulum. The evidence that the endoplasmic reticulum is a continuum of the cell membrane is compelling in the case of muscle cells (see Chapter IX), but less so at present with regard to secretory or other cell types.

Thus all membranes *may* possess a common *basic* structure and possibly origin. This does not mean, however, that the plasma membrane as well as the membranes of various subcellular systems have not evolved both morphological and functional specificity (e.g., the liver smooth endoplasmic reticulum possesses an NADPH-oxidase system specialized for drug and steroid metabolism; the mitochondrial cristae are specialized for energy liberation). Table 2.1 lists, for example, several types of membranes exhibiting differences in lipid content. Therefore, while the origin and basic structure of all membranes *may* be similar, distinct structural and *chemical* differences have evolved which may confer organ and tissue specificity.

A. Criticism of the Unit-Membrane Theory

Quite recently it has become obvious that different membranes exhibit a wide spectrum of variation in chemical content, enzymic activity, permeability characteristics, metabolism (e.g., synthesis and catabolism of membrane) and function. Furthermore, significant variations in fine structure and in dimensions of the lipoprotein layer have been observed.

If all membranes were in fact similar, the logical expectation would be that the protein-to-lipid ratios might be similar. This does not appear to

TABLE 2.1

Differences in Lipid Content Among
Various Membranes[a]

Membrane	Phospholipid (mg/100 mg protein)	Sterol content (moles/mole phospholipid)
Red blood cell	16.4	0.89
Liver plasma membrane	41.9	0.26
Muscle sarcolemma	27.8	0.24
Brain mitochondria	45.8	0.51
Liver mitochondria	28.2	0.11
Muscle mitochondria	11.0	0.15
Liver endoplasmic reticulum	30.1	0.27

[a]Adapted from L. A. E. Ashworth and C. Green, *Science* **151**, 210 (1966).

be true, however (Table 2.2). Furthermore, distinct differences in the surface area occupied by monomolecular films of protein and lipid components of the so-called unit membranes, have been found.

The argument against the universality of the Robertson model gains strength when chemical data are analyzed. While the major phospholipids of myelin are the *cerebrosides,* these are completely absent in other membranes, either animal or bacterial (Table 2.3). The neutral lipids of most membranes (plasma) are made up of cholesterol and cholesterol esters; however, bacterial plasma membranes and mitochondrial membranes have little or none of this lipid. Sphingomyelin is a most important constituent of the erythrocyte plasma membrane and of myelin (to a lesser

TABLE 2.2
Protein and Lipid Ratios of Animal
and Bacterial Membranes[a]

Membrane	Protein/lipid ratio
Myelin	0.43
Erythrocyte	2.5
Bacillus megaterium	5.4
Streptococcus faecalis	3.4
Mycoplasma laidlawii	4.1

[a]Data from E. D. Korn, *Science* **153**, 1491 (1966).

extent), but it is not present in mitochondrial or bacterial membranes or in endoplasmic reticulum. Bacterial membranes contain no steroids. The lipid composition of bacterial membranes in general, is very limited (refer to Table 2.3 for data). Moreover, fatty acid composition of individual phospholipids of different membranes varies widely. Gram-negative bacteria contain mostly saturated and monounsaturated C_{16} and C_{18} fatty acids. No polyunsaturated fatty acids are found.

There are a minimum of seven classes of phospholipids. Each class contains members with hydrocarbon chains of different length and numbers of double bonds. The multiplicity of phospholipid structures may be important in defining different types of membrane functions.

Sjöstrand (1963) has presented clear electron microscopic data (using mouse kidney and pancreas fixed with osmium and $KMnO_4$) on the differences among plasma membranes, mitochondrial membranes, and smooth endoplasmic reticulum. The dimensions appear to vary considerably, from 50 to 60 Å for mitochondrial inner or outer membrane and 90 to 100 Å for plasma membrane. Furthermore, the ultrastructure of at least one of the mitochondrial layers may consist of *globular subunits* rather than "unit-membranes." Robertson in fact has also observed the subunit structure but has interpreted them as optical artifacts. Two recent reports lend strong evidence in favor of a membrane sub-unit "micellar" structure for at least some biological membranes. In one case, the outer segment membranes of the frog retina were viewed as both unfixed, unstained pellets and as negatively stained preparations. Both showed a square array of spherical particles. The unit cell size was 70 Å and the particles had a nonpolar core about 40 Å in diameter (Blasie

TABLE 2.3

Differences in Specific Phospholipids and
Galactolipids in Membranes (percent of total)[a]

	Myelin[b]	RBC[b]	E. coli[b]	Endoplasmic[c] reticulum	Mitochondria[c]
Cholesterol	25	25	0	6	5
Phosphatidylserine	7	11	0	0	0
Phosphatidylinositol	11	23	0	64	48
Sphingomyelin	6	18	0	0	0
Cerebroside (galactolipid)	21	0	0	0	0
Cephalins	18	21	—	21	25
Lecithins	10	33	0	42	40

[a]E. D. Korn, *Science* **153**, 1491 (1966) and D. F. Parsons, *Proc. 7th Can. Cancer Res. Conf., Ontario, 1966* p. 193 (1967).
[b]Represent plasma membranes.
[c]Represent subcellular membranes.

et al., 1965). In the other case, a newer technique involving quick frozen-etched specimens which purportedly afford better preservation, revealed that most cellular membranes may be arranged in part as an extended bilayer and partly as globular or micellar subunits. The freeze-etching splits membranes and exposes the inner membrane faces. Using this technique, Branton found that myelinated membranes are unique in that the fracture faces appear smooth and devoid of subunit particles whereas other membranes reveal the characteristic micellar appearance (Branton, 1966a, b).

Even among membranes which appear to have similarity in substructure there are specific differences both in function and in chemical content. This is indicated in Table 2.4, in which the inner and outer membranes of mitochondria are compared with the smooth endoplasmic reticulum.

TABLE 2.4
Some Structural and Functional Characteristics
of Mitochondrial Membranes and Endoplasmic Reticulum
("Smooth" Microsomes)[a]

	Mitochondria		Smooth endoplasmic reticulum
	Inner membrane	Outer membrane	
Thickness (Å)	55	55	55
"Ultrastructure"	Globular with project-ing subunits (IMS)[b]	Globular without IMS	Globular without IMS
Density	1.21	1.13	1.13
Cardiolipin (% of total lipid)	21.5	3.2	0.5
Monoamine oxidase	0	Present	0
Cytochrome oxidase (μmole/gm)	0.24	0	0
Cytochrome b_5 (μmole/gm)	0.17	0.51	0.79
Permeability	Small molecules	Large molecules	Small molecules
Osmotic response	Responds	None	Slight
Response to ATP	Contracts or swells	No effect	No effect

[a]Data from D. F. Parsons, *Proc. 7th Can. Cancer Res. Conf., Ontario, 1966* p. 193 (1967), and C. A. Schnaitman and J. W. Greenawalt, *J. Cell Biol.* **31**, 100A (1966); C. A. Schnaitman, V. G. Erwin, and J. W. Greenawalt, *J. Cell Biol.* **32**, 719 (1967).
[b]IMS = inner membrane subunit.

It is perhaps not surprising that that variation in chemical composition and structure in membranes exists when one considers the numerous different functions of membranes. Some of these are outlined in Table 2.5. Myelin, however, appears to have only *one* biological function, that of an electrical insulator. It is a multilayered structure which surrounds axons and appear to consist of a proliferation of the plasma membrane of the Schwann cell that surrounds axons. Most of the evidence supporting the unit membrane theory was derived from biophysical studies of myelin and of lipid and protein model systems. A more modern depiction of the generalized cell is shown in Fig. 2.8B.

B. The "Subunit" Concept

As suggested in the previous section, most membranes may consist of repeating subunits, which possess the machinery for function. Each type of membrane would be made up of these highly structured particles, arranged in compact, multi-enzymed "containers." Most of the evidence for this has been derived from studies of the inner mitochondrial membrane, where discrete so-called "elementary particles" (referred to by Green and Fernández-Móran) or inner membrane subunits (IMS) (Parsons and Chance) have been observed (Fig. 2.9). It was first thought that the elementary particles, which were revealed by negative staining techniques, consisted of three parts: a spherical or polyhedral head piece, 80 to 100 Å in diameter; a cylindrical stalk, 45 − 50 Å long and 30 − 40 Å in width; and a base, 40 × 100 Å. The subunit along with "structural protein" appeared to define the membrane. As shown in Fig. 2.10, the *original* concept of the mitochondrial subunit suggested that the entire electron transport enzyme complex was structurally located within the unit. However, while the subunits are localized to membranes possessing electron transport and phosphorylation, the complexes (Fig. 2.9) are not situated as originally pictured. The spherical knoblike structure has now been identified as an oligomycin-sensitive ATPase of mitochondria and may represent the terminal step of phosphorylation. This will be discussed in greater detail in Chapter III.

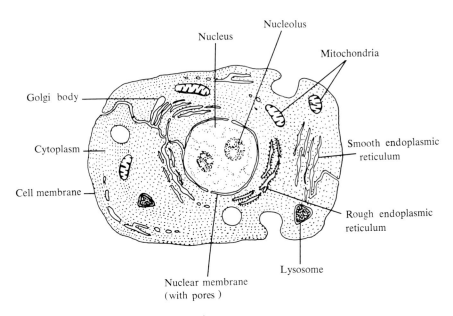

Fig. 2.8B. A more modern generalized cell. *Note:* There is no clear evidence at this time that the nuclear membrane connects with the extracellular space.

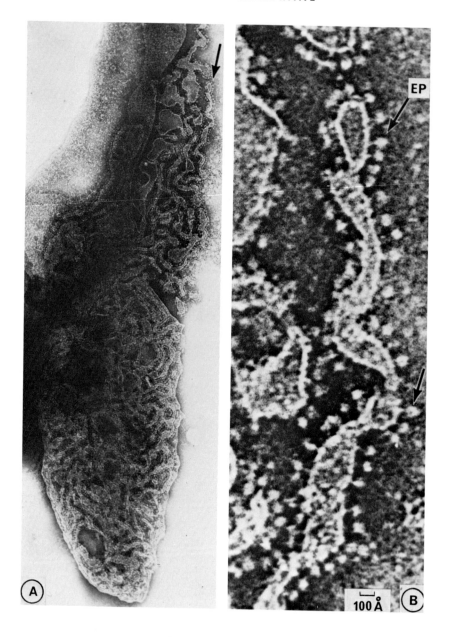

Fig. 2.9. Elementary particles of mitochondria. (A) Negatively stained beef heart mito-chondrion; isolated in 0.5 M sucrose, and prepared by surface spreading on 1% potassium phosphotungstate at pH 7.2 without prior fixation. Partly intact whole mount of flattened mitochondrial membranes showing regular arrangement of repeating particulate com-ponents. × 62,000. (B) Profile view of cristae with arrays of elementary particles (EP) in enlarged segment of (A). Polyhedral head pieces of the elementary particles are attached

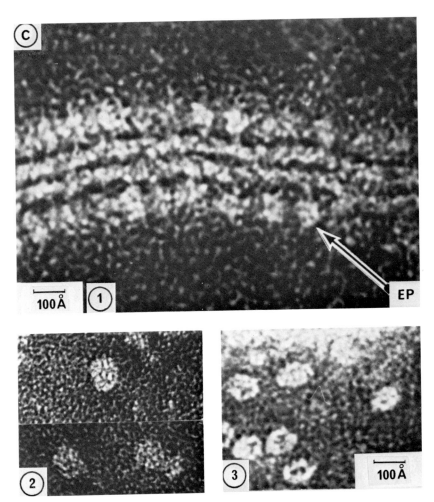

by stalks to the continuous dense outer layers of the cristae. × 420,000. From H. Fernán-dez-Morán, T. Oda, P. V. Blair, and D. E. Green, *J. Cell Biol.* 22, 63 (1964). (C) High resolution electron micrographs of negatively stained: (1) Membrane segment (cristae) from isolated beef heart mitochondria with paired arrays of "elementary particles" (EP) (showing dense substructure) attached to central membrane layer; (2) isolated particles that contain complete electron transfer chain; (3) reconstituted electron transfer particles. × 1,000,000 (indicated by scale of 100 Å). Courtesy of Dr. H. Fernández-Morán, Univ. of Chicago.

Complex	Constituent	Approx. Mol. Wgt.
I	NADH-coenzyme Q reductase	550,000
II	Succinic-coenzyme Q reductase	208,000
III	Coenzyme Q (reduced)-cytochrome c reductase	200,000
IV	Cytochrome oxidase	426,000
	Total:	1,384,000

TABLE 2.5

Multiple Functions of Membranes[a]

Function	Membrane
1. General and selective diffusion small molecules and ions	
2. Active transport	
3. Control influx and efflux of ions and substrates products between cell compartments	All
4. Phagocytosis, pinocytosis	
5. Cell adhesion and mobility	
6. Carry surface antigens	Plasma membrane
7. Limit organ growth	
8. Electrical insulator	Myelin
9. Generate nervous impulses	Nerve plasma membrane
10. Conduct nervous impulses	Sarcoplasmic reticulum
11. Convert light to electrical impulses	
12. Convert light to phosphate bond energy in ATP	Retinal rod disc membrane
13. Convert oxidation chemical energy to phosphate bond energy in ATP	Chloroplast membrane Inner membrane of mitochondria
14. Move secretory products to outside of cell	Endoplasmic reticulum and Golgi membranes
15. Wall off autolytic enzymes	Lysosomal membrane

[a]Taken from D. F. Parsons, *Proc. 7th Can. Cancer Res. Conf., Ontario, 1966* p. 193 (1967).

Subunit membrane structures have also been isolated or visualized from chloroplasts, from "microsomal" membranes (Park and Pon, 1963; Green and Perdue, 1966), and from plasma membranes (Fig. 2.11). Green and others have suggested that most membranes may be built of repeating subunits consisting of structural protein[1] associated with phospholipids and other compounds. Green has in fact defined membranes as "vesicular or tubular systems with a continuum being made up of nesting,

[1]This represents an inert insoluble protein which presumably forms the "backbone" for the membrane. According to Green, the molecular weight is approximately 22,500 and it is insoluble at neutral pH. One mole of structural protein binds with one mole of each of the cytochromes. Similar to the structural protein of tobacco mosaic virus, the N-terminal amino acid of the mitochondrial structural protein appears to be N-acetyl serine. It polymerizes readily and complexes strongly with phospholipid. Addition of small amounts of phospholipid to isolated structural protein causes the latter to aggregate into two-dimensional sheets and vesicles. There is some evidence that this protein comprises membranes other than mitochondria (myelin, erythrocyte ghosts, liver plasma membranes, chloroplasts, "microsomes").

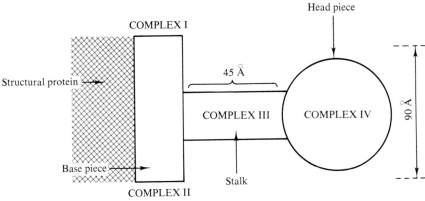

Fig. 2.10. The original membrane elementary particle of Green (This attractive postulation is no longer completely consistent with the data; see text.) Green has now modified his concept, suggesting that each "base piece" contains *one* of the four complexes (see above). This is consistent with the "oxysome" hypothesis discussed Chapter III.

lipoprotein-repeating units." Each membrane would possess its own specific structural unit as pictured below:

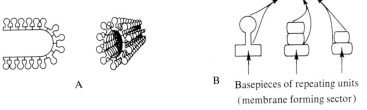

SCHEME 1. (A) Representation of a membrane as a fused continuum of repeating particles. (B) Diagrammatic representation of different possible forms of repeating units.

Thus, newer techniques of microscopy, e.g., negative staining electron microscopy and X-ray diffraction analysis, along with chemical studies, have revealed information which suggests the necessity for a revision of the original Davson-Danielli model. The main structural element of most membranes therefore may not be lipid but rather protein. There is no question, however, that the variation and complexity of the phospholipid component of membranes have direct bearing on the specific function. Recently, selective ion transport has been demonstrated in simple "artificial" phospholipid micelles (Bangham *et al.*, 1965; Mueller and Rudin, 1967). The "membrane" is formed by the addition of sphingomyelin to a mixture of α-tocopherol, chloroform, and methanol. The addition of minute amounts of a proteinaceous material from *Aerobacter cloacae* effects a measurable resting potential (about -50 mV). The subsequent addition of protamine sulfate causes fully developed action potentials.

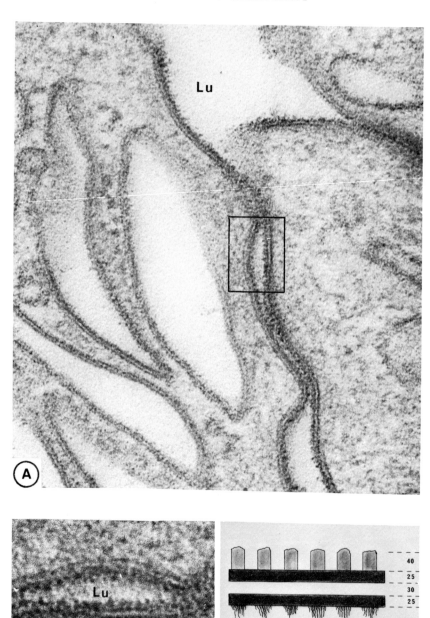

IV. CHEMICAL CONSTITUENTS OF MEMBRANES

With the development of more sophisticated biochemical and electron microscopic techniques, information is being derived which clearly indicates that the cell and subcellular membrane are multiphasic, biologically active, chemical structures, which are intimately linked with the physiological and biochemical processes of the entire cell. Some of the more important constituents of the membranes from various cell types are presented in Tables 2.6 − 2.10 and in Scheme 2. The chemical complexity of the cell membrane illustrates the functional importance of this dynamic structure. Particularly characteristic of all cell membranes is the structurally oriented ouabain-sensitive ATPase system, associated with active transport mechanisms. This is more fully discussed in Chapter VII. Associated with this important enzyme complex are possible phosphatido-peptide intermediate compounds.

The gangliosides are another class of specifically oriented compounds present in abundance in brain and liver but probably characteristic of most cell membranes.

A. Gangliosides

The gangliosides are classified as glycosphingolipids, containing the alkaloid sphingosine, saturated fatty acids, glucose, galactose, N-acetylgalactosamine and N-acetylneuraminic acid (sialic acid) in a definitive structural organization, pictured in Fig. 2.12. The fatty acids consist mainly of stearic with lesser amounts of palmitic and arachidic acids. Four or five major components and several minor ones contain the "prosthetic group" of the molecule, N-acetylneuraminic acid, ubiquitously distributed in nature. The latter was first isolated by Blix in 1936 from bovine submaxillary gland, hence the name *sialic acid* (Fig. 2.13A). The acid is usually found complexed with hexoses or N-acetylhexosamines of gly-

---

Fig. 2.11. A unit-membrane showing "elementary-like particles." (A) "Normal" urinary bladder epithelial cell showing the plasma membrane along free surface and invaginations. Unit membrane structure is evident throughout but is best resolved where the membrane is oriented normal to the section (insert). × 250,000. (B) Area of insert, shown in (A). Granular subunits are periodically spaced along the luminal aspect of the outer lamella of the unit membrane. The three-layered unit membrane is approximately 80 Å wide. The granular component noted in relation to the outer leaflet adds 40 Å to the width of the plasma membrane for a total width of 120 Å. (C) Diagram illustrating a conceptual and highly speculative reconstruction of the fine structure in this plasma membrane. Courtesy of Dr. J. J. Ghidoni, Department of Pathology, Baylor University College of Medicine, Houston, Texas.

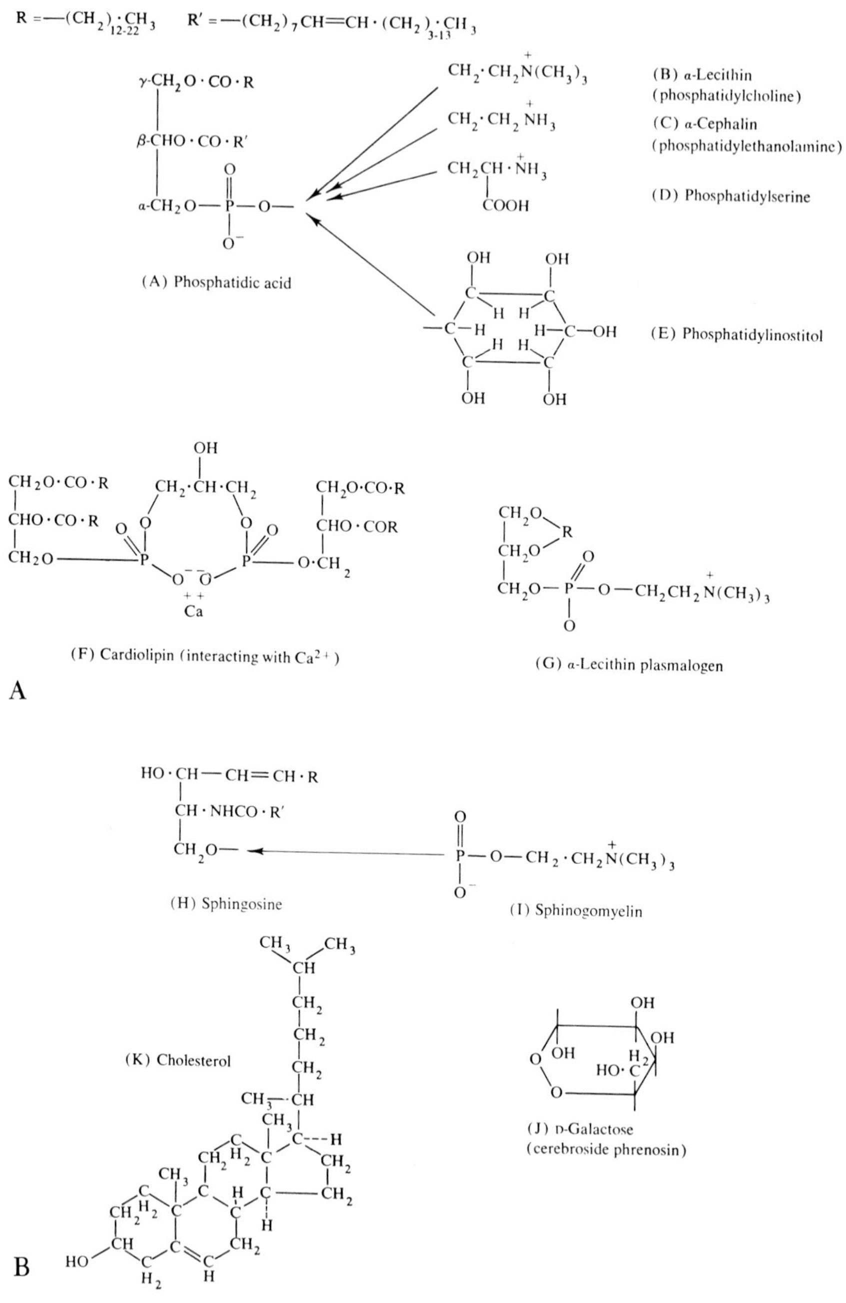

$R = -(CH_2)_{12-22} \cdot CH_3$ $R' = -(CH_2)_7 CH = CH \cdot (CH_2)_{3-13} \cdot CH_3$

$\gamma\text{-}CH_2 O \cdot CO \cdot R$

$\beta\text{-}CHO \cdot CO \cdot R'$

$\alpha\text{-}CH_2 O - \overset{O}{\underset{O^-}{P}} - O -$

(A) Phosphatidic acid

$CH_2 \cdot CH_2 \overset{+}{N}(CH_3)_3$ — (B) α-Lecithin (phosphatidylcholine)

$CH_2 \cdot CH_2 \overset{+}{N}H_3$ — (C) α-Cephalin (phosphatidylethanolamine)

$CH_2 CH \cdot \overset{+}{N}H_3$
$\quad\quad |$
$\quad\quad COOH$ — (D) Phosphatidylserine

(E) Phosphatidylinostitol

$CH_2O \cdot CO \cdot R$
$CHO \cdot CO \cdot R$
$CH_2O -$

$CH_2 \cdot \overset{OH}{\underset{}{CH}} \cdot CH_2$

$CH_2O \cdot CO \cdot R$
$CHO \cdot COR$
$O \cdot CH_2$

Ca^{++}

(F) Cardiolipin (interacting with Ca^{2+})

CH_2O
CH_2O $\searrow R$
$CH_2O - \overset{O}{\underset{O}{P}} - O - CH_2 CH_2 \overset{+}{N}(CH_3)_3$

(G) α-Lecithin plasmalogen

A

$HO \cdot CH - CH = CH \cdot R$
$\quad\quad |$
$\quad\quad CH \cdot NHCO \cdot R'$
$\quad\quad |$
$\quad\quad CH_2O -$

(H) Sphingosine

$\overset{O}{\underset{O^-}{P}} - O - CH_2 \cdot CH_2 \overset{+}{N}(CH_3)_3$

(I) Sphinogomyelin

(K) Cholesterol

(J) D-Galactose (cerebroside phrenosin)

B   HO

SCHEME 2. (A) Membrane lipids (phosphatidic acid derivatives). (B) Membrane lipids (sphingosine and cholesterol derivatives).

TABLE 2.6
Some Important Constituents of Cerebral Cortical Membranes[a]

Water	81
Protein (including phosphoprotein and ouabain-sensitive ATPase)	10
Cholesterol	1
Phospholipids (lecithin, phosphatidylserine, phosphoinositides, sphingomyelin)	4
Gangliosides	2

[a]Taken from H. McIlwain, "Chemical Exploration of the Brain." Elsevier, Amsterdam, 1963.

TABLE 2.7
Some Important Constituents of Protoplast Membranes[a]

Protein (including phosphoprotein and ouabain-sensitive ATPase)	68
Lipids	18.5
Carbohydrates	10
RNA	1
Hexosamine	1
Cytochrome c	Small amount

[a]Expressed as percent of dry weight of *Bacillus megaterium*.

TABLE 2.8
Chemical Composition of Isolated Liver Plasma Membranes[a]

Mg protein	2.6
Phospholipid phosphate	0.9
Phospholipid choline	0.5
Cholesterol	0.4
Hexosamine	0.2
Sialic acid (gangliosides)	0.03

[a]Taken from P. Emmelot, C. J. Bos, E. L. Benedetti, and Ph. Rümke, *Biochim. Biophys. Acta* **90**, 126 (1964b). Expressed as amount per micromole of membrane-bound phosphate.

coproteins and other macromolecules and imparts the acidic nature to the molecule ($pK = 2.6$) (Fig. 2.13B).

The gangliosides are found in membranous structures and are capable of complexing with cationic compounds, particularly basic proteins or other macromolecules. Both the lipid and the sialic acid moieties of the molecule are of importance in the binding process. At one time the gangliosides were considered high molecular weight polymers, e.g., 100,000 to 250,000. However, this was probably due to a micellar aggregation; the minimum molecular weight is probably much lower and may even be in the 1,000 – 3,000 range.

A number of different types of gangliosides are presently recognized and more are being discovered. The structures may vary slightly and attention is being directed toward the possible relationship between structural differences and disease states.

In the biological construction of sialic acid, glucose is an important source of the carbon chains. The incorporation of sialic acid into macromolecules, e.g., ganglioside, is thought to involve intermediate acetylneuraminic acid substances of high metabolic activity. The metabolism of the formed ganglioside is apparently slow, at least in brain, where neuraminidase activity is less than 1 μmole sialic acid liberated per gram tissue per hour. The function(s) of gangliosides in membranes therefore

TABLE 2.9

Qualitative Enzyme Content of Liver Plasma Membranes[a]

Mg^{2+} (or Ca^{2+})-ATPase
Ouabain-sensitive ATPase (Na$^+$, K$^+$-ATPase)
5'-Mononucleotidase
Glucose-6-phosphatase
NADH-cytochrome c reductase
Esterase (nonspecific)
RNase (alkaline and acidic)
NAD-pyrophosphatase
Acid phosphatase-(nitrophenylphosphate)
Acid phosphodiesterase
Alkaline phosphodiesterase
Acid phosphophosphatase-glycerolphosphate
DPNase
NADase
p-Nitrophenylphosphatase (Mg-dependent and Mg-independent alkaline and a "neutral" activity; also an acid and a K$^+$-activated alkaline and neutral activity)
Acetylphosphatase (neutral pH)—also a K$^+$-activated activity
Ribonuclease
Leucyl-β-naphthylamidase (in $5-60$ Å globular knobs)
Inosine diphosphatase
NAD-nucleosidase
Adenyl cyclase
ATP-pyrophosphatase (alkaline)
Acetylcholinesterase
Lipase
Phosphatidylinositol kinase

[a]Taken from P. Emmelot, E. L. Benedetti, and Ph. Rümke, *Symp. Electron Microscopy, Modena, Italy, 1963* p. 253 (1964a), and P. Emmelot, personal communication (1968).

TABLE 2.10

Some Important Constituents of the Human Erythrocyte Membrane[a]

Proteins	Lipids
Phosphoprotein	Free cholesterol
Stromatin	Phosphatidylcholine
Elinin (lipoprotein)	Phosphatidylethanolamine
Acetylcholinesterase	Phosphatidylserine
Ouabain-sensitive ATPase	Sphingomyelin
	Phosphatidylinositol
	Phosphatidic acid
	Cholesterol esters

[a] The red blood cell membrane = 1 – 4% dry weight of the whole red blood cell. The lipid to protein ratio is 1.6:1.8.

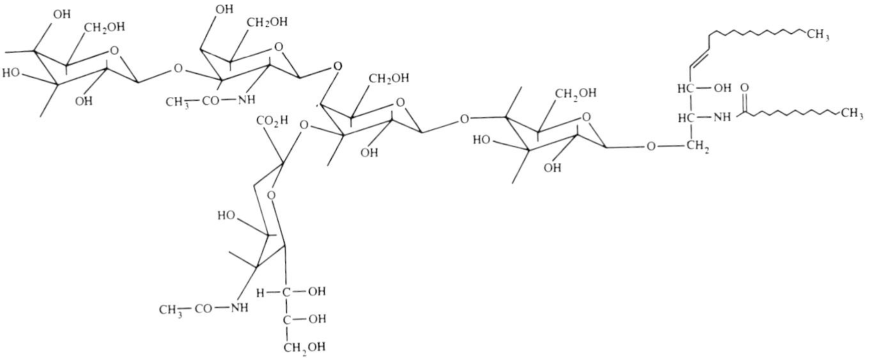

Fig. 2.12. Proposed structure of a monosialoganglioside. Taken from R. Kuhn and H. Wiegandt, *Chem. Ber.* 96, 866 (1963).

probably is not associated with the continual synthesis and destruction of acidic groups, but rather with the entire formed structure itself.

These compounds have been implicated in ion transport mechanisms, particularly in neural tissues where the content is quite high (Table 2.11). McIlwain has suggested that membrane "pores" may be lined with electronegative charges conferred by sialic acid compound molecules (see Chapter VII).

N-Acetylneuraminic acid (NANA)-containing glycoprotein is abundant on the erythrocyte surface, conferring an overall negative charge. Incubation with neuraminidase specifically removes NANA from the erythrocyte surface, greatly reducing the electrophoretic mobility of the cell.

A number of viruses and bacteria are rich in neuraminidase. Removal of
NANA from the erythrocyte membrane by this enzyme, which, for
example, is located in the coat of the influenza virus, may be the basis for
the hemagglutination phenomenon first noticed by Burnet. The initial
requirement for the entrance of many viruses into cells may be the removal of sialic acid. It is well known that erythrocytes low in or depleted
of NANA exhibit increased fragility, altered antigenicity, and shortened
survival time.

The positioned ganglioside is of importance, therefore, in cellular
cohesion. The ganglioside also possesses specific "receptor" characteristics (see p. 93 for "serotonin receptor" theory) and may be involved in
cation transport (Chapter VII). The sialic acid moiety may also provide
appropriate receptor sites for the attachment of viruses to cells, thus
playing an important role in the susceptibility of these cells to various
disease states.

B. The "Sodium-Pump" Adenosinetriphosphatase Associated with Membranes

Another important constituent or system which is present in membranes, particularly the plasma membrane and possibly the endoplasmic
reticulum, is a phosphotransferase or phosphohydrylase complex, measurable by its rapid hydrolytic rate of ATP, called for convenience an
adenosinetriphosphatase (ATPase). It should be emphasized that the
enzymic-like breakdown of ATP is the end result of what probably is a

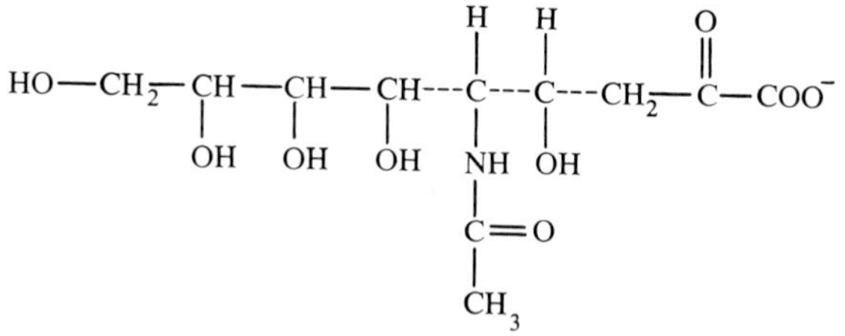

$$HO-CH_2-CH-CH-CH---\overset{H}{\underset{OH}{C}}---\overset{H}{\underset{NH}{C}}---CH_2-\overset{O}{C}-COO^-$$

with OH, OH, OH, NH, OH and
$$C=O$$
$$CH_3$$

Fig. 2.13A. N-Acetylneuraminic acid (sialic acid = acidic component, "prosthetic group" of gangliosides).

N-Acetylneuraminic acid — galactose — glucose — N — stearyl — sphingosine
galactose

Fig. 2.13B. Possible sequence of ganglioside.

long series of reactions. Indeed, the various ATPases, of which there are many distributed in nearly every subcellular constituent, are usually described as possessing intermediate reactions involving a high-energy, labile constituent which transfers the phosphate moiety from the ATP to an acceptor molecule. Regardless of the somewhat crude nature of the ATPase present in membranous fractions, the enzyme exhibits properties which appear to be quite specific. For example, 2,4-dinitrophenol, which stimulates ATPases associated with mitochondria, does not affect the membrane enzyme even in high concentrations. Magnesium ions are required for activity and of particular significance, the addition of "physiological" concentrations of Na^+ and K^+ effects a marked stimulation of the ATP hydrolysis. Furthermore, this stimulation is completely abolished by ouabain, a cardiac glycoside that specifically inhibits active transport processes. The latter property and a number of other important considerations provide evidence that this specific ATPase system is associated with the active transport process and is indeed part of the sodium pump" (see Chapter VII).

Since the endoplasmic reticulum (or the sarcoplasmic reticulum of the muscle cell (see Chapter IX) may be contiguous with the cell membrane as well as with possibly some subcellular membranes (Fig. 2.8A and B), it is possible that the transport ATPase system is located in all membranous systems which maintain or possess active transport processes. Indeed, the enzyme has been found in the "microsomal" fraction of a number of cells and its properties appear to be quite similar to those of the ATPase associated with the plasma membrane.

TABLE 2.11

Subcellular Distribution of Ganglioside in Liver, Ehrlich Ascites Cells, and Brain[a]

	Percent		
	Ehrlich ascites cells	Brain (gray matter)	Liver
Plasma membrane nuclei	−	−	64
Nuclei	3	11	0
Mitochondria	17	28	6
Endoplasmic reticulum	70	55	21
Supernatant	1	6	9

[a]Taken from Wallach and Eylar (1961); McIlwain (1963); Emmelot et al. (1964b); Kuhn and Wiegandt (1963).

V. FUNCTION OF MEMBRANES

A. Transport Mechanisms

As discussed in the preceding sections of this chapter, many membranous systems appear to share structural similarities, e.g., trilaminar nature, structural protein mesolayer, micellar character of the phospholipid and possibly paired arrays of elementary-like particles. While the various cellular organelles possess membranes with specialized functions, at least one very fundamental process appears to be characteristic of all membranes, namely, ion regulation. Ion concentration is intimately involved in fluid regulation, excitability of muscles and nerves, activation of various enzymes, protein synthetic processes, pH, and contractility. With few exceptions all cells possess a process which maintains a concentration gradient of ions, particularly Na^+ and K^+, the intracellular concentration of the latter, in general, being very much higher than the former. Tables 2.12 and 2.13 give examples of approximate concentrations and fluxes of ions in a number of cell types and in Fig. 2.14, both intracellular and extracellular values for Na^+, K^+, and Cl are shown. These data are approximate and vary considerably depending upon a number of factors, the most important of which is the method of determination. However, the important point is that K^+ is at least 40 times more concentrated inside the cell than outside, a fundamental characteristic of almost all cells. The intervening barrier of a membrane thus serves to effectively "compartmentalize" ions. What is less apparent is *how* the membrane selectively discriminates among ion species at various intervals during cellular activities. For example, it is known from the studies of Hodgkin (1958) that the resting potential of a nerve fiber or muscle cell is a

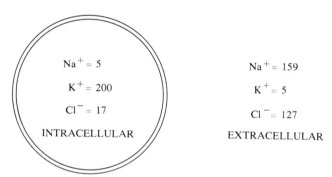

Fig. 2.14. Intra- and extracellular ionic concentrations in resting cat heart. Values are in milliquivalents per kilogram of cell water and are adapted from Hecht (1961) and Page (1962).

TABLE 2.12

Ionic Content of Various Tissues (Fresh)

Tissue	(mmoles/kg cell water)[a]				
	(Extracellular) Na$^+$	(Intracellular) K$^+$	(Total) Ca^{2+}	(Extracellular) Cl$^-$	Reference[b]
A. Heart muscle					
1. Cat papillary	43	164	–	17	1
2. Guinea pig atrium	–	–	1.70	–	2
B. Brain					
1. Guinea pig	40.4	114.2	2.17	38.5	3, 8
2. Mouse	42.4	105.5	–	28.7	4
3. Rat	47.2	99.8	–	–	5
4. Dog	51.8	95.4	2.1	36.3	6
C. Skeletal muscle					
1. Dog	29.2	98.1	1.9	19.4	
2. Frog	42	92	2.78	–	7

[a] These figures are very approximate and are in general based upon substances which presumably locate exclusively in the extracellular space (e.g., inulin, mannitol). Subtracting the total concentration of cation from the calculated extracellular concentration yields the intracellular concentration. The complexity of the extracellular space and errors in measurement techniques lend doubt to the quantitative accuracy of published data. For example, Page (1965) found that the intracellular sodium concentration of cat papillary muscles, using mannitol as the extracellular tracer, was about one eighth of that found using inulin as the determinant of extracellular space. Differences in potassium and chloride were also found. These observations suggest that certain extracellular structures can discriminate between mannitol and inulin according to size, by a process called "molecular sieving." Mannitol has a diameter of about 8 Å compared to 30 Å for inulin. In addition to these complexities, the presence or absence of calcium ions may be critical in extracellular space (and, pari passu, intracellular measurements) determinations. The absence of calcium in papillary muscles, e.g., significantly alters the extracellular space, a phenomenon which may be due to the presence of complex polyelectrolytes present in the "ground substance" in which heart muscle cells are imbedded. Calcium produces rather marked configurational alterations of certain polyanionic compounds (Katchalsky, 1964), a process which may have relevance in basic structural orientations.

[b] Key to references:
1. Page and Solomon (1960); Page (1962).
2. Winegrad and Shanes (1962).
3. Lolley (1963).
4. Timiras et al. (1954)
5. Katzman and Leiderman (1953).
6. Mannery (1954).
7. Bianchi (1963).
8. Keesey et al. (1965).

TABLE 2.13

Ionic Fluxes Associated with Excitation in Various Tissues

Tissue	$\mu\mu$moles/cm²/impulse			Reference[d]
	K⁺ Efflux	Na⁺ Influx	Ca²⁺ Influx	
Frog heart	20	15	–	1
Dog heart	6 – 9	11 – 18	–	1
Rat skeletal	16	19	–	1
Squid nerve	3	3.5	–	1
Guinea pig atrium (at rest)			0.02[a]	2
Guinea pig atrium (during contraction)			0.55[a]	2
Dog papillary muscle (during contraction)			0.92[b]	3
Guinea pig atrium (at rest)			0.50[c]	4
Guinea pig atrium (60 beats/min.)			0.89[c]	4

[a] $\mu\mu$moles/cm²/sec (solution contains 2.5 mM Ca²⁺).
[b] $\mu\mu$moles/cm²/beat (solution contains 5 mM Ca²⁺).
[c] mmoles/kg (solution contains 2.3 mM Ca²⁺).
[d] Key to references:
  1. Hecht (1961); Page (1962).
  2. Winegrad and Shanes (1962).
  3. Langer and Brady (1963).
  4. Grossman and Furchgott (1964).

"potassium diffusion potential" due to the unequal distribution of K⁺. The resting potential is that potential difference recorded in millivolts, between the outside and the inside of a cell prior to excitation. Measurements are done with the aid of microelectrodes.

The rising phase of the action potential approximates an Na⁺ diffusion or equilibrium potential (Fig. 2.15). The membrane during the resting state of the cell (inexcitable state) is presumably "permeable" to K⁺ and relatively "impermeable" to Na⁺. When the electrical activity of the cell becomes manifest in the form of the "spike" of the action potential [phase (b), Fig. 2.15], a marked increase of Na⁺ permeability may be measured and approximately 15 $\mu\mu$moles/cm²/beat of Na⁺ move into the cell.[2] During this interval the permeability to K⁺ has decreased to a min-

[2] This value applies to the ventricular fiber of frog.

In heart muscle, the energy required by the "Na pump" to remove this amount of Na⁺ from the cell and to move K⁺ back into cell if a 1:1 coupling exists (actually probably a 3:2 is more nearly correct) is 8×10^{-2} cal/gm or 13% of total energy liberated (Johnson, 1957).

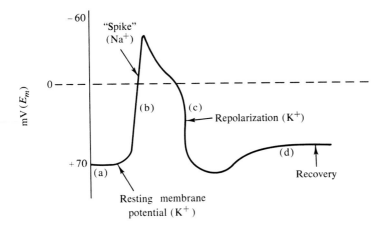

Fig. 2.15. A typical action potential of nerve. The amplitudes are relative to the cell interior which is set at zero; E_m's, therefore, are "transmembrane potentials." Phase (a) is the resting membrane potential due to K^+ differential; (b) is the "spike" or rising phase of the action potential, due to Na^+ influx; (c) is the repolarization phase due to K^+ efflux; (d) is part of the recovery phase which involves active transport. The time scale is in milliseconds.

imum. This situation is apparently reversed during subsequent phases of electrical activity; i.e., the Na^+ and K^+ are "pumped" against their respective concentration gradients. These phasic changes in permeability have been known for years and have been ascribed to various possible mechanisms, some of which will be discussed in subsequent sections. (Table 2.13 presents a compilation of ion fluxes in a number of tissues associated with the process of excitation.)

There are, essentially, three general mechanisms by which differential concentrations and movements of ions can be effected (all requiring the intervening membrane barrier):

1. PASSIVE DIFFUSION OR LEAK   [see Figs. 2.16(a) and (b)]

The ions move down their electrical or chemical gradients. Energy is not expended although a reversible combination with a specific chemical substance in the membrane is probable.

The supposition of the existence of pores or channels within the membrane is implicit. These pores may alter in size or shape, perhaps through the action of hormones. The "pore concept" is discussed on page 89.

The following sections present various aspects of the diffusion process:

a. SIMPLE DIFFUSION. A passage of substances through aqueous pores in the cell membrane with penetration resulting from random molecular

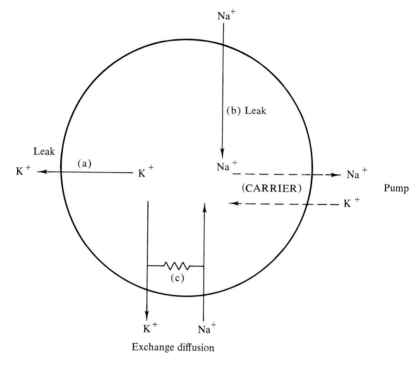

Fig. 2.16. Diagrammatic representation of membrane transport.

motion without an interaction with another molecular species. The process occurs rather infrequently because usually some interaction with a molecular entity exists. An example of this mechanism is the simple passage of water across membranes.

b. SOLVENT DRAG. The penetrating substance is swept through "aqueous pores" in the membrane as a result of the bulk flow of water. The movement of thiourea across the toad bladder is consistent with this process. Osmotic diffusion (Franck and Mayer, 1947) is similar to this mechanism. The solute circulates in the cell by an anomalous osmosis produced by an electrical current.

c. DIFFUSION RESTRICTED BY MEMBRANE CHARGE. Small anions such as chloride may pass through positively charged aqueous channels which presumably exclude cations. This aspect is discussed on page

d. DIFFUSION RESTRICTED BY A LIPID BARRIER. In this process, the penetrating molecule enters the cell, provided the molecule has the appropriate solubility characteristics to "dissolve" first in the lipid phase of the membrane and subsequently in the aqueous phase located on the other side of the membrane. The mechanism does not require the presence of aqueous pores in the membrane, and has been experimentally documented by the correlation of lipid-water partition coefficients with membrane penetration rates, in the case of many nonelectrolytes. It is obvious that the membrane in a system of this type must have solvent properties which are quite different from those of the cellular phases. In living systems, the cellular phases consist of aqueous solutions, and therefore substances of a nonpolar or a lipoid character would be appropriate as structural constituents of such membranes. This "solubility" theory was originally promulgated by Overton (1902) to explain permeability and action of anesthetic gases in membrane systems. More recently, Rosenberg (1948) has used this concept to explain mechanisms of active transport, a topic which will be discussed in detail in a separate chapter (Chapter VII).

e. FACILITATED DIFFUSION ("Mediated Transport"). The transported molecule presumably binds reversibly with a carrier molecule or substance present in the membrane in a process that may be initiated or facilitated by certain catalysts called "hydrolases" or "permeases." The complex formed oscillates between the surfaces, i.e., the outer and inner sections of the membrane, releasing or picking up molecules on either side. Because of the thinness of the membrane, the motion necessary for such a membrane carrier would be very slight and could be accounted for in terms of thermal movement or some type of reversible "molecular deformation." An example is the transport of glucose in erythrocytes.

None of the above processes directly requires metabolic energy nor leads to an accumulation of substances against a concentration gradient without the intervention of some type of intracellular binding.

f. PINOCYTOSIS. The cell membrane can under certain circumstances develop invaginations. These result from a "pinching off" of specific areas of the cell membrane, forming a type of "intracellular vesicle" (Fig. 2.17).

Pinocytosis was first described by Lewis in 1931 as an engulfment of fluid droplets by the active movement of undulating membranes at the periphery of cells flattened against a tissue culture vessel. The movements of the membrane projections were observed to entrap the droplets which then appeared as clear vacuoles among the contours of the undulating membrane. The folds project upward from the cell surface. Presumably the vacuoles moved inward toward the cell center and became smaller as the contents were "withdrawn." This surface activity has been observed particularly in capillary endothelial cells, wherein flask-shaped vesicles about 800 Å in diameter are formed by *invaginations* of the cell

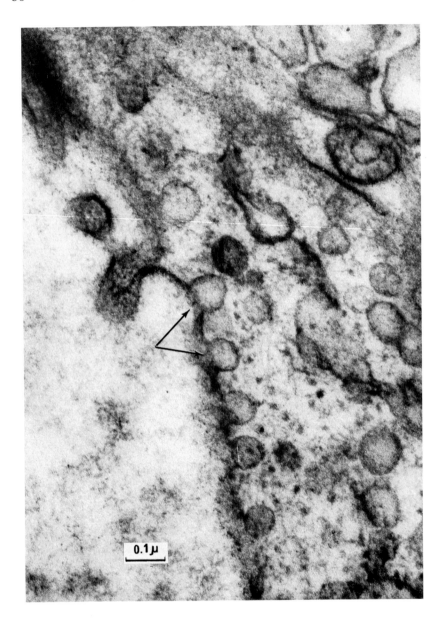

Fig. 2.17. Pinocytotic vesicles. Pinocytotic vesicles along the luminal surface of an endothelial cell of a capillary in rat jejunum. × 124,400. Osmium fixed, stained with lead and uranyl acetate. Courtesy of Mr. G. Adams, Department of Pathology, Baylor University College of Medicine, Houston, Texas.

surface, i.e., "micropinocytosis" (see Palade, 1956). Palade has suggested that these vesicles may detach from the luminal surface, migrate across the cell, and discharge their contents at the basal surface (Fig. 2.18) thus functioning as a means for the transfer of fluid across the capillary wall. A mucopolysaccharide coating on the surface of the membranes may have the required properties for binding of large ions or molecules, prior to pinocytotic formation.

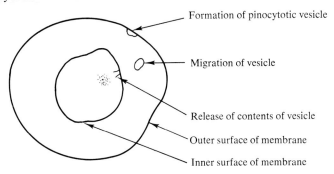

Fig. 2.18. Diagrammatic representation of pinocytosis.

Pinocytosis has been observed in ameba plasmolemma, erythroblast membrane, oocytes of insects, sinusoidal surfaces of liver cells, Kupffer cells, Purkinje cells in cerebellum, neuroglia, nerve cell bodies, synaptic endings, and in canalicular invaginations at the base of the brush border in the proximal convoluted tubule of the kidney. Whenever micropinocytosis involves a special localized differentiation of the plasma membrane, a selective absorption of macromolecules may occur. Pinocytosis might be considered as an "active deformation of the cell surface" for the purpose of selective absorption of certain macromolecules. Originally the term pinocytosis referred to entrapment by projecting folds but at present is more generally considered to describe uptake of material by membrane surface changes.

The giant freshwater ameba, *Chaos chaos*, has been employed in pinocytocic studies to a great extent. A number of polyelectrolytes, particularly some basic proteins, *induce* the pinocytotic cell surface change by apparently binding with anionic sites in the membrane. The binding reaction is a passive one but the uptake or engulfment aspect may be active, i.e., showing an energy requirement. Another process which is somewhat similar to pinocytosis and also incompletely understood in terms of transport mechanisms is *phagocytosis*. Phagocytosis is seen particularly in leukocytes and involves the engulfing of relatively large aggregates.

2. EXCHANGE DIFFUSION

This process, originally proposed by Ussing (1952), is depicted in Fig. 2.16 (c). Substances, ions in this case, can cross the membrane only in a complexed form. For each molecule transported to one side of the membrane, a similar molecule must be transported back. Therefore, no net transfer is observed. Exchange diffusion requires the presence of a "carrier" which may be produced in one of the solutions in contact with the membrane, (for example, the interior or the intracellular portion of the membrane), and consumed in the other, or extracellular portion of the membrane. On the other hand, the carrier may be a mobile one which is not consumed or produced, but oscillates or rotates back and forth from the exterior of the membrane to the interior of the membrane in a type of "ferry boat" arrangement. An important criterion of exchange diffusion is that no energy is expended. Therefore, it is essential that the flux of a complexed ion should be the same in both directions and of course the complexed carrier and ion should be confined to the membrane phase. Also apparent is the *one to one* exchange arrangement. In the case of sodium and potassium ions, sodium would be transported inward by combining with some sort of carrier in the membrane, while a simultaneous transfer of potassium with the same carrier occurs outwardly. The passive movement of sodium and potassium down their potential gradients in nerve and muscle tissue requires a "carrier" of some type since simple diffusion rates alone are too slow to account for the potential differences observed. Exchange diffusion may function as a possible mechanism.

3. ACTIVE TRANSPORT

Probably the most important of membrane functions is the process of active transport (Fig. 2.16 "pump"). Adequate descriptions are attributed to a number of individuals, particularly Rosenberg, Ussing, and Wilbrandt. Basically, active transport is a process characterized as a transfer of matter, which for energetic reasons cannot take place spontaneously from a lower to a higher chemical or electrical potential, or both. Substances that can pass faster along concentration gradients than would be expected from simple physicochemical considerations would be excluded by this narrow strictly thermodynamic definition. For example, as mentioned above, inward movement of sodium across an axon membrane occurs much faster than would be defined by simple diffusion mechanisms. Hence, even this aspect of transport involves some type of membrane-located "carrier" mechanism (see Section V,A). In addition, a transfer of a substance may occur against its chemical potential gradient by a type of intracellular binding which would not require energy. Consequently, a broader and perhaps better definition of active transport is as follows: The transfer or movement of substances across a membrane which requires work or the expenditure of energy whether the energy is used to overcome a potential difference, a concentration difference, or a combination of both. An important aspect of active transport then is the necessity for the utilization of chemical energy derived from cellular metabolism. We will reserve further and more detailed discussion of this important field for Chapter VII.

B. The Pore Concept

Utilizing intracellular microelectrode techniques and osmotic gradient methods,[3] it has been estimated that the membrane which separates the cytoplasm from the extracellular spaces possesses pores with diameters greater than 3.5 Å but less than 8 Å. It should be emphasized that these studies assume an idealized water-filled cylindrical pore which traverses the membrane from inside to out, and that the pores discriminate between water-soluble nonelectrolytes *by size only.* In other words, this is a measure of *passive diffusion* only.

The pore radius should, in general, be only two to three times that of the solvent molecule. A better term for pore size as measured by this procedure is *"equivalent pore radius."* Equivalent pore radii of a variety of tissues have been examined (Solomon, 1961) in erythrocytes, squid axons, *Necturus* kidney slices, intestines, and cultured HeLa cells. In general, it appears that 4 Å is characteristic of the pore in the resting or "passive" state. While the *pore concept* defines a passive system, i.e., movement of materials without the expenditure of energy or work, probably two main factors control the fluxes: *size of pore* and *charge density within the pore.*

[3]The tissue is first preincubated in a hypotonic, high K^+ solution in which Cl^- is replaced by an anion which does not presumably diffuse, causing a loss of intracellular endogenous Cl^-. Mannitol or inulin is added to the medium to raise the osmotic pressure, which causes a rapid movement of water out of the cells, resulting in crenation, which increases intracellular K^+ concentration. When the external K^+ is kept constant, the rise in internal K^+ can be measured by an electrode. This system represents a method for studying the entry of substances through the membrane. If the molecule is a slowly penetrating or nonpenetrating one, the elevation in voltage due to the unequal K^+ distribution remains for several hours; if the molecule is rapidly penetrating, the voltage rise will fall toward the control level rapidly (due to the movement of water back into the cell, thus returning the internal K^+ levels to the control). If the molecular radii of the test species is known, the pore size in the membrane can readily be estimated (Page, 1962a,b).

The pore size implies that exclusion of one or more species is based upon the radius of the ion and of the pore. It has often been suggested that hormones may play a role in the control of the pore opening. Pitressin, for example, has been shown to significantly increase pore size in a number of systems.

Another factor which may be vital in regulating the size of membrane pores is calcium. The action of hormones on membranes, for instance, may be moderated or mediated by calcium ion *within* the membrane. Curran and Gill (1962) suggest that calcium exerts its action on the outside of the membrane, causing a "nonspecific tightening" in some region, leading to a decreased permeability to Na^+ and to Cl^-. Thus, calcium may control sodium movements through a direct membrane effect, or through hormone mediation. The importance of calcium in muscular events is discussed in detail in Chapter IX.

Pore size may also be controlled by polybasic substances which may "open up bridges between inner and outer membranes." Interactions between basic and acidic compounds present in membranes may regulate movements of ions, an hypothesis which is expanded in the chapter on active transport (Chapter VII).

The second characteristic of the pore alluded to above is the charge density. Certain fixed negative charges may exist within the walls of the pore, which can attract ions of opposite charge by electrostatic forces, while repelling negatively charged ions. When the pore is "widely opened" [position (A)], free diffusion of ions occurs since the fixed pore charges do not appreciably influence the charge density of the pore. When the pore becomes smaller [position (B)], for example, under the influence of calcium or a hormone, the fixed charges become *moderately* effective in limiting the rate of diffusion of ions of like sign, e.g., Cl^-, SO_4^{2-}, or PO_4^{3-}. As the pore becomes still smaller [position (C)] the pore charges

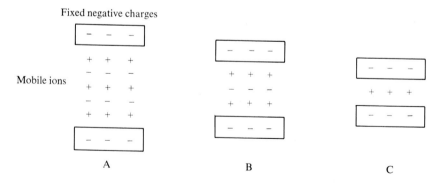

SCHEME 3. The pore concept.

effectively allow only the positively charged species to pass, e.g., K^+ or Na^+.

Whether Na^+ or K^+ is "permitted" to pass may depend upon the relative diameter of their hydrated shells. It is generally agreed that K^+ is smaller than Na^+ and although estimations of the exact size vary, a reasonable estimate is about 5 Å for Na^+ and 4 Å for K^+ (Joseph et al., 1961; Page, 1962). The fixed negative charges within the pore membrane may be due to acidic polyelectrolytic compounds, gangliosidic in nature (see p. 73).

C. Cell Contact

Another important function of membranes is concerned with cell interaction or contact. Minute quantities of basic polyelectrolytes can cause strong agglutination of bacterial, plant, and animal cells that is preceded by an adsorption of the polyelectrolytes to the cell's surface and which leads to numerous biological changes such as inhibition of growth, lowering of toxicity, and hemolysis of erythrocytes (De Vries et al., 1954). These adsorption phenomena have been extensively studied during the past 12 years and in general it has been found that the data can be fitted to a Langmuir isotherm. The uncharged polymer molecule is adsorbed with a large number of its segments in direct contact with the surface and it is often impossible to remove this layer by washing with pure solvent. The adsorption of charged polyelectrolytes follows a similar pattern, although the adsorption is not as strong. Figure 2.19A shows a typical adsorption isotherm for polylysine on red blood cell membranes.

The polyions are arranged partially perpendicular to the cell surface in an orientation which is due to the polarization effects. The adsorbed molecules which protrude from the surface into the solution may use their free ends to attract other surfaces and to bind one cell to another, thus producing the phenomenon of "agglutination" or formation of clumps. The surfaces of the cells which are part of the clumping do not actually touch but rather adhere to one another through "macromolecular bridges" formed between one cell surface and another (Fig. 2.19B). The forces exerted by these macromolecular bridges are strong and profoundly affect the cell surface. For example, polylysine produces a change in the normal bidiscoidal shape of erythrocytes which then assume a shape allowing maximal contact between cells and the highest utilization of the interaction energy. The distance between cells in an agglutinated colony corresponds approximately to the length of the fully stretched macromolecules or polyelectrolytes. The polyelectrolytes, in this case, polylysine or histones, literally "glue" the membranes together so powerfully that they cause a merging of the surfaces. According to Katchalsky (1964),

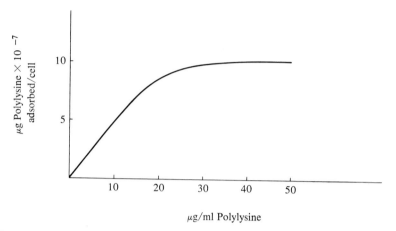

μg/ml Polylysine

Fig. 2.19A. Adsorption isotherm for polylysine bromhydrate on erythrocyte "ghosts." Taken from A. Katchalsky, *Biophys. J.* **4,** 9 (1964).

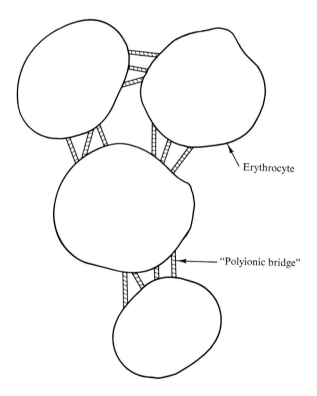

Fig. 2.19B. Diagram of agglutination phenomenon.

the polyelectrolytes which interact with acidic compounds in membranes may also join distant parts of the surface of the membrane and open up larger "holes" in the cell membrane. These holes apparently are the cause of hemolysis observed in the treatment of erythrocytes with higher concentrations of basic compounds.

The interactions between polyelectrolytes and membrane constituents also may be of importance in the defense mechanism of the organism. When *Bacillus anthracis* penetrates mammalian tissues, the organism produces huge quantities of polyglutamic acid, an acidic polyelectrolyte, which acts as a "spreading factor," opening up channels between cells (Bloom and Blake, 1948), which facilitates bacterial penetration. The released polyglutamic acid may be counteracted by a polybasic tissue factor which combines and neutralizes the acidic polyelectrolyte. It may be that other polyelectrolytes exist in nature which play a role in biological defense mechanisms. Heparin, for example, functions as an endogenous anticoagulant factor by virtue of its acidic composition.

D. Protein Synthesis and Growth

The biochemical processes of protein synthesis will be discussed in detail in Chapter V. It was seen that ribosomes are usually membrane associated when they carry out their synthetic activities in the living cell (Hendler, 1962). What is the importance of this type of system? Why are the ribosomes not considered to exist free in the cytoplasm when protein synthesis takes place? A number of investigators have found that the membrane represents the major site of incorporation of amino acids into proteins in a variety of systems (Hendler, 1965). An appreciable fraction of the total cellular RNA is bound to membranes (Schlessinger, 1963). Membrane-bound ribosomes are at least five times more effective than cytoplasmic ribosomes for the incorporation of amino acids. Therefore, it appears that a *membrane constituent* may be a limiting factor in protein synthesis. Loss of membrane integrity is frequently associated with impairment of protein synthesis (Dubin *et al.*, 1963). The membrane therefore plays a part in the control of the activity of the attached particles. The mechanism is at this time unknown.

The importance of membrane structure with respect to drug action has long been recognized. In 1902, for example, Overton suggested that the lipoidal nature of the membrane played a significant role in the mechanism of action of anesthetic substances. Recently, Woolley and his collaborators (1964) have postulated that sialic acid (see Chapter IX) present in the cell or subcellular membranes functions as a specific receptor for an important chemical present in brain and other tissues, known as serotonin (5-hydroxytryptamine). This reiterates the vital nature of acidic and basic compounds present in membrane-bound systems and suggests an ubiquitous importance in pharmacological action (see Chapter VII). A wide variety of pharmacological agents are believed to exert their effects through alterations in "membrane permeability." These include chlorpromazine, digitalis, and catecholamines.

VI. CONCLUSION

The prime importance of *structural organization* in biological processes cannot be overemphasized. As we will learn from subsequent chapters, specific orientation of enzymes, cofactors, substrates, and products of reactions is a vital part of cellular function. The membrane represents the basic structural element upon which the arrangement of these "cellular chemicals" occurs. Since the morphological pattern of the plasma membrane is ubiquitously dispersed throughout the cell, a knowledge of the complexity of this structure can be with some reservation, extended to subcellular components. The material discussed in this chapter defines the relationship between the chemical nature and structure of the membrane, and their importance in cellular activity.

GENERAL REFERENCES

Ashworth, L. A. E., and Green, C. (1966). *Science* 151, 210.
Bianchi, C. P. (1963). *J. Cellular Comp. Physiol.* 61, 255.
Danielli, J. F., and Davson, H. (1934). *J. Cellular Comp. Physiol.* 5, 495.
Danielli, J. F., and Harvey, E. N. (1934) *J. Cellular Comp. Physiol.* 5, 483.
Emmelot, P., Benedetti, E. L., and Rümke, Ph. (1964a). *Symp. Electron Microscopy, Modena, Italy, 1963*, p. 253. Tipografia S. Pio, Roma, Italy.
Emmelot, P., Bos, C. J., Benedetti, E. L., and Rümke, Ph. (1964b). *Biochim. Biophys. Acta* 90, 126.
Fernández-Morán, H., Oda, T., Blair, P. V., and Green, D. E. (1964). *J. Cell Biol.* 22, 63.
Goodwin, T. W., and Lindberg, O. (1961). "Biological Structure and Function," Vol. 1. Academic Press, New York.
Grossman, A., and Furchgott, R. F. (1964). *J. Pharmacol. Exptl. Therap.* 145, 162.
Hecht, H. H. (1961), *Am. J. Med.* 30, 720.
Katchalsky, A. (1964). *Biophys. J.* 4, 9.
Katzman, R., and Leiderman, P. H., (1953). *Am. J. Physiol.* 175, 263.
Kuhn, R., and Wiegandt, H. (1963). *Chem. Ber.* 96, 866.
Langer, G. A., and Brady, A. J. (1963). *J. Gen. Physiol.* 46, 703.
Lolley, R. N. (1963). *J. Neurochem.* 10, 665.
McIlwain, H. (1963). "Chemical Exploration of the Brain." Elsevier, Amsterdam.
Mannery, J. F. (1954). *Physiol. Rev.* 34, 334.
Page, E. (1962). *J. Gen. Physiol.* 46, 201.
Page, E., and Solomon, A. K. (1960). *J. Gen. Physiol.* 44, 327.
Robertson, J. D. (1959). *Biochem. Soc. Symp. (Cambridge, Engl.)* 16, 3.

Timiras, P. S., Woodbury, D. M., and Goodman, L. S. (1954). *J. Pharmacol. Exptl. Therap.* 112, 80.
Wallach, D. F. H., and Eylar, E. H. (1961). *Biochim. Biophys. Acta* 52, 594.
Winegrad, S., and Shanes, A. M. (1962). *J. Gen. Physiol.* 45, 371.

SPECIFIC REFERENCES

Bangham, A. D., Standish, M. M., and Watkins, J. C. (1965). *J. Mol. Biol.* 13, 238.
Blasie, J. K., Dewey, M. M., Blaurock, A. E., and Worthington, C. R. (1965) *J. Mol. Biol.* 14, 143.
Bloom, W. L., and Blake, F. G. (1948). *J. Infect. Diseases* 83, 116.
Bloom, W. L., Watson, D. W., Cromartie, W. J., and Freed, M. (1947). *J. Infect. Diseases* 80, 41.
Branton, D. (1966a). *J. Cell Biol.* 31, 15A.
Branton, D. (1966b). *Proc. Natl. Acad. Sci. U.S.* 55, 1048.
Bungenberg de Jong, H. G. (1949). *Colloid Sci.* 2, 232 and 335.
Curran, P. F., and Gill, J. R. (1962). *J. Gen. Physiol.* 45, 625.
De Vries, A., Stein, Y., Stein, O., Feldman, J., Gurevitch, J., and Katchalski, E. (1954). *Proc. 4th Intern. Cong. Soc. Hematol., Paris, 1954* Vol. VII-8, p. 385.
Dubin, D. T., Hancock, R., and Davis, B. D. (1963). *Biochim. Biophys. Acta* 74, 476.
Franck, J., and Mayer, J. E. (1947). *Arch. Biochem.* 14, 297.
Gesner, B., and Thomas, L. (1966). *Science* 151, 590.
Gorter, E., and Grendel, F. (1925). *J. Exptl. Med.* 41, 439.
Gorter, E., and Grendel, F. (1926). *Koninkl. Ned. Akad. Wetenschap., Proc.* 29, 315.
Gough, A. (1924). *Biochem. J.* 18, 202.
Green, D. E., and Perdue, J. F. (1966). *Proc. Natl. Acad. Sci. U. S.* 55, 1295.
Hendler, R. W. (1962). *Nature* 193, 821.
Hendler, R. W. (1965). *Nature* 207, 1053.
Hodgkin, A. L. (1958). *Proc. Roy. Soc.* B148, 1.
Johnson, J. A. (1957). *Am. J. Physiol.* 191, 487.
Joseph, N. R., Engel, M. B., and Catchpole, H. R. (1961). *Nature* 191, 1175.
Katchalski, E., and Sela, M. (1958). *Advan. Protein Chem.* 13, 243.
Katz, B. (1962). *Proc. Roy. Soc.* B155, 455.
Keesey, J. C., Wallgren, H., and McIlwain, H. (1965). *Biochem. J.*, 95, 289.
Korn, E. D. (1966). *Science* 153, 1491.
Lewis, W. H. (1931). *Bull. Johns Hopkins Hosp.* 49, 17.
Mueller, P., and Rudin, D. O. (1967). *Nature* 213, 603.
Nevo, A., De Vries, A., and Katchalsky, A. (1955). *Biochim. Biophys. Acta* 17, 536.
Overton, E. (1902). *Arch. Ges. Physiol.* 92, 115.
Page, E. (1962a). *J. Gen. Physiol.* 46, 201.
Page, E. (1962b). *Circulation* 26, 582.
Page, E. (1965). *Ann. N. Y. Acad. Sci.* 127, 34.
Palade, G. E. (1956). *J. Biophys. Biochem. Cytol.* 2, Suppl., 85.
Park, R. B., and Pon, N. G. (1963). *J. Mol. Biol.* 6, 105.
Parsons, D. F. (1967). *Proc. 7th Can. Cancer Res. Conf., Ontario, 1966*, p. 193.
Rosenberg, T. (1948). *Acta Chem. Scand.* 2, 14.
Schlessinger, D. (1963). *J. Mol. Biol.* 7, 569.
Schnaitman, C. A., Erwin, V. G., and Greenawalt, J. W. (1967). *J. Cell Biol.* 32, 719.
Schnaitman, C. A., and Greenawalt, J. W. (1966). *J. Cell Biol.* 31, 100A.

Sjöstrand, F. S. (1963). *J. Ultrastruct. Res.* 9, 561.

Sjöstrand, F. S., Andersson-Cedergren, E., and Karlsson, U. (1964). *Nature* 202, 1075.

Solomon, A. K. (1961). *Membrane Transport Metab., Proc. Symp. Prague, 1960*, p. 94. Academic Press, New York.

Ussing, H. H. (1952). *Advan. Enzymol.* 13, 21.

Ussing, H. H., and Levi, H. (1949). *Nature* 164, 928.

Woolley, D. W., and Gommi, B. W. (1964). *Nature* 202, 1074.

Chapter **III**

MITOCHONDRIA

I. INTRODUCTION

The cell acquires its usable energy by converting carbohydrates, fats, and proteins, in a series of enzymic reactions in the cytoplasm, to pyruvate (see Appendix) which, through another series of stepwise reactions occurring in the mitochondria, is then oxidized, and the energy obtained is conserved in the form of adenosine triphosphate (ATP). The "energy" in ATP is released and utilized in the maintenance and function of the cell. An understanding of the functional nature of the mitochondrion will afford an insight into the remarkable efficiency of cellular energetics.

The name, "mitochondria" (Greek: *mitos,* filament; *chondros,* granule) is applied to the filamentous and granular cytoplasmic bodies observed by Benda in 1897.

The original observations and description of these cytoplasmic organelles are generally credited to Altmann, who in 1894 used a special fixation and staining procedure which revealed the presence of what he termed "bioblasts." There is evidence, however, that what we now know to be mitochondria were described many years ago by Henle (1841), who observed granules in the sarcoplasm of striated muscle, and by Aubert (1853), who found unusually large and most abundant granules in insect muscle. Moreover, Kölliker, in an extensive study (from 1857 to about 1888) of these "interstitial granules," found them to be widely distributed in many different species of animal and even described a method of separation of these granules from insect muscle. He also recorded the effects of a hypotonic saline medium on the microscopic appearance of the structures. These studies were continued by Retzius (1890's), who called these specific bodies "sarcosomes," in order to distinguish them from other cellular inclusions, particularly fat bodies.

97

II. MORPHOLOGY AND DISTRIBUTION

Morphological studies of mitochondria were facilitated with the advent of dark field and phase microscopy and particularly with the introduction of a supravital staining procedure by Michaelis in 1900. The coloration produced with special dyes such as Janus green B, crystal violet, or toluidine blue depends upon the integrity of mitochondria, since the reaction between mitochondria and dye requires the presence of an intact electron transport system.

Using these techniques, it has been shown that the shape and mobility of mitochondria vary considerably in different animal cells. For example, mitochondria in the adrenal cortex take the form of short rods or granules, while in the intestine they are usually filamentous. Occasionally, mitochondria may clump together and coalesce into what are called chondriospheres; these are typically seen in scurvy.

While the distribution of mitochondria in the cell is somewhat uniform, some differences have been noticed. In kidney cells, for example, they are aggregated next to blood capillaries and adjacent to the tubules. In some cells they may move freely and in others they have a more or less permanent position. In some striated muscle cells, mitochondria are shaped as rings or braces around the I-band of the myofibrils (Chapter IX). Occasionally, mitochondria congregate around a fat droplet, in order to expedite the acquisition of an important substrate (Fig. 3.1A).

Both the size of the mitochondrion and the density of matrix vary considerably from tissue to tissue and even in the same tissue in particular stages of development (see Fig. 3.1B).

During cellular division, mitochondria may be found around the spindle, and after division of the cell they are distributed in almost equal numbers between the daughter cells (Chapter VIII). Presumably, the structure and the distribution of mitochondria are intimately related to the energy requirements of the particular cell.

The development and perfection of electron microscopic techniques have produced tremendous advances in our understanding of the structure and function of the mitochondrion. From the extensive and definitive studies of cytologists and biochemists, a composite picture has evolved (Figs. 3.1 and 3.2).

Structure

The highly organized nature of mitochondria promulgated the hypothesis that these organelles are complete "operational" units. The

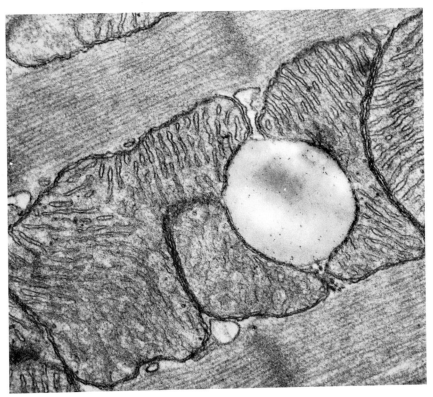

Fig. 3.1A Mitochondria surrounding a lipid droplet. Mouse heart mitochondria, *in situ;* fixed in 6% glutaraldehyde; postfixed in 1% osmium tetroxide in phosphate buffer and observed in a Siemens Elmiskop × 75,000. Courtesy of Dr. Federico Gonzales, Department of Anatomy, Northwestern University, Chicago, Illinois.

presence of a series of enzyme systems (about 70) associated with the mitochondria is entirely consistent with this concept.

The illustration shown in Fig. 3.2 is a presentation of our current knowledge of mitochondrial structure. An outer membrane of about 60 Å in width envelopes the mitochondrion. Separating this structure from the inner membrane is a light space of about 80 Å. The inner membrane, also 60 Å thick, sends infoldings or rugaelike projections (wrinkles or folds) into the matrix. These complex intramitochondrial structures were termed *cristae mitochondriales* by Palade who first observed them in 1954. The cristae appear to be continuous, in some places at least, with the inner membrane and are of similar dimensions and structure as the membrane itself. The outer space is found not only between the two mem-

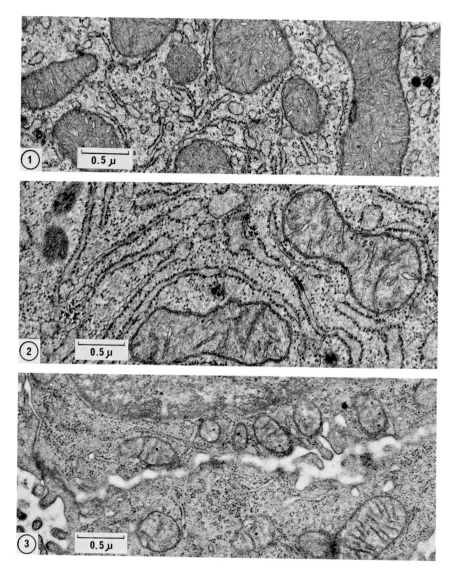

Fig. 3.1B. Electron micrographs of liver and tumor mitochondria, *in situ*. (1) Control rat liver; portion of cytoplasm from hepatocyte. The mitochondria have tubular cristae and moderately dense matrices. × 40,500. (2) Fetal rat liver; portion of cytoplasm from hepatocyte. Mitochondria have mixed tubular and platelike cristae with loose matrices that appear less dense than in control liver. × 40,500. (3) Mouse mammary tumor; portion of cytoplasm from tumor cell. Mitochondria have mixed tubular and platelike cristae with loose matrices that appear less dense than in control liver. × 40,500. Two percent glutaraldehyde fixation; postfixed in 1% osmium tetroxide. Courtesy of Drs. E. Rabin and Z. Blailock, Department of Pathology, Baylor University College of Medicine, Houston, Texas.

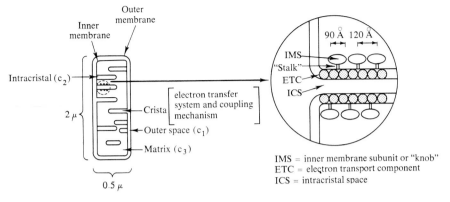

Fig. 3.2. Structure of the mitochondrion. IMS = inner membrane subunit or "knob"; ETC = electron transport component; ICS = intracristal space.

branes but in the core of the cristae. The matrix, which is continuous within the inner chamber and bounded by the inner membrane, is generally homogeneous but occasionally contains small granules of high density and finely filamentous material which is probably DNA. The matrix is semisolid in character and appears to be of a lipoprotein nature. The mitochondrion is a compartmentalized organelle consisting of outer, inner, and possibly intracristal spaces (labeled c_1, c_2, and c_3 in Fig. 3.2). We will discuss later the possible relationship between the compartments and functional activity.

Recent studies using high resolution techniques of electron microscopy have revealed that each of the mitochondrial membranes is quite complex. While the structures may correspond to the "unit membrane" hypothesis of Robertson, recent evidence (data obtained by negative staining techniques) suggests a globular or micellar arrangement, with projecting subunits (see Chapter II). The ultrastructure of the mitochondrial membranes is particularly significant since this represents the first intact system in which the lipoprotein framework is assigned specific functional biochemical details.

III. PROPERTIES OF MITOCHONDRIA

A. History and General Aspects

While it is true that the very early cytologists predicted a relationship between mitochondria and cellular oxidation, Kingsbury is credited with the suggestion that the mitochondria are centers for cellular respiration. He based this hypothesis on the observation that lipid solvents reduced the rate of respiration and dissolved the mitochondria.

The first definitive biochemical studies were performed by Batelli and Stern who in 1912 found that a frog muscle mince rapidly oxidized citric, succinic, malic, and fumaric acids, and by Warburg who in 1913 found that the insoluble fraction of cellular extracts contained respiratory enzymes. Warburg's recognition of the association of the terminal respiratory activities of the cell with intracellular particles profoundly influenced subsequent investigations of cellular energetics. The initial observations and theories of the early biochemists were largely ignored until Keilin in 1929 succeeded in isolating the first integrated multienzyme system from muscle cell particles, the succinic and cytochrome oxidation systems.

The oxidation of foodstuffs in the body proceeds in a highly controlled manner. A mole of glucose, for example, when completely oxidized to CO_2 and H_2O, yields about 690,000 cal. Since thermodynamic concepts are independent of mechanism and molecular pathways, one gram-molecular weight of glucose (180 gm) when completely oxidized always yields 690,000 cal regardless of the molecular mechanisms or conditions. A simple release of this amount of energy within the cell in the form of heat (as is found in a heat engine or in a calorimeter) would be incompatible with life. In addition, heat energy can perform work only when there exists a temperature differential, i.e., when it flows from a warmer region to a cooler one. Since there is practically no temperature differential in the living organism, the cell is unable to function as a type of heat engine. Therefore, in order to obtain an efficient recovery of the energy in the "fuel," the processes of cellular oxidation must proceed in a stepwise controlled manner. It is known that glucose is anaerobically oxidized to pyruvate by a series of about 12 enzymes located in the cytoplasmic matrix of the cell via the Embden-Meyerhof pathway (see Appendix).

Even before the complete mechanism for anaerobic glycolysis was elucidated, however, it was realized that the primary energy-generating sites must be *aerobic* in nature. Culminating years of extensive studies, Krebs in 1937 proposed a scheme for the terminal pathway of oxidation in animal tissues. Using pigeon breast muscle, he suggested that citric acid was formed by an enzymic reaction between oxalacetic acid and pyruvic acid and that subsequent oxidation of this 6-carbon tricarboxylic acid to CO_2 and H_2O took place in a cyclic series of reactions which he called "the citric acid cycle" (Fig. 3.3). Subsequent studies, using ^{14}C-labeled intermediates, confirmed the validity of the scheme and localized the citric acid cycle to the mitochondrion. The theoretical considerations invoked by Ogston in 1949 and the experimental studies of Ochoa and Stern proved that citric acid is in fact the primary product of the initial condensation steps in Krebs cycle oxidations (Fig. 3.3).

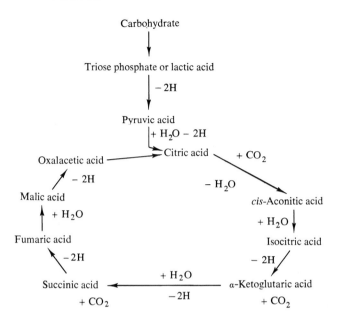

Fig. 3.3. The originally constructed citric acid cycle. Krebs, 1937; see H. A. Krebs, *Chem. Pathways Metab.* 1, 109 (1954).

It is known that the cycle represents the terminal oxidative pathway for fatty acids, ketone bodies, proteins, and carbohydrates. The key intermediate in the metabolism of these body fuels is the formation of an activated acetate (acetyl coenzyme A) which occurs in the mitochondria and is an absolute requirement for the initial condensation reaction. The discovery of the function of active acetate in intermediate metabolism represents one of the most important biochemical chapters in this century and one which has had a profound influence on a great number of biochemical and physiological disciplines.

The observation, in 1945, by Lipmann, that coenzyme A (CoA) was required for the acetylation of sulfanilimide in pigeon liver preparations represents the starting point in this remarkable chapter. Nachmanson and Berman, a year later, demonstrated a CoA requirement for the synthesis of acetylcholine from choline and acetate in the brain. The structure of CoA is shown in Fig. 3.4. Panthothenic acid is an essential feature of the coenzyme and a thiol represents the prosthetic group of the molecule.

Active acetate is an S-acetylated CoA. A simplified scheme depicting the formation of acetyl-CoA from pyruvic acid is shown in Fig. 3.5. The reaction requires NAD^+, thiamine (vitamin B_1) in the form of a pyrophosphate (thiamine pyrophosphate), Mg^{2+}, and lipoic acid (thioctic

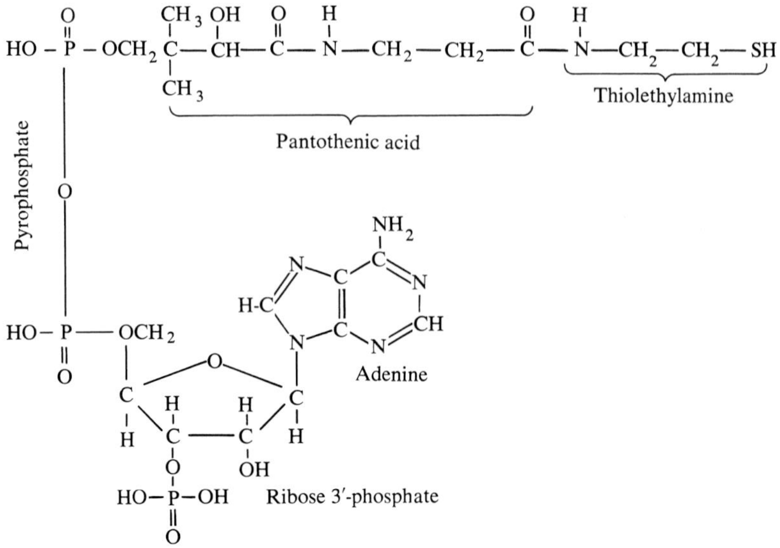

$$HO - \overset{\overset{\displaystyle O}{\|}}{P} - OCH_2\ \overset{\overset{\displaystyle CH_3}{|}}{\underset{\underset{\displaystyle CH_3}{|}}{C}} - \overset{\overset{\displaystyle OH}{|}}{CH} - \overset{\overset{\displaystyle O}{\|}}{C} - \overset{\overset{\displaystyle H}{|}}{N} - CH_2 - CH_2 - \overset{\overset{\displaystyle O}{\|}}{C} - \overset{\overset{\displaystyle H}{|}}{N} - CH_2 - CH_2 - SH$$

Thiolethylamine

Pantothenic acid

Pyrophosphate

$$HO - \overset{\overset{\displaystyle O}{\|}}{P} - OCH_2$$

Adenine

Ribose 3'-phosphate

$$HO - \overset{\overset{\displaystyle O}{\|}}{P} - OH$$

NH$_2$

Fig. 3.4. Structure of coenzyme A.

1. Glycolysis ➤ pyruvate

2. Pyruvate + thiamine pyrophosphate (TPP) ➤ α-lactyl-2-TPP (active pyruvate)

3. Active pyruvate $\overset{-CO_2}{\longrightarrow}$ 2-hydroxyethyl-TPP (active acetaldehyde)

4. Active acetaldehyde + lipoic acid* ➤ 2-acetyl-TPP + lipoic acid (reduced)

5. 2-Acetyl-TPP + lipoic acid (reduced) ➤ Acetyldihydrolipoic acid (Ac-LA) + TPP

6. Ac-LA + coenzyme A ➤ acetyl-CoA + lipoic acid (reduced)

7. Lipoic acid (reduced) + FAD ➤ Lipoic acid (oxidized) + FADH$_2$

8. FADH$_2$ + NAD$^+$ ➤ FAD + NADH + H$^+$

to electron transport chain

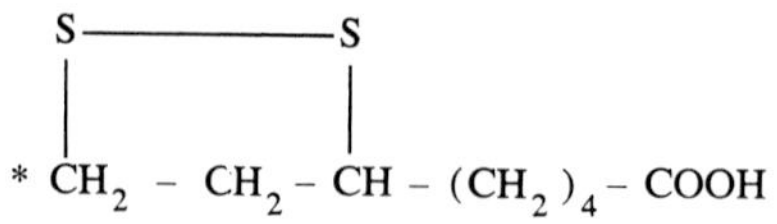

$$S \text{———} S$$
$$* CH_2 - CH_2 - CH - (CH_2)_4 - COOH$$

Fig. 3.5. Formation of active acetate from pyruvate.

acid). The thiamine reacts with the pyruvate to form an activated acetaldehyde-thiamine pyrophosphate complex which combines with oxidized lipoic acid in a reductive cleavage reaction, forming acetyldihydrolipoic acid. The S-acetyl group then is transferred to CoA, catalyzed by a thiotransacetylase. Reduced lipoic acid is then reoxidized by flavin-adenine dinucleotide (FAD). The reduced FAD shunts its electrons and protons into the electron transport chain via NAD.

Acetyl-CoA, then, is the terminal product of cytoplasmic oxidative reactions and is the first step in the succeeding oxidative reactions which are carried out in mitochondria. The stepwise release and "storage" of energy packets within the mitochondrion is of primary significance and will be discussed subsequently.

The various enzyme complexes involved in the citric acid cycle are shown in Fig. 3.6 and below. The points at which energy is transferred from NADH, which is formed from NAD^+ during the stepwise oxidations, to ADP via the electron transport chain are defined as A through D.

Step	Enzyme complex involved	Coenzyme	ATP formed[a]
A	Isocitrate dehydrogenase	NADH	3
B	α-Ketoglutarate dehydrogenase	NADH(NADPH)	4
C	Succinate dehydrogenase	$FADH_2$	2
D	Malate dehydrogenase	NADH	3

[a]Moles of ATP formed per mole of coenzyme oxidized via the electron transport chain.

B. Isolation and Investigative Techniques

The elucidation of the mechanism of energy release and oxidation would have been impossible without adequate procedures for separating mitochondria from the myriad of cellular constituents. The isolation methods have evolved from the initial studies of Bensley and Hoerr in 1934 who discovered that mitochondria were insoluble in saline solutions. They devised the procedure of differential centrifugation (see Chapter I) to isolate large granules from liver cells. The granules actively oxidized glutamic and succinic acids and gave a positive reaction for cytochrome oxidase activity by the "NADI" reagent (α-naphthol + p-phenylenediamine). Although numerous procedures have subsequently been developed (density gradient, column chromatography, gel filtration, sectioning procedures, etc.), the Schneider scheme is still the most widely used preparative method and is discussed in detail in Chapter I. A more recent adaptation for the preparation of large amounts of mitochondria

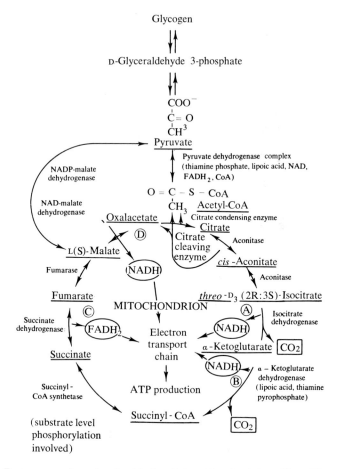

Fig. 3.6 Enzyme complexes involved in the citric acid cycle. Adapted from "Intermediary Metabolism," Gilson Medical Electronics, Madison, Wisconsin, 1964. See page 105 for explanation of letters A-D.

is described in Table 3.1. The ultrastructure of typical isolated mitochondrial preparations appears in Figs. 3.7A and 3.7B. Note the abundance of cristae and size of heart mitochondria as compared to the liver preparation. The characteristic heterogenicity of size and shape of mitochondria is apparent. The isolated mitochondria are comparable to *in situ* sections (Fig. 3.1B) of mitochondrial-containing tissues. The *in vitro* liver preparations [Fig. 3.7A(1,2)] represent mitochondria in two different functional states (IV and III, respectively). This will be discussed in a later section.

Most of the early observations on the mechanisms of mitochondrial metabolism were made using the Warburg constant volume respirometer

TABLE 3.1

A Rapid Preparative Method for Large Quantities of Calf Heart Mitochondria[a]

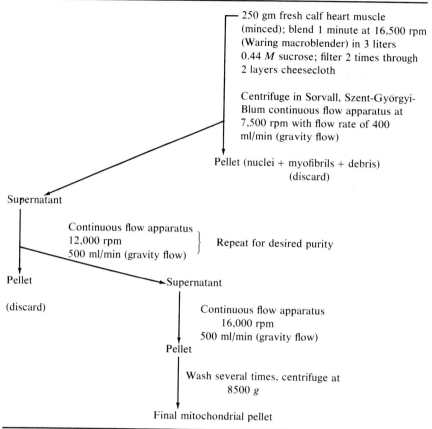

250 gm fresh calf heart muscle (minced); blend 1 minute at 16,500 rpm (Waring macroblender) in 3 liters 0.44 *M* sucrose; filter 2 times through 2 layers cheesecloth

Centrifuge in Sorvall, Szent-Györgyi-Blum continuous flow apparatus at 7,500 rpm with flow rate of 400 ml/min (gravity flow)

Pellet (nuclei + myofibrils + debris) (discard)

Supernatant

Continuous flow apparatus 12,000 rpm 500 ml/min (gravity flow) } Repeat for desired purity

Pellet

(discard)

Supernatant

Continuous flow apparatus 16,000 rpm 500 ml/min (gravity flow)

Pellet

Wash several times, centrifuge at 8500 *g*

Final mitochondrial pellet

[a]*Note:* For preparation of small amounts of heart mitochondria from laboratory animals (rat, guinea pig, and rabbit), employ the Schneider procedure (Chapter I) using a medium containing 0.18 *M* KCl, 10 m*M* EDTA, 0.5% bovine serum albumin at pH 7.2. A shearing instrument (Polytron; Brinkmann Instrument Co.) is most convenient and rapid (Safer and Schwartz, 1967).

(Fig. 3.8). The mitochondrial suspension [usually in a buffered 0.25 *M* sucrose medium containing a small amount (0.5−1 m*M*) of ethylene-diaminetetraacetic acid (EDTA) to help maintain mitochondrial stability] is added to a buffered saline medium containing inorganic phosphate and ATP, in the main compartment. The EDTA functions by chelating undesirable heavy metals and possibly Ca^{2+}. A folded piece of filter paper saturated with a 10−20% KOH solution is placed in the center well in

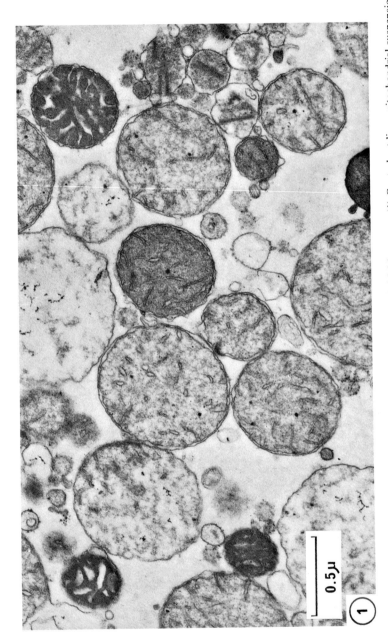

Fig. 3.7A. Electron micrographs of liver mitochondria, *in vitro*, isolated in 0.25 *M* sucrose. (1) Control rat liver mitochondrial suspension, fixed after 15-minute incubation in state IV (presence of substrate (glutamate) and inorganic phosphate, absence of phosphate acceptor, ADP). The majority of mitochondria are similar to those seen in tissue (orthodox conformation). × 50,000.

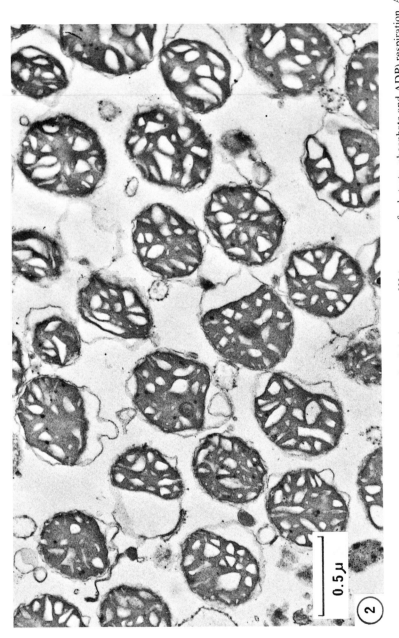

Fig. 3.7A. (2) Control rat liver mitochondrial suspension, fixed during state III (presence of substrate, phosphate and ADP) respiration. All mitochondria have condensed morphologic appearance. × 50,000. Two percent glutaraldehyde fixation, etc. Courtesy of Drs. E. Rabin and Z. Blailock, Department of Pathology, Baylor University College of Medicine, Houston, Texas.

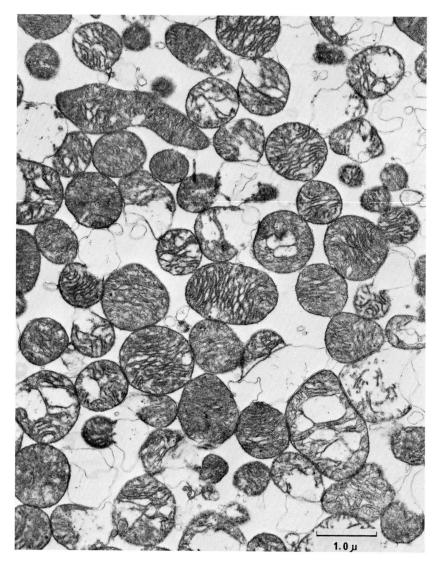

1.0 µ

Fig. 3.7B. Electron micrograph of mitochondria from normal rabbit heart. *Isolation Medium:* 180 mM choline chloride, 20 mM tris chloride, 1% bovine serum albumin, 10 mM EDTA at pH 7.4. This medium yields heart mitochondria possessing optimum respiratory control and P/O ratios. Heart mitochondria contain more cristae than mitochondria from other tissues. The cristae are narrow, platelike, and closely packed together. The matrices appear moderately dense in this freshly isolated preparation. Two percent glutaraldehyde fixation, postfixed in 1% osmium tetroxide. × 27,000. Courtesy of Drs. E. Rabin and Z. Blailock, Department of Pathology, Baylor University College of Medicine, Houston, Texas.

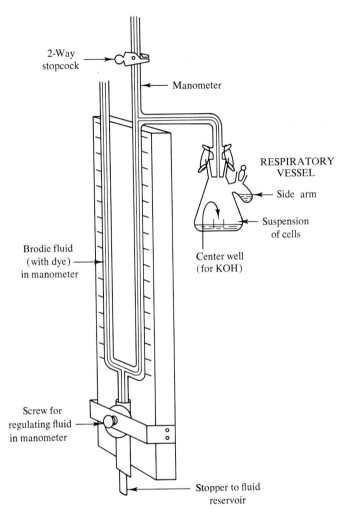

2-Way
stopcock

Manometer

RESPIRATORY
VESSEL

Side arm

Suspension
of cells

Brodie fluid
(with dye)
in manometer

Center well
(for KOH)

Screw for
regulating fluid
in manometer

Stopper to fluid
reservoir

Fig. 3.8. A Warburg manometer with flask.

order to adsorb the CO_2 produced during the reaction. The vessel is attached to the manometer and placed in a constant temperature water bath.
A shaking mechanism is activated and incubation is continued for various
time intervals. The pressure changes are directly proportional to oxygen
consumption according to Boyle's law; with the aid of formulas, the
amount of oxygen taken up by mitochondria is easily calculated (Umbreit
et al., 1964).
Frequently, various inhibitors and activators are placed in the side

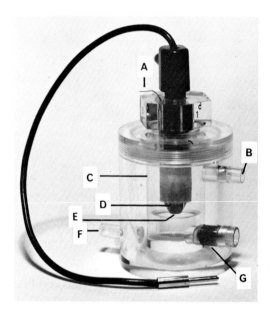

Fig. 3.9. Oxygen electrode. (A), Silver-silver oxide anode; (B), water outlet; (C), inner chamber (3.3 ml); (D), polyethylene or Teflon membrane; (E), platinum cathode; (F), water inlet; (G), rubber diaphragm through which materials are injected.

bulb and may be conveniently tipped in at appropriate intervals. In manometry, it is highly important to maintain adequate and optimal conditions (Umbreit *et al.*, 1964).

Within recent years a sensitive polarographic procedure for measuring oxygen consumption in isolated mitochondrial suspensions has been developed which obviates many of the difficulties encountered with manometry (Fig. 3.9). This method depends upon the fact that molecular oxygen is reduced at a platinum cathode. At a voltage of about -0.6 V, current (measured in microamperes) is directly proportional to the concentration of oxygen in the solution. Changes in current due to the uptake of oxygen by mitochondria may be conveniently registered on a suitable recorder.

The main advantages of this type of measurement over the Warburg procedure include multiple studies on one sample, greater sensitivity, and more accurate kinetic measurements.

C. Biochemical and Morphological Correlates

In this section we will discuss the mechanism of mitochondrial oxidative phosphorylation as related to structure. It will be recalled that the

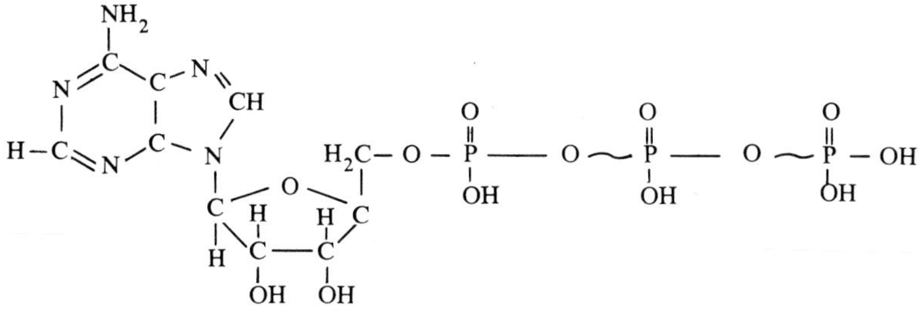

Fig. 3.10. Adenosine 5′-triphosphate (ATP. The two terminal phosphate moieties are "energy rich" (shown as "~").

energy residing in the acetate molecule (derived from cytoplasmic oxidations) is released in a series of cyclic reactions with the formation of CO_2 and H_2O. What happens to this energy? At the time Krebs was formulating the citric acid cycle, Kalckar in Denmark and Belitzer in Russia independently recognized that the incorporation of inorganic phosphate into a specific compound, adenosine triphosphate, accompanied the oxidation of glucose by homogenates of kidney muscle. ATP had already been identified as the energy source of muscle a few years earlier. Kalckar also found that adenosine diphosphate (ADP) and inorganic phosphate disappeared concomitantly during the oxidative processes and suggested that biological oxidation is coupled to the phosphorylation of ADP to form ATP (Fig. 3.10).

ATP "pushes" endergonic reactions to completion by virtue of its high phosphate-transfer potential, usually called "high-energy bonds." The latter term is inadequate since it implies that there is a "concentration" of energy between phosphate bonds (P ~ P) which tends to make the terminal phosphate spontaneously discharge. This is actually not true. In order to remove the terminal phosphate group, energy would have to be put into ATP. Bond energy technically is energy which has to be inserted into the molecule in order to break a bond between two atoms. A much better term used to describe the energy-rich nature of ATP is *high group-transfer potential,* defined as the change in chemical potential $(\Delta F°)$, which occurs when one mole of a substituent group of a donor molecule, (e.g., ATP) is transferred to a standard acceptor molecule, usually H_2O, under standard conditions of pressure, temperature, and pH. The unit of measure of potential is calories per mole of transferred group. When such a reaction occurs under these conditions in the case of ATP hydrolysis, the $\Delta F° = -7000$ calories. The negative sign indicates that the reaction (i.e., the reaction between ATP and H_2O) yields 7000 cal per mole. The ease with which ATP can be hydrolyzed is used by the organism to "drive" reactions which require energy (Klotz, 1957).

As already described, the sequential breakdown of foodstuffs and the participation of soluble enzymes in the cytoplasm leads to the formation of the energy-laden acetyl coenzyme A. The terminal oxidative events occurring in mitochondria liberate and subsequently store this energy in the form of ATP. What is the nature and function of the complex intramitochondrial reactions?

There are three main aspects of oxidative phosphorylation and these are probably intimately associated with the structure of the mitochondrion: (1) electron transport, (2) the citric acid or Krebs cycle, and (3) coupled oxidative phosphorylation.

1. ELECTRON TRANSPORT CHAIN

It will be recalled that the mitochondrion is a highly organized organelle, consisting of a series of inner and outer membranes with a semifluid matrix between the membranes (Figs. 3.1, 3.2, and 3.7). The inner membrane in some areas infolds into the matrix as cristae. The cristae are directly related to oxidative activity, possessing the electron transport system and the mechanism for coupling the transfer of electrons to the phosphorylation of ADP into ATP. About 30% of the membranes consists of lipid, 50% of a structural protein which is not concerned with oxidative reactions, and the remainder consist of active enzymic protein involved in metabolic activities.

The *respiratory or electron transport chain* is a highly organized structured complex present in the inner membranes. A currently popular view of the specifically arrayed electron transport chain is shown in Fig. 3.11.

In this composite scheme, the metabolites derived from carbohydrates, fats, and proteins are oxidized via the citric acid cycle donating stepwise, $2H^+$ and 2 electrons initially to either NAD^+, in the case of NAD^+ dehydrogenase-dependent substrates, such as α-ketoglutarate, isocitrate, and malate or to a flavin adenine dinucleotide (FAD)-dependent flavoprotein in the case of succinate. From there, the electrons are transferred through a sequence of carriers: flavoproteins, a quinone, a number of cytochromes, and finally to oxygen. The prosthetic group of the cytochromes is Fe^{3+}; the electrons reduce the iron ($2Fe^{3+} + 2e^- \rightarrow 2Fe^{2+}$) while the protons ($2H^+$) probably pass into solution. The terminal cytochrome electron acceptor (cytochrome oxidase) is either a complex with two different prosthetic groups attached to the same protein called "cytochrome oxidase complex" (cytochrome aa_3), or it is a single cytochrome containing two prosthetic groups attached to different proteins. The terminal electron and proton acceptor in the chain is molecular oxygen.

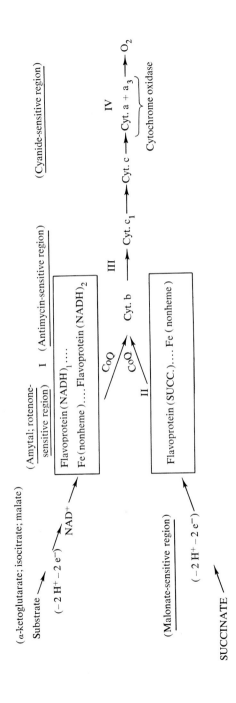

Fig. 3.11. A current view of electron transport chain. *Note:* The importance and position of coenzyme Q (CoQ) is still in doubt. Four "complexes" have been isolated, designated I – IV in approximate positions above. I = NADH-cytochrome c reductase complex (contains NADH dehydrogenase, nonheme iron. CoQ_{10}, cytochrome b, and cytochrome c_1); II = succinic-CoQ reductase; III = reduced CoQ-cytochrome c reductase; IV = reduced cytochrome c oxidase system.

Thus the hydrogen atoms and electrons do not react with oxygen in a single step but rather are transferred in a sequential manner to a system of intermediate carriers in a highly efficient process.

In recent years two possible intermediates have been implicated in the electron transport chain, coenzyme Q and vitamin K. Coenzyme Q (CoQ), otherwise known as ubiquinone, represents a class of compounds with the general formula diagrammed in Fig. 3.12. All of the ubiquinones have the same absorption spectra with a maximum at 275 mμ, a minor peak at 407 mμ, and an inflection at 320 mμ. Upon reduction to the quinol, the maximum shifts to 290 mμ and the inflection and minor peaks disappear. The ubiquinones are widely distributed in nature (which accounts for their name) and are located primarily in the mitochondrial portion of the cell in high concentration.

The position of CoQ in the respiratory chain scheme is believed to be somewhere between cytochrome c and flavoprotein (Fig. 3.11), i.e., in the antimycin A-sensitive region. The main unresolved problem with regard to CoQ is whether or not it is an *obligatory* component of the main electron transport pathway or is on a side pathway and functions not as an intermediate in oxidative phosphorylation, but in some other system, perhaps involving ion transport (see Chapter VII).

Another quinone, vitamin K, has also been implicated in oxidative phosphorylation. It has been shown that when mitochondria, particularly from liver, are exposed to ultraviolet light, a marked depression of P/O ratios (amount of ATP formed/oxygen consumed) occurs in the presence either of NAD^+-dependent substrates or succinate. Addition of vitamin K and cytochrome c restores the P/O ratios to normal, at least in the NAD^+-substrate oxidations. This observation, plus the ease with which

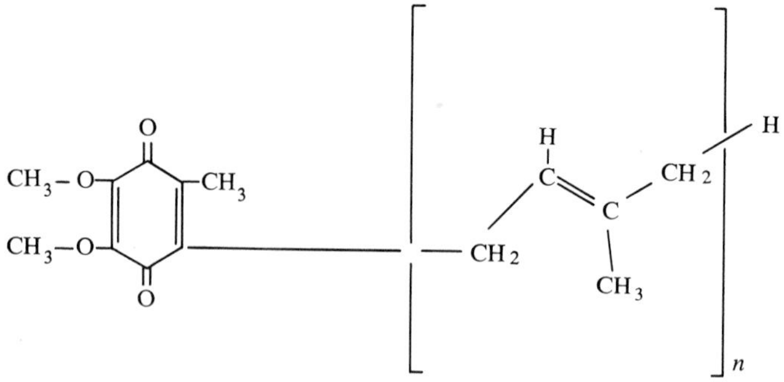

Fig. 3.12. Coenzyme Q (ubiquinone), n = the number of isoprenoid residues in the side chain and may be from 6 to 10 in the naturally occurring quinones. These are designated as CoQ_{10}, etc.

vitamin K undergoes oxidation and reduction, suggests an involvement in oxidative phosphorylation. While vitamin K as such has not been found in mitochondria, an enzyme which catalyzes the reduction of a closely related substance, menadione (2-methyl-1, 4-naphthoquinone), has been observed in mitochondria. The enzyme is designated as pyridine nucleotide menadione reductase. In addition, a vitamin K_1 reductase and vitamin K_1,H_2-cytochrome c reductase have been isolated from liver mitochondria. The natural electron carrier for these enzymes however, is still unknown, but continuing investigations may yet reveal some definitive role for vitamin K as well as other quinol compounds in mitochondrial activity.

Summarizing the known electron transport chain constituents, there are at least five participating oxidation-reduction components: *flavin, heme, nonheme iron* (Fe_{NH}), *copper,* and possibly *coenzyme Q.* There are probably at least three different flavoproteins, one associated with succinate dehydrogenase (f_S) and two with NADH dehydrogenase (f_{N_1}, and f_{N_2}) The flavoproteins have been isolated and their properties determined. The prosthetic group of both f_S and of f_N is FAD. However, the FAD of f_S is attached to the apoenzyme presumably by a strong peptide linkage, while those of the f_N's are relatively weak. In addition to the flavoproteins, there are at least four cytochromes, cytochrome a, b, c, and c_1 possessing similar prosthetic groups, probably an iron-protoporphyrin IX. The previously mentioned cytochrome oxidase complex appears to contain a copper protein in addition to the iron-porphyrin linkage.

The highly organized nature of the mitochondrion has been revealed by subfractionation procedures using the detergent compounds, digitonin, deoxycholic acid, and Triton X-100, or ultrasonic vibration, mechanical disintegration, and hypoosmotic solutions. The resulting fragments from these treatments can be separated and purified by centrifugation procedures. Studies of this type have suggested that the mitochondrion consists of a large number of orderly recurring structural units, each of which contains an array of respiratory carriers in a specific ratio. One such experimental approach is illustrated in Fig. 3.13. Membrane particles produced by digitonin treatment reveal relatively intact assemblies of enzymes capable of electron transport and coupled phosphorylation. The particles however, are devoid of substrate-level enzymic apparatus as well as all of the Kreb cycle intermediates. A single liver mitochondrion, according to Lehninger, probably contains $5000-10,000$ "respiratory assemblies" evenly distributed in the membrane. Most of the membrane mass is made up of these arrays of enzymically active molecules, so that the basic structural plan of the mitochondrial membrane is that of layers of highly oriented lipid and protein molecules.

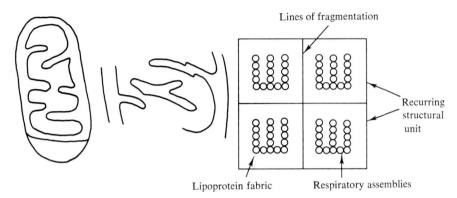

Another procedure which has been used extensively to fragment and study morphological and biochemical aspects of mitochondria is sonic oscillation. The particles obtained from this treatment, designated ETP_H by Green and his co-workers (electron transport particle from heart), are capable of coupling the oxidation of NADH of or succinic acid to the synthesis of ATP (Fig. 3.14A). Similar to the "Lehninger particle" described above, the ETP_H is unable to carry out the complete citric acid oxidative reactions. The rates of oxidation and of phosphoric acid esterification by ETP_H are three to four times higher than the parent mitochondrion, presumably due to the loss of endogenous control mechanisms (see Section III, D). In addition, ETP_H is able to readily oxidize externally added NADH, whereas intact mitochondria cannot. In fact, the insensitivity of a respiring mitochondrial preparation to external NADH is indicative of intact membranes. The bound form of NADH is detached during sonication and the sites for oxidation of added NADH become exposed and reactive.

Recently, in a combined morphological and biochemical study, the ETP was fragmented into an "elementary particle" (EP) which is $80 - 100$ Å

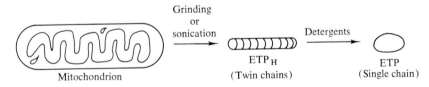

Fig. 3.14A. Subparticles of mitochondria obtained by sonication and detergent treatments. Drawn from data by D. E. Green. *Advan. Enzymol.* **21**, 73 (1959); Comp. Biochem. Physiol. **4**, 81 (1962).

in diameter (Green, 1962; Fernández-Morán, 1962). These particles were thought to be the basic structural subunits of mitochondria (see Chapter II). The molecular weight of the EP "headpiece" was estimated at approximately 500,000−750,000. It was postulated that together with a "stalk" and a cylindrical base piece, these units would accommodate the entire enzymic system for electron transport.

Phosphotungstic acid staining combined with surface spreading and electron microscopy has recently afforded a more detailed examination of the inner mitochondrial membrane. The "knobs" or EP's alluded to above are diagrammed in Fig. 3.2 and are seen to consist of a globular portion (IMS) approximately 90 Å in diameter and a stalk which is attached to the membrane. Complete removal of the head pieces by ultrasonic vibration still leaves the entire electron transport activity intact; the content of

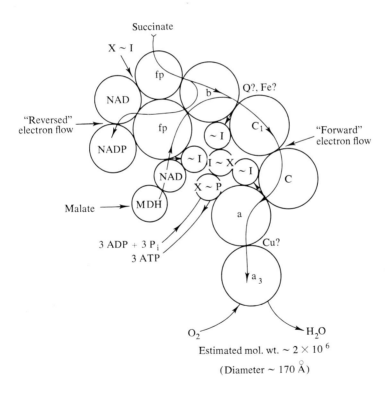

Fig. 3.14B. A possible oxysome arrangement. A schematic diagram of the sites by which oxidative phosphorylation controls electron transfer in the respiratory chain or *oxysome* (the letter symbols refer to cytochromes, Q to ubiquinones, fp to flavoprotein; X and I are hypothetical intermediates of oxidative phosphorylation). Taken from B. Chance, *J. Gen. Physiol.* **49**, 163 (1965).

cytochromes was in fact slightly increased (Chance *et al.*, 1964). Furthermore, recent studies indicate that the IMS may represent the terminal
step in oxidative phosphorylation, i.e., formation of ATP and not the
entire electron transport function, as suggested by Green. Enzymic activity appears to be identical with an oligomycin-sensitive ATPase
(Racker and Conover, 1963; Racker, 1967). Thus IMS may, in fact, be
coupling factor F_1 (see pp. 131-132). Chance, in a series of quantitative
kinetic studies, has been able to determine the sequence in which oxidations of the various carriers occur upon the addition of oxygen to anaerobic mitochondria (i.e., cytochromes a_3, a, c, b; flavoprotein; and NADH;
see Fig. 3.11). This information would then allow for an analysis of the
sequence in which these proteins interact. Chance has concluded that the
entire function of electron transport is accomplished with one each of the
cytochromes; $2-3$ molecules of flavoprotein and $10-20$ molecules of
NADH. He refers to this as a basic unit for a respiratory assembly — the
oxysome (see diagram below and Fig. 3.14B). Thus, the repeating subunit,

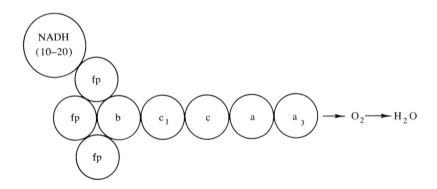

the diameter of which is about 170 Å, according to Chance consists of the
various electron transport carriers bound together in specific molecular
proportions. As we shall discuss later, phosphorylation is closely associated with, and in fact controls the rate of, electron transport. The binding of the carriers then would presumably involve, according to Chance,
"inhibitory ligands" which maintain the endogenous rate of electron
transport at a low level. When energy demand becomes manifest, the
electron transport chain would be "released" or "derepressed" so that the
rate of oxidation could increase substantially. The carriers of the electron transport chain have very high turnover times; the half-times for
the oxidation of the reduced cytochromes are indicated below:

Cytochrome	Half-time (msec)
b	30.0
c	3.3
a	2.1
a_3	1.4

Electron transfer in the "resting steady state," however, is only $10-20$ electrons per second per carrier. The potentiality of the electron transfer chain is therefore extremely high. Chance has formulated a possible binding arrangement of the respiratory-phosphorylation subunit, as diagrammed in Fig. 3.14B. The "inhibitory ligands" referred to above are indicated in the figure as "$\sim$ I" at each of three sites: between NAD and fp; cytochromes b and c_1; and cytochromes c and a. "I $\sim$ X" can be utilized in the phosphorylation of ADP, in the reversal of electron transport causing NAD^+ reduction by succinate and subsequent transhydrogenation to NADPH, and in active cation transport. These processes will be discussed in detail in later sections of this chapter.

The respiratory assembly subunits are located, as indicated in Fig. 3.2, specifically in the inner membrane and cristae. They are not part of the outer membrane system.

The reader is reminded that data on the substructure of mitochondria as related to function are still in the formative stages and any theory presented should be regarded with interest but should not be tacitly accepted. The procedures employed for microscopic examination are constantly being reevaluated. Likewise, data acquired from biochemical studies are also in a reinterpretative state, as can be attested by the number of symposia on structure and function of mitochondria.

Structural protein (SP) accounts for about 50% of the total protein of the ETP. It is water insoluble, colorless, and is devoid of redox groups, and has been isolated lipid free. It is believed that stable, hydrophobic bonds or regions in the SP contribute to the formation of complexes with lipid material. The structural protein may form the coat of the structural unit within which both lipid and cytochromes are embedded.

In summary, then, the composite picture of the mitochondrial membrane includes thousands of compact and/or interrelated respiratory enzyme assemblies or groups, aligned in regular array, possibly held together by a structural protein coat. The highly structured and oriented electron transport chain is believed to consist of two intercommunicating chains, one for the oxidation of succinate and the other for the oxidation of NADH (Fig. 3.11). Coenzyme Q or cytochrome c or both *may* be the mobile links between each segment, moving in all directions through the lipid phase. They *may* be considered as the molecular instruments by which electrons in the mitochondrion can move to the various components of the chains.

While the ideas presented in the preceding section are provocative and revealing, it still is not possible to formulate a totally acceptable molecular explanation for electron transport. The combined morphological and biochemical approach, as described, should lead to the solution of this basic and highly complex problem.

2. THE CITRIC ACID CYCLE AND AUXILIARY ENZYMES

Another important aspect of mitochondrial function is represented by the citric acid cycle (Fig. 3.6). The enzyme and cofactors associated with these reactions may be located in the outer membrane of the mitochondrion (although there is some serious doubt about this; see Quagliariello *et al.*, 1967) and are external to the highly structured electron transport chain. An enormous amount of work has been done in elucidation and characterization of the mechanisms of substrate oxidation (including fatty acid oxidation) and probably more is known about these enzymes than any other aspect of mitochondrial function. Some of the enzyme complexes involved are depicted in Fig. 3.6. The mechanisms of each will not be discussed; the reader is referred to the reference section. In addition to the citric acid cycle, the outer membrane contains a number of enzymes which are unrelated to electron transport. These are associated with the synthesis of material necessary for the maintenance of mitochondrial activity and include various phospholipids, fatty acids, phosphoproteins and possibly cytochromes. Net synthesis of mitochondrial cofactors has not actually been demonstrated although very active incorporation of amino acids into mitochondrial proteins, probably structural in nature, occurs, particularly in mitochondria from heart muscle.

The outer membrane also contains monoamine oxidase (MAO) (Schnaitman, Erwin, and Greenawalt, 1967). This enzyme is apparently localized exclusively to the mitochondria and, in addition to functioning in the metabolism of catecholamines, it may also be involved in control of respiratory activity. The deamination of tyramine catalyzed by MAO may exert a regulatory effect on the oxidation of succinate (Gorkin and Krivchenkova, 1966).

The mechanisms by which the citric acid and the fatty acid cycles, the auxiliary enzymes, and the electron transport chains are architecturally arranged with respect to one another are not known. However, it is recognized that a very precise molecular orientation does exist. Both sterioisomerism and the geometric placement of enzymes and enzyme systems are of paramount importance in the regulation and maintenance of mitochondrial activity.

Recently, specific mitochondrial DNA and RNA have been found, suggesting that the mitochondrion may contain the entire or part of the apparatus for protein synthesis and self-perpetuation (see Section IV, C).

3. OXIDATIVE PHOSPHORYLATION

The third and most interesting of the basic mitochondrial functions is oxidative phosphorylation, i.e., the coupling of electron transport to phosphate esterification. While the details are still not completely understood, a number of important observations and concepts are worth considering.

It has been established, for example, that there are at least three sites along the electron transport chain where phosphorylation of ADP takes place. These are probably between NADH and flavoprotein, between cytochrome b and c, and between cytochrome c and cytochrome oxidase. The sites are designated 1, 2, and 3 in Fig. 3.15.

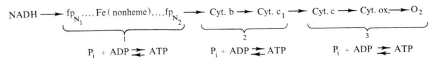

Fig. 3.15. Proposed coupled phosphorylation sites.

The use of inhibitory and uncoupling agents, artificial electron donors and acceptors, and exchange reactions, as well as studies of ATPases have led to a plausible mechanism for the coupling phenomenon between electron transport and the synthesis of ATP.

a. INHIBITORS OF ELECTRON TRANSPORT (see Table 3.2 and Figs. 3.11 and 3.16). There are at least four sites along the electron transport chain where various inhibitors can act.

i. *The cytochrome oxidase region.* It has been shown that cyanide, azide, and carbon monoxide act at this point, possibly by binding the necessary copper ions.

ii. *Coenzyme Q-cytochrome c region.* Antimycin A, BAL (2,4-dimercaptopropanol), the naphthoquinone antimalarials (e.g., Atabrine), and 2-alkyl-4-hydroxyquinoline oxides act here by binding an unknown factor or factors.

iii. *$NADH_2-fp_N$ area.* Amytal and rotenone act in this region by interacting with an unknown factor.

TABLE 3.2

Summary of Inhibitors at Specific Sites

Compound	Effects and/or possible sites

ELECTRON TRANSPORT
INHIBITORS

Amytal — Inhibits, primarily, aerobic oxidation of NADH-linked substrates; also succinate; not overcome by 2,4-DNP; acts between NADH and flavoprotein; also inhibits phosphorylation mechanisms (energy transfer probably at the NADH-flavin site)

Rotenone — Relatively specific for the NADH-flavin-linked site; possibly between fp_{N1} and fp_{N2}

Cyanide — Inhibits cytochrome oxidase (complexes Cu?)

Antimycin A — Blocks unknown factor between cytochromes b and c

British anti-lewisite (BAL) — Blocks distal to the antimycin-sensitive factor

TRUE UNCOUPLERS OF
OXIDATIVE PHOSPHORY-
LATION

2,4-Dinitrophenol (2,4-DNP) — Probably acts by discharging or hydrolyzing or complexing with a high-energy intermediate; inorganic phosphate not required for action; the intermediate attacked is one formed before P_i is esterfied; may combine with some lipid component; stimulates ATPase and oxygen consumption of intact mitochondria

Carbonyl cyanide-
(m-chlorophenylhydrazone)
(mCl-CCP) — Similar to 2,4-DNP

Arsenate — Site is obscure but probably hydrolyzes an intermediate distal to the 2,4-DNP-sensitive sites

Dicoumarol — Similar to 2,4-DNP

Thyroxine
(10^{-6} M and higher) — Similar in some respects to 2,4-DNP; probably acts primarily on structure rather than on energy-intermediate

OXIDATIVE PHOSPHORY-
LATION INHIBITORS

Oligomycin — Inhibits only phosphorylating electron transport in intact mitochondria; inhibits ATPase and oxygen consumption; inhibits arsenate but not 2,4-DNP-uncoupled respiration; the inhibition of oxygen consumption by oligomycin is relieved by 2,4-DNP but not arsenate; site is separate from electron transport and from the portion of energy transduction that is 2,4-DNP-insensitive

Aurovertin — Similar to oligomycin in some respects; differs, however, on mitochondrial swelling; site is probably slightly different from oligomycin

TABLE 3.2 (*continued*)

Compound	Effects and/or possible sites
Guanidines	Probably acts at an NADH-flavin couple; similar to Amytal; does not inhibit 2,4-DNP-induced ATPase
Parathyroid hormone	Stimulates a vitamin D-dependent release of Ca^{2+}, non-phosphorylative oxidation, increases uptake and efflux of ions; site unknown
Valinomycin	Stimulation of phosphate (or arsenate) and potassium-dependent respiration; action may be mediated via ion transport since it markedly increases K^+-influx
Histone (below 50 μg/mg mitochondrial protein)	Stimulates ATPase and phosphorylating (or nonphosphorylating) respiration and K^+ efflux; inhibits exchange reactions; octylguanidine but not aurovertin or oligomycin inhibits histone-induced changes; stimulates oligomycin-inhibited respiration; does not stimulate octylguanidine-inhibited respiration; causes energy-dependent swelling; site unknown

iv. Succinate dehydrogenase. Oxalacetate and malonate act at this point and unlike the above, are competitive inhibitors.

There are a number of other inhibitory agents, the mechanisms of which are not as clear as those mentioned above. Among these are atractylate (acts probably at an energy transfer site involving ADP), guanidine, methylene glycol, and progesterone (acts possibly between NADH and a flavin). It should be emphasized that none of the inhibitors mentioned with the *possible* exception of rotenone really have absolute specificity.

In addition to the inhibitors of electron transport, a great number of so-called true uncoupling agents, i.e., substances that dissociate oxidation from phosphorylation, have been studied. These include 2,4-dinitrophenol (DNP), thyroxine, dicoumarol, pentachlorophenol, gramicidin, and the carbonyl phenylhydrazones (Heytler's reagents). The proposed sites of the various inhibitors will be discussed in a subsequent section. Table 3.2 and Fig. 3.16 give summaries of information concerning most of the relatively specific inhibitors.

b. ELECTRON ACCEPTORS AND DONORS. The use of ferricyanide as an electron acceptor in intact mitochondrial suspensions showed that one phosphorylation occurs between cytochrome c and oxygen. It is established that ferricyanide is reduced (i.e., accepts electrons from NADH or NADH-linked substrates) in the cytochrome c region. Thus if α-ketoglutarate or glutamate serves as an electron donor, in the presence of ferricyanide in an anaerobic environment (either azide, cyanide, or N_2 is

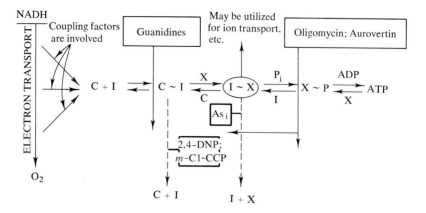

Fig. 3.16. Intermediates in oxidative phosphorylation and agents that act on phosphorylation mechanisms. *Note:* It is assumed that uncouplers act by hydrolyzing or complexing with preexisting high-energy intermediates or by preventing formation of the intermediates. I and X are energy transfer intermediate carriers. Dotted lines represent activation; solid lines extending from boxes represent inhibition. C represents a reduced or oxidized carrier such as NADH, CoQ or cytochromes. This scheme is called the chemical hypothesis and is opposed to the chemiosmotic concept (see pp. 129-131).

used to depress cytochrome oxidase activity), the P/2e ratio (this is analogous to P/O ratio under aerobic conditions) is 2, instead of the usual 3. This is adequate evidence for the occurrence of at least two phosphorylating sites between NADH and cytochrome c (Figs. 3.11 and 3.15).

Ascorbic acid, p-phenylenediamine or tetramethyl-p-phenylenediamine (TMPD), a redox dye, plus ascorbic acid, donate electrons nonenzymically to cytochrome c (Fig. 3.17). Consequently, the step or steps between cytochrome c and oxygen may be studied. Investigations of this type have revealed that one phosphorylation of ADP occurs between cytochrome c and oxygen. Some of the artificial acceptors and donors serve as convenient bypasses, and are valuable in studies of discrete areas of the electron transport chain.

In summary, our present knowledge indicates that with NADH-dependent substrates (those having no substrate-level phosphorylation) the maximum P/O ratio (micromoles of inorganic phosphate esterified to microatoms of oxygen consumed) is 3, with two phosphorylations taking place between NADH and cytochrome c and one phosphorylation occurring in the cytochrome oxidase region. When succinate is utilized as substrate, however, the NADH dehydrogenase site is bypassed (Fig. 3.11) and the maximum P/O ratio is 2, implying that one phosphorylation occurs between NADH and cytochrome b or CoQ.

c. EXCHANGE REACTIONS AND ADENOSINETRIPHOSPHATASES
(ATPases). There are four basic chemical reactions which are observed
either in intact respiring mitochondria or in certain types of fragmented
mitochondrial preparations:

1. 2,4-Dinitrophenol (DNP), a potent uncoupling agent, stimulates the breakdown of ATP into ADP and inorganic phosphate.

2. An ATP-inorganic phosphate ($^{32}P_i$) exchange reaction. The terminal phosphate of ATP very rapidly becomes labeled upon incubation with $^{32}P_i$ in the presence of a mitochondrial suspension. This reversible reaction is completely depressed by 2,4-DNP and by oligomycin.

3. An ATP-^{32}ADP exchange reaction, which is also inhibited by 2,4-DNP and by oligomycin.

4. A P_i-$H_2{}^{18}O$ exchange reaction, which is inhibited by both 2,4-DNP and oligomycin.

The effect of 2,4-DNP is particularly significant because there is correlation between the uncoupling of oxidative phosphorylation induced by relatively low concentrations of 2,4-DNP, and the resulting stimulation of oxygen consumption and of ATPase activity of mitochondria. This suggests an interrelationship between ATPase activity, "tightness" of coupling mechanism and respiration, and has bearing on the concept of mitochondrial respiratory and cellular control. Before pursuing the exchange reactions, a brief discussion of the ATPase concept follows.

For some time it was recognized that intact mitochondria possess enzymic activity for hydrolysis of ATP (Lardy and Elvehjem, 1945) and that this activity, called adenosinetriphosphatase, varied according to the method of isolation, age, and source of the mitochondria. Freshly isolated and washed mitochondria exhibit little or no ATPase activity. If the mitochondria are aged by incubation at 37°C for 15 − 30 minutes, or by freezing and thawing or by detergent treatment, the ATPase activity

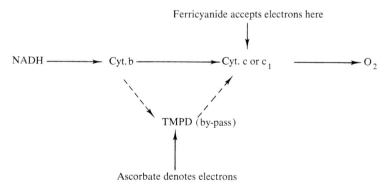

Fig. 3.17. Artificial electron acceptors and donors.

becomes quite active. The ATPase activity in freshly isolated mitochondria is therefore termed "latent." The addition of 2,4-DNP to this preparation greatly stimulates enzyme activity.

ATPase may or may not be a single enzyme and may represent merely a series of reactions which ultimately result in the release of inorganic phosphate from ATP. Recently, however, a mitochondrial ATPase has been isolated and characterized (Penefsky et al., 1960) from heart muscle, which appears to be identical to a part of the mitochondrial coupling mechanism, i.e., the enzymic site or sites responsible for ATP hydrolysis and the coupling of phosphorylation to electron transport may be on the same protein(s). These investigators suggest that the ATPase activity as such may be an aberrant one which is only observed following some disruption of the mitochondria. It is of interest that this enzyme system is uniquely cold labile. When the mitochondria are intact, this system may function as a *phosphate transfer agent* at the terminal phosphorylation step. The ATPase may also represent an important coupling factor (see p. 132).

We return now to our discussion of the exchange reactions in mitochondria. The ATP-^{32}P$_i$ reaction can occur in the absence of net electron transport; ADP, inorganic phosphate, and ATP are essential requirements. Furthermore, the reaction proceeds most rapidly when the carriers in the chain of electron transport are in the fully oxidized state.

The ATP-^{32}ADP exchange on the other hand, occurs in the absence of inorganic phosphate and loses its sensitivity to 2,4-DNP upon aging. As a result of the above-cited studies, Lehninger has proposed the following scheme for oxidative phosphorylation:

$$(1)\ \text{Carrier} + X\ \xrightarrow{2\ e^-}\ \text{Carrier} \sim X$$
$$\text{(reduced)} \qquad\qquad \text{(oxidized)}$$

$$(2)\ \text{Carrier} \sim X + P_i\ \rightleftharpoons\ \text{Carrier} + P \sim X$$
$$\text{(oxidized)} \qquad\qquad\quad \text{(reduced)}$$

$$(3)\ P \sim X + ADP\ \rightleftharpoons\ ATP + X$$

The carriers may be the various intermediates in the electron transport chain, such as NADH, etc. The X represents an unknown intermediate which forms a high-energy compound first with the carrier and then with inorganic phosphate, whereupon the phosphate is esterified with ADP to form ATP. This scheme is similar to the expanded one shown in Fig. 3.16, in which *two* intermediates, X and I, function.

Dinitrophenol uncouples oxidative phosphorylation by acting at a point proximal to the uptake of phosphate, namely (see also Fig. 3.16),

$$\text{Carrier} \sim X + 2{,}4\text{-DNP} \longrightarrow \text{Carrier} + X$$

While this scheme for oxidative phosphorylation has been slightly modified by several investigators, the basic tenet, i.e., the presence of *carrier molecules* and combination with *intermediates* to form high-energy compounds which subsequently lead to the synthesis of ATP, is accepted by most investigators, with the exception of Mitchell in England who advocates a chemiosmotic mechanism.

On the basis of arsenate, 2,4-DNP, and oligomycin[1] effects, at least two unknown intermediates are postulated, X and I (Fig. 3.16). Efforts have been expended without success for many years in attempts to elucidate the nature of these compounds.

Chemical vs. chemiosmotic theory of oxidative phosphorylation. It may be of interest at this point to compare the main aspects of the traditional chemical hypothesis of oxidative phosphorylation with that of the chemiosmotic theory. The former is based on the original concept of Lipmann (1946) which involves the binding of an electron carrier to an energy carrier by a bond which becomes "energy rich" after the redox reaction. The reduction of the respiratory carrier in some way facilitates its phosphorylation and subsequent oxidation increases the energy of the bond:

$$A \xrightarrow{+e^-} Ae^- \xrightarrow{P_i} Ae^- - P \xrightarrow{-e^-} A \sim P \xrightarrow{ADP} ATP + X$$

This scheme is now revised in only one way; phosphate does not enter into the formation of the primary "energy-rich" bond but rather at a later stage:

$$A \xrightarrow{+e^-} Ae^- \xrightarrow{+I} Ae^- - I \xrightarrow{-e^-} A \sim I \xrightarrow{X} X \sim I \xrightarrow{P_i} P_i \sim X$$
$$\qquad\qquad\qquad\qquad\qquad\qquad\qquad\qquad\qquad + \qquad\quad +$$
$$\qquad\qquad\qquad\qquad\qquad\qquad\qquad\qquad\qquad A \qquad\quad I$$
$$\xrightarrow{ADP} ATP + X$$

A scheme similar to this was first proposed by Slater in 1955 (see Slater, 1956).

As indicated above, through the use of oligomycin, dinitrophenol, arsenate and other relatively specific agents, the concept of nonphosphorylated and phosphorylated "energy-rich" intermediates became

---

[1]Oligomycin is a toxic antibiotic studied originally by Lardy *et al.* (1958). It is a relatively specific inhibitor of the transfer reaction between inorganic phosphate and a high-energy intermediate (see Table 3.2 and Fig. 3.16).

popular, although none has ever been isolated. The compound "Ae⁻ — I"

popular, although none has ever been isolated. The compound "$Ae^- - I$" should be stable and, therefore, theoretically is isolatable. Again, this has not, to date, been accomplished. Accordingly, Mitchell has formulated a hypothesis which does *not* necessarily require the intermediary compounds postulated. The liberation of H^+ during the oxidation of substrates or NADH, fpH_2, etc., serves as a motive force for the phosphorylation of ADP as follows:

    1. Oxidation yields protons and electrons which separate and accumulate on either side of the inner membrane.

    2. The proton accumulation facilitates the liberation of *"phosphorylium"* ion in the active center of an ATP synthetase. This latter enzyme is located on the membrane, and makes contact with the pool of accumulated H^+. The hydroxyl ions formed in the phosphorylium reaction neutralize the H^+ to form metabolic water; the phosphorylium ion then reacts with ADP to yield ATP:

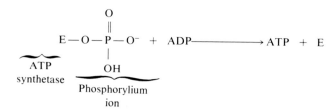

Mitchell has recently introduced into his scheme a number of hypothetical *coupling* intermediates, which are formed by the electrical membrane potential.

    The chemiosmotic hypothesis of Mitchell is seen to be radically different from the so-called traditional chemical coupling concept. Based upon a *vectorial* migration of charges ("protonmotive force") through the mitochondrial membrane, the theory obviates the necessity for the existence of "energy-rich intermediates."

    While this hypothesis is attractive, experimental support is scanty. The promulgators of the original hypothesis for oxidative phosphorylation argue vigorously against the chemiosmotic concept. Chance, e.g., has utilized sensitive recording procedures along with a dye, bromothymol blue, that presumably is specific for intramitochondrial regions and detects intravesicular acidification. He concludes that the rate of proton formation is over a thousand fold *slower* than the initial rate of oxidation of reduced carriers. That is, the so-called "hydrogen or proton-pump" of Mitchell cannot be coupled to electron transport. Movements of H^+ appear to be a consequence of a primary cation pump (see Section IV, E, and Chapter VII) (Chance and Mela, 1967). This information would tend to support the chemical mechanism (presence of intermediary "energy-rich" compounds) for oxidative phosphorylation. Solution of this important aspect of mitochondrial function must await further exploration, perhaps dependent upon the development of newer techniques of "intermediate" isolation.

d. COUPLING FACTORS AND INTERMEDIATES. *i. Protein coupling factors.* It should be recalled that there are three sites at which electron transport is linked to phosphorylation (Fig. 3.15). At each of these sites, phosphorylation of ADP to ATP occurs according to a proposed mechanism (Fig. 3.16). The close association between electron transport, i.e., oxidation, and phosphorylation, requires certain physiological coupling factors (Fig. 3.16). Extensive isolation and characterization studies have revealed at least six important factors, the properties of which are summarized in Table 3.3. The exact position and function of these and other coupling proteins are still not known, nor is it yet known whether the three sites of phosphorylation (Fig. 3.15) share the same energy transfer enzymes or require specific factors.

Other protein coupling substances have been suggested in the past. The "C"-factors (I and II) of Lehninger (1962) are notable among these and have been identified as glutathione peroxidase and catalase, respectively. The function of these, however, is obscure, but deserve further study. A more complete discussion of C-factors as possibly related to contraction and swelling of mitochondria may be found in Section IV, A.

ii. High-energy phosphorylated intermediates. As already discussed, intermediary substances may be formed or activated prior to the final formation of ATP during the process of oxidative phosphorylation. These substances are distinct from the protein coupling factors which purportedly link electron transport to phosphorylation (see above).

Recently, a labile protein-bound phosphohistidine was implicated as a high-energy intermediate, as follows:

$$\text{inorganic phosphate} \longrightarrow \text{phosphohistidine} \longrightarrow \text{ATP}$$

However, it is now clear that the rate of incorporation of $^{32}P_i$ into phosphohistidine proceeds *maximally* at only about 1.5% of the rate of the incorporation into ATP, making it highly improbable that phosphohistidine is an intermediate in oxidative phosphorylation. However, it is of importance in the substrate level phosphorylation reaction catalyzed by succinyl-CoA synthetase (Fig. 3.6). In this reaction, it will be recalled that α-ketoglutarate is oxidized to succinate with the formation of four energy-rich phosphate bonds, one of which represents *substrate level phosphorylation*. The latter term indicates the formation of 1 ATP in a

TABLE 3.3

Protein Coupling Factors Involved
in the Coupling of Oxidation to Phosphorylation[a]

Factor	Properties
F_1	Cold-labile ATPase activity; required for oxidative phosphorylation, $^{32}P_i$ATP exchange and reversal of electron transport[b]. Complexes with ADP and is related to ADP-^{32}ATP exchange. May be identical with IMS (see Fig. 3.2)
F_2	Required for phosphorylation and $^{32}P_i$-ATP exchange
F_3	Stimulation of $^{32}P_i$-ATP exchange
F_4	Required for oxidative phosphorylation, $^{32}P_i$-ATP exchange and reversal of electron transport[b].
F_0	Conferral of oligomycin sensitivity on ATPase
ATPase inhibitor factor	Soluble; naturally occurring, may function in control of mitochondrial activity

[a]This table is adapted from E. Racker and T. E. Conover [*Federation Proc.* 22, 1088 (1963)] and from various papers in the *Journal of Biological Chemistry*, cited in the reviews (Racker and Conover, 1963; Racker, 1967).
[b]See Section IV, B.

reaction not catalyzed by oxidative phosphorylation mechanisms. Hence this reaction is not sensitive to 2,4-dinitrophenol. Phosphohistidine is an energy-rich intermediary compound in this substrate level formation of ATP.

The most important criterion that should be fulfilled by any proposed intermediary substance is that the *initial* rate of $^{32}P_i$ incorporation should be the same as or greater than the optimal rate of formation of ATP. This has not as yet been fulfilled by any of the proposed compounds.

D. The Influence of Mitochondrial Metabolism in Control of Cellular Respiration

Interest in respiratory control mechanisms began with the discovery by Pasteur over 100 years ago that oxygen inhibited the rate of glycolysis in a fermentation system. Since that time, it has clearly been demonstrated that aerobically respiring mitochondrial systems markedly inhibit the rate of glucose uptake and lactic acid formation in Embden-Meyerhof glycolytic reactions. This is the well-known Pasteur effect which received a good deal of attention from biochemists, particularly after Warburg proposed that an alteration in the Pasteur effect might be characteristic of cancerous tissue. With the acquisition of definitive knowledge con-

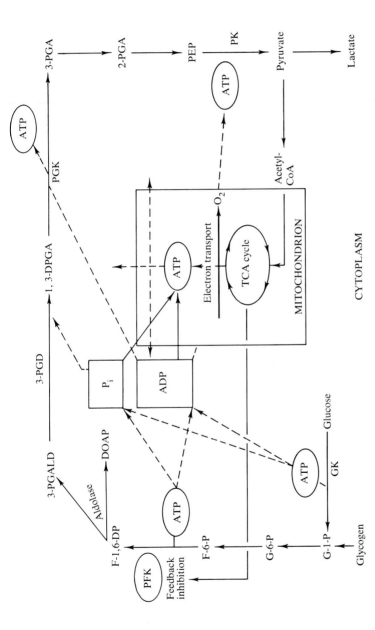

Fig. 3.18. Metabolic interrelationships in the cell. 3-PGALD, phosphoglycericaldehyde; 3-PGD, phosphoglycericaldehyde dehydrogenase; 1,3-DPGA, diphosphoglyceric acid; PGK, phosphoglycerokinase; PEP, phosphoenol pyruvate; PK, pyruvate kinase; GK, glucokinase; PFK, phosphofructokinase; F-6-P, fructose-6-phosphate; DOAP, dihydroxyacetone phosphate; F-1,6-DP, fructose 1,6-diphosphate; G-6-P, glucose 6-phosphate; G-1-P, glucose 1-phosphate; 3 PGA, 3-phosphoglyceric acid.

cerning enzymic pathways for glucose metabolism as well as the realization of the highly structured and compartmentalized nature of mitochondria, a number of theories have been proposed for metabolic controlling mechanisms which are based upon the Pasteur phenomenon.

Mitochondria and the soluble cytoplasm are compartmentalized in the cell. The distinct separation of enzymic activities conveys a type of *competition* for common intermediary substances which therefore may be rate limiting. A convenient diagrammatic representation is presented in Fig. 3.18.

It may be seen that a competition for ADP and/or inorganic phosphate between anaerobic cytoplasmic reactions and aerobic mitochondrial processes would result in a diminution of glucose consumption and lactic acid formation. Since ADP and phosphate are obligatory requirements for the phosphoglycerate kinase (PGK) step as well as for the highly compartmentalized mitochondrial oxidative phosphorylation, one may vision a restriction of glycolysis as a result of limiting available concentrations of ADP and/or phosphate.

Elegant and sensitive procedures have been devised by Britton Chance for the purpose of studying respiratory controlling factors (see Chance and Williams, 1956). The methodology, essentially, embodies the use of a double-beam (two-monochromator) spectrophotometer, and a split-beam recording instrument. This allows for extremely rapid spectrophotometric scanning techniques which have yielded very clear and concise spectra of respiratory chain intermediates in the intact cell as well as in isolated mitochondrial suspensions. The procedure has enabled Chance to obtain recordings of absorbancies of the cytochromes, flavoproteins, and pyridine nucleotides at all wavelengths throughout the entire spectrum. Minute changes representing the differences of optical densities between anaerobic and aerobic phases have yielded an extensive amount of information, which previously was unobtainable by other procedures.

Rapid and minute variations in respiration measured by a vibrating platinum electrode using a polarographic procedure (see Section III,A) coupled with the above technique showed that the intracellular concentration of ADP is rate limiting and is an important control of respiration (see Table 3.4).

Phosphofructokinase (PFK) has been found to be extremely sensitive to ATP, as well as to ADP and AMP. In addition, citric acid, a Krebs cycle intermediate, markedly inhibits PFK activity and this inhibition may be reversed by fructose 1,6-diphosphate. It is possible that the PFK step in glycolysis is another key metabolic controlling device. Balances between citric acid, fructose diphosphate, and ATP in the cell may function as a type of servomechanism working in conjunction with ADP and inorganic phosphate ratios (Fig. 3.18).

The rate-limiting importance of ADP in mitochondrial electron transport may be conveniently demonstrated by the addition of either ADP or an ADP-producing enzymic system

$$\text{glucose} + \text{ATP} \xrightarrow{\text{hexokinase}} \text{glucose 6-phosphate} + \text{ADP}$$

to a mitochondrial suspension in the presence of oxygen and substrate. If mitochondria are freshly prepared and "tightly" coupled, the *increase* in oxygen consumption will be anywhere from 10−20 times the control value, depending upon the tissue and the source. The structure of the mitochondrion appears to be directly associated with respiratory activity (see Section IV, A). Chance has defined various states of mitochondrial activity (Table 3.4).

The procedure known as the "respiratory-control method" or "acceptor-response procedure" is a much more sensitive and rapid indicator of mitochondrial metabolism than the standard P/O measurements (see below):

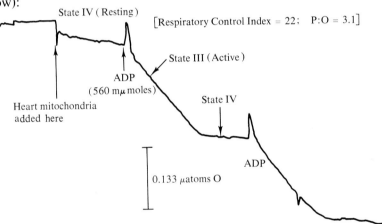

State IV (Resting)

[Respiratory Control Index = 22;   P:O = 3.1]

State III (Active)

ADP
(560 mμ moles)

State IV

Heart mitochondria
added here

0.133 μatoms O

ADP

SCHEME 1. Oxygen electrode analysis of mitochondria: respiratory control. Medium: Glutamate, 5 mM; inorganic phosphate, 8.5 mM; sucrose, 0.25 M; Tris · HCl, 10mM; pH 7.4 (See Fig. 3.7A and B for morphological correlation.) See Table 3.4 for explanation of states of activity.

IV. IMPORTANT ANCILLARY FEATURES OF MITOCHONDRIA

A. Contraction and Swelling

Mitochondrial swelling, observed originally by Raaflaub (1953), has been studied in great detail. Swelling is an active process, somewhat analogous to muscle relaxation (Chapter IX), and therefore is probably dependent upon energy-rich intermediate activity. There are at least two types of swelling, large-amplitude and low-amplitude; the former may be only partially energy-dependent while the latter is completely dependent upon a source of energy. Inhibitory substances such as Amytal, Rotenone, antimycin A, cyanide, or nitrogen, can completely prevent low-amplitude mitochondrial swelling. Swollen mitochondria can be made to contract (and simultaneously extrude water) upon the addition of ATP in the presence of Mg^{2+}. The reversal of the swelling by $ATP + Mg^{2+}$ can occur in the absence of oxidative processes but does require the presence of at least a portion of the coupling machinery, since oligomycin prevents the antiswelling action of ATP. The same antibiotic does not affect swelling. Consequently, it is apparent that swelling and contraction operate through different processes or at least through different portions of the same enzymic process.

The contraction and swelling of mitochondria may be intimately linked with volume changes. The foldings of the inner membrane system or cristae [Figs. 3.1, 3.2, and 3.7A(1,2)] may correlate to some extent with changes in the *state* of mitochondrial activity. Mitochondrial states are defined according to the relative concentrations of substrate and phosphate acceptor, ADP, and respiratory rate (Table 3.4). Mitochondria in state 3, for example, exhibit an irregularly folded inner membrane with a decrease in the volume of the matrix. Transition to state 4 results in an increase in volume of the matrix with a return of the regular pattern of cristae [Hackenbrock, 1967: Fig. 3.7A(1,2)]. Hence, functional activity of mitochondria may be associated with ultrastructural characteristics of the inner organelle membrane system. It is conceivable that control of the configuration of mitochondria by hormones such as thyroxine, or by polyelectrolytes or by other endogenous agents may exert influences on cellular energetics.

METHOD OF MEASUREMENT OF THE SWELLING-CONTRACTILE CYCLE IN MITOCHONDRIA. There are two types of swelling-contractile phenom-

TABLE 3.4

Mitochondrial States[a]

State	O_2	ADP	Substrate	Respiration
1	>0	Low	Low	Slow
2	>0	High	Low	Slow
3	>0	High	High	Fast
4	>0	Low	High	Slow
5	Zero (N_2)	High	High	Zero

[a]Taken from B. Chance and G. R. Williams, *Advan. Enzymol.* **17**, 65 (1956).

Several more recent "states" have been suggested: (a) State 6, induced by the addition of low concentrations of cations in the absence of phosphate and presence of a substrate. The inhibited rate of respiration (greater inhibition than that observed in state 4) is relieved by addition of phosphate. (b) State 7, specific inhibition of the energy-linked reversal of electron transport (NAD reduction) involving hyperbaric oxygen (Chance and Schoener, 1966).

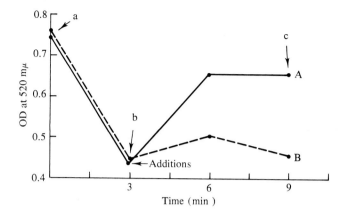

Fig. 3.19A. Large-amplitude mitochondrial swelling and contraction. Medium: 0.125 M KCl, 20 mM, Tris · Cl (pH7.4), 1 mM CaCl$_2$, 0.8 mg mitochondrial protein per milliliter (final volume is 3.0 ml). Additions: Curve A, 10 mM ATP, 6 mM MgCl$_2$, 4 mg/ml BSA; Curve B, 30 μg f$_{2a}$ histone, 10 mM ATP, 6 mM CaCl$_2$, 4 mg/ml BSA.

 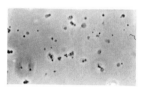

Fig. 3.19B. Phase micrographs of contracted and swollen mitochondria. Left: Aliquot taken at Point a or Point c of Fig. 3.19A. Right: Aliquot taken at Point b.

ena, one involving small changes in optical density (due to light scattering which appears to correspond to mitochondrial volume changes), called low-amplitude or "phase II swelling" (Chance and Packer, 1958; Packer, 1960, 1963) and the other involving a large-amplitude change called "phase II swelling" (Price et al., 1956; Lehninger, 1965). Both types are depicted in Fig. 3.19A and C.

The swelling agent (1 mM calcium in this case) is added to liver mitochondria, which is suspended in a buffered saline medium, at "a" (Fig. 3.19A). The decrease in optical density at 520 mμ represents the large-amplitude swelling. At point "b," 8 mM ATP + 3 mM MgCl$_2$ is added. The rapid increase in extinction represents the contractile phase. This process may be repeated several times. The line (----) represents the addition of an agent which inhibits the contractile phase, 30 μg of f$_{2a}$, a histone obtained from rat liver. The phase micrographs [Fig. 3.19(B)]

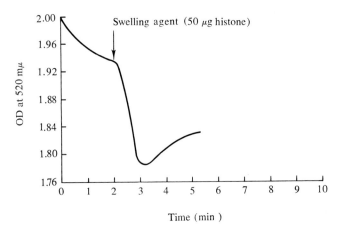

Fig. 3.19C. Low-amplitude mitochondrial swelling and contraction. Medium: 0.124 M KCl, 12.5 mM Tris·P$_i$ (pH 7.4), 2.5 mM Tris·succinate, 1.26 mg mitochondrial protein per milliliter.

clearly show the nature of the swelling process. In low-amplitude swelling, a greater amount of protein is usually used; the swelling agent effects a drop in optical density which under some conditions may reverse (Fig. 3.19C). It should be emphasized that the relationships between light-scattering changes, swelling and contraction and actual morphological alterations are still incompletely understood.

One of the disadvantages of the optical density type of volume change measurement is the inability to distinguish between membrane disintegration and extreme degrees of swelling. A method has recently been introduced which depends not upon volume change but upon a unique property of the membrane, its electrical resistance. The Coulter particle counter has been useful in measuring permeability changes with or without membrane disruption (Gebicki and Hunter, 1964).

The resemblance between mitochondrial swelling and contraction and muscle contraction and relaxation has led Lehninger to propose a uniform mechanistic hypothesis, as shown in Fig. 3.20. Two plausible explanations are represented: I. An alteration of membrane state occurs through ATP-driven changes in shape of protein molecule. The "mechanoenzyme" may be an intermediate in energy coupling (E) whose configuration depends upon the binding of P$_i$, ADP, or ATP. II. In this scheme, an independent membrane protein (possibly "phosphoprotein") may be activated by ATP to yield mechanical changes. These effects may be of fundamental significance in all active membrane systems.

1. "CONTRACTILE PROTEIN" IN MITOCHONDRIA

The demonstration of a muscle-type contractile-swelling cycle in isolated mitochondria supplied the impetus to seek a contractilelike protein in mitochondria. It should be mentioned that the presence of contractile elements in systems other than muscle has been known for some time. Hoffmann-Berling in 1956 extracted a contractile protein from sarcoma cells. Contractile proteins have also been found in blood platelets and in the tail sheath of bacteriophage. Spindle formation in the mitotic apparatus exhibits the properties of a contractile system (see Chapter VIII). The first demonstration of a contractile protein in mitochondria was reported by the Ohnishi's in 1962 who revealed some of the detailed properties of this protein such as Mg^{2+}-dependent ATPase activity, syneresis, "contraction" and "relaxation," of a glycerinated preparation, all of which strongly suggest actomyosin-like characteristics. The *in vitro* incorporation of amino acids into this protein can be demon-

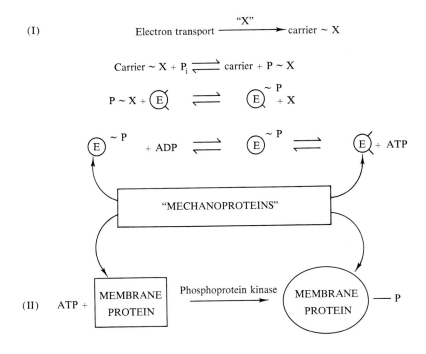

Fig. 3.20. The mechanoenzyme concept. From A. L. Lehninger, *in* "Biological Structure and Function" (T. Goodwin and O. Lindberg, eds.), Vol. 2, p. 45. Academic Press, New York, 1961.

strated. Moreover, both incorporation of labeled amino acids and ATPase activity of the protein are inhibited by puromycin and by actinomycin D. The possible presence of a contractile protein in mitochondrial membranes has important consequences. First, it further defines the fundamental similarity of all membrane-bound systems, cellular and subcellular. Second, it has implications in the mechanism of active ion transport in mitochondria (see Chapter VII). Third, it may have an important relationship to the coupling and uncoupling factors in oxidative phosphorylation. Fourth, it may explain the mechanism of mitochondrial motility, which has been postulated for years.

B. Reversal of Electron Transport in Mitochondria

Under certain conditions, the presence of succinate can cause a reverse flow of electrons in mitochondria, resulting in a reduction of NAD^+. This is illustrated in Fig. 3.21A. This is an ATP-requiring reaction which, of course, indicates that energy is necessary to drive the reaction to the left. The reduction of NAD^+ can also occur when electrons are "fed" into the chain at the cytochrome c level, by tetramethyl-p-phenylenediamine (TMPD) plus ascorbate or by various quinones plus ascorbate. The reaction is extremely sensitive to Amytal, antimycin A, thyroxine, and other respiratory chain and energy-transfer inhibitory substances and proceeds via the same carriers which convey electrons to oxygen. Recently, it has been suggested that a new cytochrome, cytochrome b_{555}, may be involved as a nonphosphorylated intermediate in the reversal action (Chance and Schoener, 1966). The physiological importance of this phenomenon, while still unclear, nevertheless offers extremely interesting and plausible possibilities. For example, a mechanism of active transport postulated in the early 1950's involved electron transport and possibly pyridine nucleotide (see Chapter VII). Even more

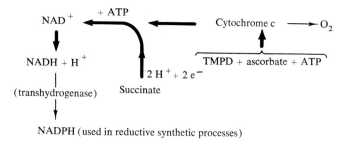

Fig. 3.21A. Reversal of electron transport (reductive transphosphorylation).

attractive is the recent development of the concept of (NAD^+/NADH) and ($NADP^+$/NADPH) ratios and their involvement in a variety of reductive synthetic processes as well as in catabolic activities. The backward flow of electrons described herein could conceivably regulate these parameters.

It is becoming more and more apparent that the classic concept of the sole function of the mitochondrion being an energy-generating machine needs to be altered. The rapidly accumulating evidence strongly suggests that high-energy intermediates are functioning in energy-linked *reductive* processes which bear importance in such physiological processes as general synthetic mechanisms, transport of ions across membranes and possibly, maintenance of membrane integrity.

C. DNA and RNA in Mitochondria

While it has been known for years that mitochondria are capable of actively incorporating amino acids into protein (Siekevitz, 1952), net protein synthesis has never been demonstrated. Early studies have also shown that inorganic phosphate could be incorporated into a crude mitochondrial fraction of liver, which may have been RNA. Hence the concept that cellular organelles like the mitochondrion might be capable of genetically directing their own synthetic activities has gained popularity in the past few years. "Cytoplasmic heredity" theories are not new however, originating perhaps with Meves in 1908. The existence of extranuclear DNA has been discussed for several years because of the known association of mitochondria with the nuclear envelope (see Chapter IV). The tacit assumption that any DNA present in mitochondria is due to nuclear contamination has hampered the recognition that small amounts might in fact be specific and significant.

In recent years, studies of extranuclear DNA were initiated in attempts to uncover the mechanism of morphogenesis of plant and animal mitochondria as well as plant plastids. The latter are cytoplasmic organelles which function in the photosynthetic process. Do these cellular structures originate *de novo* or from preexisting organelles or structures? The results of these experiments support the latter view. Using relatively specific staining techniques and electron microscopy, Swift *et al.* (1964) and Nass and Nass (1963) have observed the presence of intramatrical fiberlike material which appears to possess the characteristics of DNA. For example, DNase treatment of mitochondria results in a partial to complete disappearance of the fibers, without destroying the integrity of the mitochondria. DNA has been isolated, intact, from mitochondria of a variety of organisms. The mitochondrial DNA is partially resistant to

DNase treatment and exhibits a buoyant density significantly different from nuclear DNA. Furthermore, mitochondrial DNA, unlike nuclear DNA, is *renaturable* under certain conditions. It appears that mitochondrial DNA may be of the circular viral type (Nass, 1966). Incorporation of labeled precursors into mitochondrial RNA proceeds at an *in vitro* rate which exceeds that found in nuclei. Table 3.5 presents some information on content of and incorporation of precursors into nucleic acids present in isolated mitochondria. A typical RNA pattern from mitochondria appears in Fig. 3.21B. It is of interest that heart mitochondria not only contain more DNA than liver but apparently incorporate precursors into RNA at an accelerated rate as compared with liver mitochondria. Heart mitochondria are, of course, very much more active organelles (contain an abundance of cristae) than liver and consequently a more active DNA-RNA system might be expected. Mitochondria also apparently demonstrate the presence of DNA polymerase activity, ribosome, transfer and possible messenger RNA's.

After partial hepatectomy, the specific activity of mitochondrial DNA (determined by $^{32}P_i$ injection), 4 hours after $^{32}P_i$ was six times as great as nuclear DNA (Nass, 1967).

Very recently, a double-stranded DNA has also been isolated from erythrocyte membranes.

TABLE 3.5

DNA and RNA in Mitochondria

Mitochondrial source	DNA[a] (μg/mg protein)	RNA[b] (μg/mg protein)	Incorporation rate into RNA[c] (cpm/μg RNA)
Rat liver	0.65[d]	9[d]	1.1 (UTP-^{14}C)[e]
Rat liver after DNase treatment	0.21[d]	—	—
Calf heart	3.2 [f]	15	200 (Ur-^{3}H)
Calf heart after DNase treatment	0.22	—	—
Rabbit heart	—	14	800 (Ur-^{3}H)
Rat heart	—	14	5000 (Ur-^{3}H)

[a] DNA was determined by ultraviolet absorption and by the diphenylamine reaction.

[b] RNA was determined by the orcinol method after isolation by the procedure of Schmidt and Thannhauser (1945). (see Chapters I and VIII).

[c] The label (uridine or UTP) was added to a buffered saline medium containing mitochondria and incubated for 1 hour at 37° C.

[d] Data taken from Nass *et al.* (1965).

[e] Data taken from Neubert and Helge (1965). (All other data from the authors' laboratories.)

[f] Estimation of DNA in one mitochondrion, 5×10^{-11} μg (Corneo *et al.*, 1966).

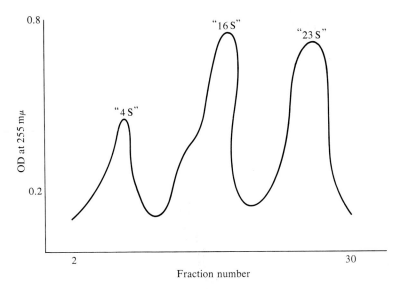

Fig. 3.21B. Sedimentation analysis of heart mitochondrial RNA. One hundred grams of dog heart was blended for 30 seconds in a medium containing $0.18\ M$ KCl, 10 mM EDTA, 0.5% BSA (bovine serum albumin), and 20 mM tris at pH 7.2 in a 10% w/v ratio. The suspension was subjected to centrifugations of 6×10^3 and 1.5×10^7 g-min. The pellet was suspended in 100 ml of the medium and centrifuged at 6×10^3 and 1.5×10^7 g-min. The pellet was suspended in 5 ml of the medium and layered over a 15% Ficoll solution. After centrifugation at 1.5×10^7 g-minute, the pellet was suspended in 5 ml and was subjected to a modified Kirby extraction for RNA. The RNA was dissolved in 1 ml of H_2O and layered over a 5–40% sucrose gradient. The gradient was centrifuged for 5.8×10^7 g-minute. The optical density of the gradient was determed by use of an ISCO optical density (at 255 mμ) analyzer.

What is the functional significance, if any, of extranuclear mitochondria-located DNA? Several rather attractive possibilities are worth mentioning. The mitochondrion may possess its own highly specialized protein-synthesizing apparatus sufficient to code for hundreds of proteins specific to the mitochondrion. Some of the genes may be of the constitutive type, i.e., they may be responsible for duplication of the organelle as well as for other biochemical activities such as specific enzyme synthesis. The DNA might also function as part of an intramitochondrial regulatory system, controlling the activity of some of the mitochondrial enzymes. It has recently been suggested that limited protein synthesis may occur of perhaps only certain enzymes located in the inner membrane system. The outer membrane constituents may be synthesized in the endoplasmic reticulum and transported to the mitochondrion. It is possible, too, that the DNA and/or RNA present in mitochondria function in a nongenetic manner. Polynucleotides have been implicated in the mechanism of oxidative phosphorylation.

It is clear from the foregoing that the generally accepted hypothesis that the nucleus is the exclusive site for cellular DNA has to be modified.

D. Histones and Mitochondria

Since the work of Smith and Conrad in 1956, it has been recognized that polycationic compounds have marked inhibitory effects on electron transport in isolated mitochondria, due to a binding action with cytochrome oxidase. The concentrations needed to produce these inhibitory effects, however, are "high," i.e., in milligram quantities. Much lower concentrations have been shown to stimulate, rather than inhibit, oxygen consumption of isolated mitochondria concomitant with a significant increase in ATPase activity, a decrease of the ADP-ATP exchange reaction, a swelling of the mitochondrial membrane, and a potassium efflux. This is diagrammatically pictured in Fig. 3.22. These observations implicate the nucleus in possible control of mitochondrial activity.

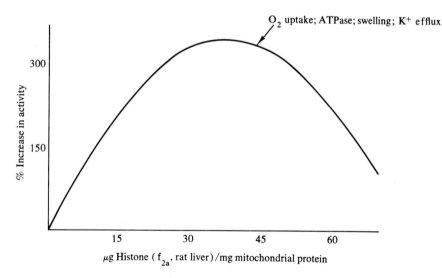

Fig. 3.22. A diagrammatic representation of the effects of histones on mitochondria. Taken from data in A. Schwartz, C. L. Johnson, and W. C. Starbuck, *J. Biol. Chem.* 241, 4505 and 4513 (1966); C. L. Johnson, C. M. Mauritzen, W. C. Starbuck, and A. Schwartz, *Biochemistry* 6, 1121 (1967).

E. Ion Transport and Related Activities in Mitochondria

Another general property of mitochondria which again emphasizes the multiphasic activities of these organelles is the ability to secrete or concentrate various ions (Na^+, K^+, Sr^{2+}, Mn^{2+}, Mg^{2+}, Ca^{2+}). This phenomenon originated from the studies of Bartley and Davies in 1952. Using mitochondria isolated from sheep kidney cortex, they calculated the ratio of a number of ions between intra- and extramitochondrial spaces, as shown in Table 3.6. They also found that the "turnover" of ^{42}K in mitochondria was quite high and suggested that the secretory

TABLE 3.6
Accumulation of Ions into Mitochondria[a]

Ion	Ratio of internal to external
H^+	2.5
Na^+	2.6
K^+	2.0
Mg^{2+}	4.5
PO_4^{3-}	6.0

[a] Adapted from Bartley and Davies (1952, 1956).

activity of some cells might in fact depend upon the mitochondria. Subsequent data confirmed the significant intracellular ionic gradients, including calcium ions, and suggested that the mitochondria are the "pumps" which enable cells to do osmotic work.

The importance of these observations lies not so much in the fact that mitochondria from a variety of plant and animal tissues can actively accumulate ions, but rather that it reemphasizes the interrelationships between electron transport, phosphorylative activities and ion transport, all having as a common basis the integrity of membranes (see Chapter II).

Ion accumulation is an energy-dependent process. Electron transport itself, however, is not an absolute requirement. The energy can be supplied by externally added ATP, the hydrolysis of which furnishes the required intermediates. Consequently, two possible energy pathways are involved: (1) an ATP-supported one which is inhibited by the inhibitor oligomycin, and (2) a substrate-supported one which appears to be insensitive to oligomycin. While it is apparent that heart mitochondria can accumulate, in the presence of inorganic phosphate, up to 1000 μmoles of calcium per milligram mitochondrial protein per minute, the functional significance still remains unclear. This massive accumulation involves an intramitochondrial precipitation of hydroxyapatite:

$$[Ca_3(PO_4)_2]_3 \cdot Ca(OH)_2$$

When large amounts of calcium accumulate, however, an uncoupling of oxidative phosphorylation and an increase in ATPase activity, mitochondrial swelling, and a stimulation of endogenous free fatty acids result. In addition, the formation of insoluble complexes (hydroxyapatite) leads to an ejection of H ions. The physiological significance of this type of reaction is questionable. However, very small amounts of calcium are transported over two times as rapidly as larger amounts—a reaction which occurs in the absence of inorganic phosphate. Accompanying this reaction is a short burst of oxygen consumption which rapidly returns to normal or control levels (Fig. 3.23). This reaction, then, is not similar to the uncoupling type induced by larger amounts of calcium but is analogous to the effects of ADP on a mitochondrial suspension (see Section III, D). The extremely rapid uptake of small amounts of calcium may involve a high-energy intermediate in the first phosphorylation site

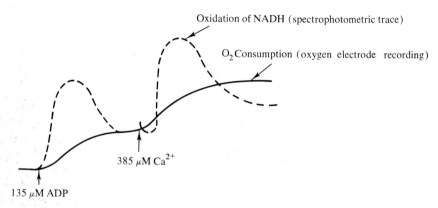

Oxidation of NADH (spectrophotometric trace)

O_2 Consumption (oxygen electrode recording)

385 μM Ca^{2+}

135 μM ADP

Fig. 3.23. Effect of low concentrations of calcium compared to ADP on mitochondrial redox activity. Taken from B. Chance, "Energy-Linked Functions of Mitochondria." Academic Press, New York, 1963.

(NADH $\sim$ I or NADH $\sim$ P) and may also involve cytochrome b and cytochrome b$_{555}$.

The effect of even lower levels of calcium on mitochrondrial respiration is depicted in Fig. 3.24. Notice the short burst of oxygen consumption after the addition of calcium and the subsequent return of respiration to the control value. Measurements of calcium binding to mitochondria, during this period, show that calcium is accumulated stoichiometrically from the medium in a ratio of 1.85 to 2.0 molecules of Ca^{2+} per energy-conserving site (per 2 electrons), a ratio which is consistent with the extra oxygen consumed.

The effects of calcium ions on mitochondria have been stressed in this section because of the intimate involvement of this cation in membrane functions in general, particularly in the process of muscle contrac-

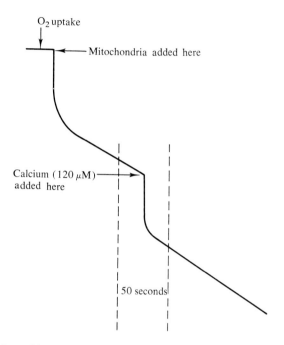

Fig. 3.24. The effect of low calcium on mitochondria. Oxygen electrode trace. Adapted from C. S. Rossi and A. L. Lehninger, *J. Biol. Chem.* **239**, 3971 (1964).

tion and relaxation (see Chapter IX). The mitochondrion serves as a membrane model and many of the effects discussed may in fact be related to general membrane activity.

As already mentioned, potassium is another important cation which is actively accumulated by the mitochondrion, and one which has received a great deal of attention in recent years. Pressman, employing a number of antibiotics, has found that one in particular, valinomycin, appears to trigger the influx of K^+ (Figs. 3.25 and 3.26B and C). The rapid uptake of K^+ is accompanied by changes in light scattering (swelling), by a significant increase in incorporation of inorganic phosphate into ATP, and by an increase in oxygen consumption. There may be specialized mitochondrial "receptors" which react with active inducers (e.g., valinomycin, gramicidin) of transport. This defines the transduction of chemical bond energy into mechanical work, the latter being either active cation transport or contraction and swelling. The cation pump may be of basic importance in mitochondrial function (Fig. 3.26A).

It should be apparent from Fig. 3.26A that an intermediate "~ X" on the main pathway of ATP production may be rate limiting, i.e., a competition exists between ion transport and ATP formation. This is conveniently

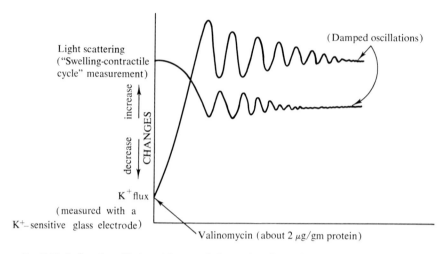

Fig. 3.25. Induced oscillations (alternate influx and outflux) of K^+ in mitochondria. Taken from B. C. Pressman, *Symp. Mitochondria at 5th Ann. Meeting Am. Soc. Cell Biol., Philadelphia, Pennsylvania* (1965).

Flux of ^{42}K into Mitochondria

Condition	μmoles of K^{42}/gm protein/min
Control	0.2
Valinomycin (30 μg/gm protein)	100
Valinomycin + 2,4-DNP (100 μmoles)	80

demonstrated, in fact, in Figs. 3.26B and C. A potassium-sensitive electrode combined with an oxygen electrode recorded changes in potassium movements and oxygen consumption in isolated heart mitochondria. Note that the addition of ADP, the phosphate acceptor, after a valinomycin-induced K^+ influx occurred, causes the K^+ to be extruded. At the exact point when ADP is completely esterified to ATP, K^+ reenters the mitochondria. These data indicate that when the equilibrium of phosphorylation reactions is directed toward the formation of ATP, ion transport becomes limited. They further suggest the presence of an intermediate(s) common to both ion transport and oxidative phosphorylation.

Oxidative phosphorylation, contraction and swelling, and cation transport in mitochondria are interdependent and are probably of prime importance in the regulation of cellular activity.

Another and very important concept which has developed as a result

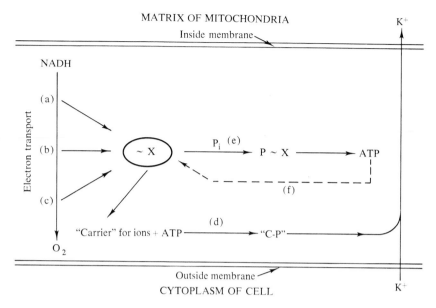

Fig. 3.26A. Cation pump in mitochondria as primary directing force. *Note:* The energy-rich intermediate ($\sim$ X) may be produced either by electron transport (a), (b), (c), or by ATP hydrolysis (f). The intermediate is subsequently involved in the formation of other compounds ("C-P"), which function in mechanical work (active cation transport, swelling-contractile cycle, etc.). The phosphorylation of the ion-transport carrier (d) is an oligomycin-insensitive step. Adapted from M. Hofer and B. C. Pressman, *Biochemistry* **5,** 3919 (1966).

of cation studies in mitochondria involves possible intracellular pH control. It is known that the active uptake of cations is accompanied by the production of hydrogen ions. Accompanying the uptake of 2 calcium ions per site, for example, is an ejection of 2 protons and an accumulation of 2 hydroxyl anions. Cation accumulation is an energy-dependent process supported either by electron transport or by ATP hydrolysis. The ejection of H^+ from the mitochondrion in response to the influx of Ca^{2+} has prompted a number of investigators to propose that the *primary* basis for mitochondrial transport is an activation of an outwardly directed hydrogen ion pump. This is an extension of a theory originally promulgated by Mitchell and is fully discussed in Chapter VII, and on pp. 129-131.

F. Mitochondrial Morphogenesis

It will be recalled that mitochondria are limited by a double membrane similar in some respects in appearance and size to those membranes

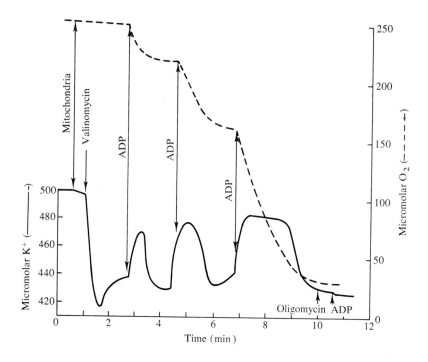

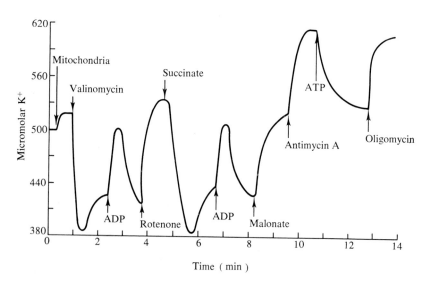

Fig. 3.26B (*top*), **C** (*bottom*). Demonstration of the competition between ion transport and ATP formation. Taken from B. Safer and A. Schwartz, *Circulation Res.* 21, 25 (1967).

associated with the surface of the cell, and subcellular organelles. It has therefore been suggested at varying times that mitochondria may arise from the cell surface membrane (plasma membrane), the nuclear membrane, endoplasmic reticulum, and practically every membrane-lined cellular inclusion. Probably the most provocative of the numerous hypotheses was the one proposed by Robertson (1959), diagrammed in Fig. 3.27. Robertson frequently observed in electron micrographs small fingerlike evaginations of cell surfaces, particularly of muscle fibers ("p" in Fig. 3.27). If these evaginations were to extend down into invaginations in the cell membrane, which also have been observed, the beginning of one of the mitochondrial bodies ("tM") would be formed. The inner membrane could then fold to become Palade's *cristae mitochondriales.* The connection or "umbilical cord" would then break or pinch off. While this was an interesting theory, there is little or no evidence to support it. Robertson has also proposed that the mitochondria may possibly be extensions of the endoplasmic reticulum, by a similar process. Modern biological approaches however, have led to more plausible theories of morphogenesis (see Section IV, C).

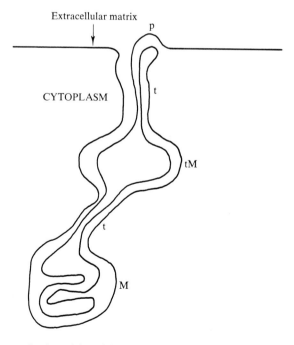

Fig. 3.27. Robertson's view of the origin of mitochondria. Redrawn from J. D. Robertson, *Biochem. Soc. Symp. (Cambridge, Engl.)* **16**, 3 (1959).

THE MITOCHONDRION

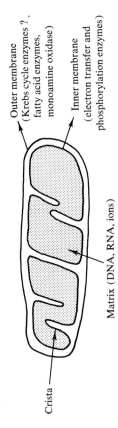

Outer membrane
(Krebs cycle enzymes ?,
fatty acid enzymes,
monoamine oxidase)

Inner membrane
(electron transfer and
phosphorylation enzymes)

Crista

Matrix (DNA, RNA, ions)

Functions:

1. Oxidative – phosphorylation (electron transport, ATPase activity, etc.)
2. Swelling – contractile cycle
3. Ion transport
4. Reversal of electron transport (reductive transphosphorylation)
5. DNA and RNA
6. Transhydrogenase reaction

THE ELECTRON TRANSPORT CHAIN

Electrons and Protons (H$^+$) Are Transported in This Direction ⟶
when energy is "stored" as ATP

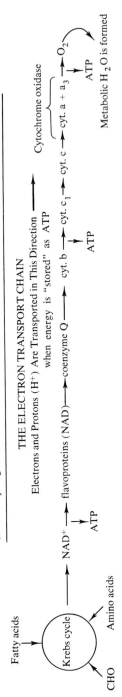

Fatty acids

CHO

Krebs cycle

Amino acids

Fig. 3.28. The mitochondrion. Note: The *exact* locations of DNA, RNA, and the Krebs cycle enzymes are unknown.

G. Energy-Linked Transhydrogenase Reactions

Another important energy-requiring process associated with mitochondria is the production of reduced nicotinamide-adenine dinucleotide phosphate (NADPH) from reduced nicotinamide-adenine dinucelotide (NADH) as follows:

$$NADH + NADP^+ + I \sim X \longrightarrow NAD^+ + NADPH + I + X$$

The "$I \sim X$" indicates that this reaction occurs with the expenditure of one "high energy" intermediate generated at one coupling site of the respiratory chain (see Fig. 3.15). ATP is not, therefore, directly involved. This may constitute a physiological regulatory device between NAD^+-linked oxidative catabolism and NADPH-linked reductive anabolism (Fig. 3.21A). Further investigations are needed to shed light on what may be a vital energy-linked mitochondrial process.

We have discussed in some detail at least three processes in mitochondria which function with the *expenditure* of high-energy intermediates generated during the process of electron transfer and phosphorylation: reversed electron transport, active ion translocation, and the transhydrogenase reaction. Our thinking concerning this complex organelle has changed considerably in the past ten years, so that we no longer look at the mitochondrion as a one-way energy liberating machine.

V. CONCLUSIONS

The past decade has been one of intensive activity concerning morphological and biochemical correlates of mitochondrial function. As a result of these endeavors, we are now closer to an understanding of the fundamental mechanism of electron transport, oxidative phosphorylation, and possible hormonal control of cellular activity. There has been, during this period, a great interest in the role of the mitochondrion in a variety of disease states. It is not inconceivable that we will soon unify the knowledge acquired from these efforts into an integrated picture of the methods by which the mitochondrion, an organelle of minute dimensions, can control and modify the very existence of the cell (summary, Fig. 3.28).

APPENDIX

Extramitochondrial Cytoplasmic Metabolism

Summaries of the various known oxidative pathways present in the extramitochondrial cytoplasmic regions of the cell are presented below:

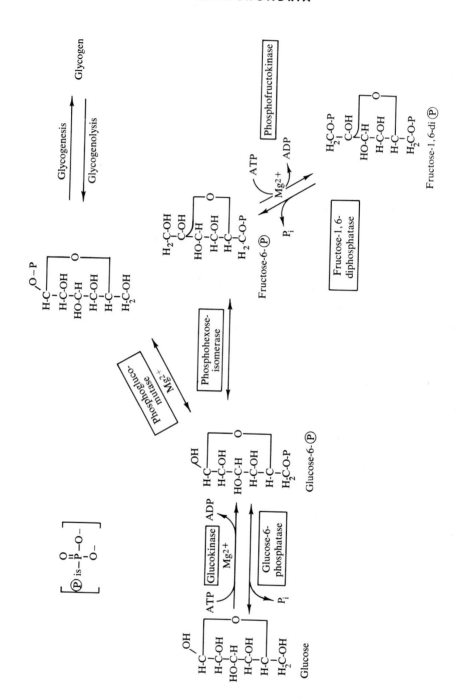

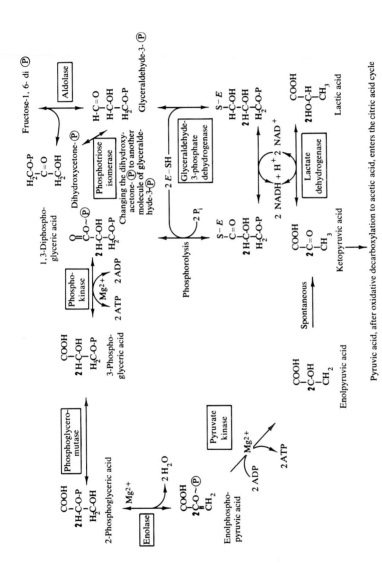

SCHEME 2. Embden–Meyerhof pathway (anaerobic glycolysis).

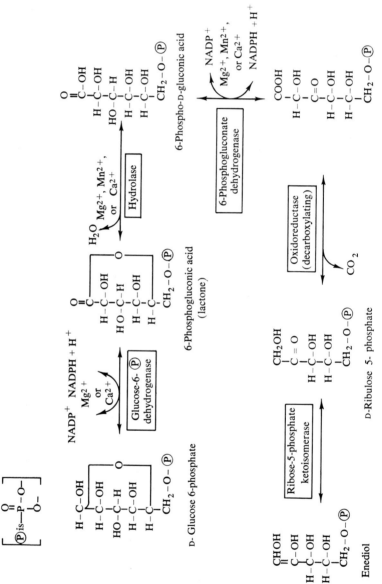

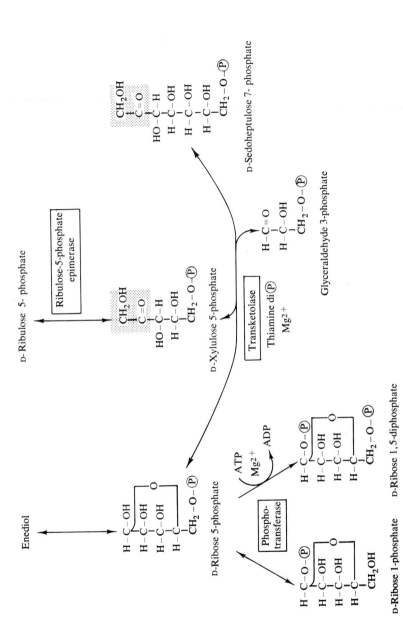

SCHEME 3. The direct oxidative pathway (hexose monophosphate shunt).

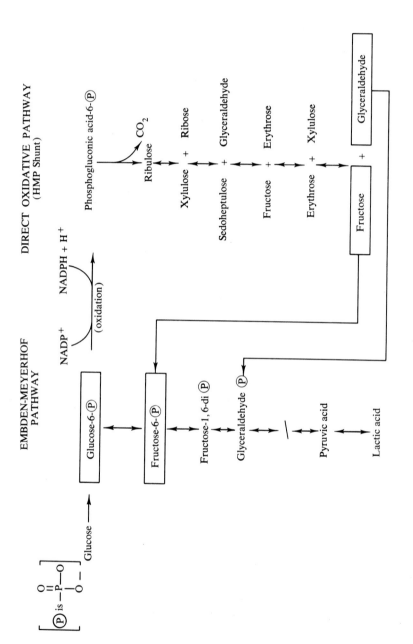

SCHEME 4. Comparison of Embden-Meyerhof pathway and direct oxidative pathway of glycolysis.

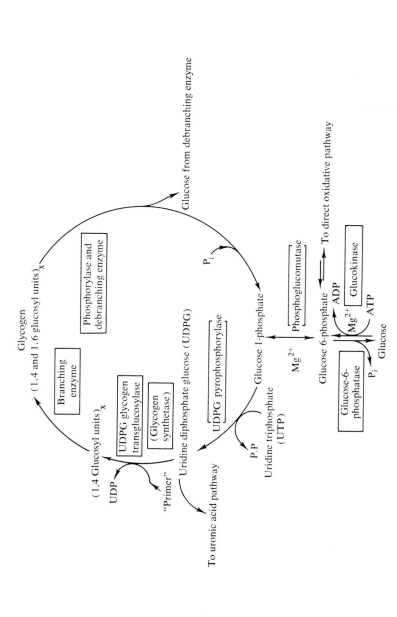

SCHEME 5. Pathway of glycogenesis and glycogenolysis. Taken from H. A. Harper, "Review of Physiological Chemistry," 10th ed. Lange Med. Publ., Los Altos, California.

GENERAL REFERENCES

Allfrey, V. (1959). *In* "The Cell" (J. Brachet, and A. E. Mirsky, eds.), Vol. 1, p. 193. Academic Press, New York.

Belitzer, V. A., and Tzibakova, E. T. (1939). *Biokhimiya* 4, 516.

Chance, B. (1963). "Energy-Linked Functions of Mitochondria." Academic Press, New York.

Chance, B. (1965). *J. Gen. Physiol.* 49, 163.

Chance, B., and Hollunger, G. (1961). *J. Biol. Chem.* 236, 1534.

Chance, B., and Mela, L. (1967). *J. Biol. Chem.* 242, 830.

Chance, B., and Schoener, B. (1966). *J. Biol. Chem.* 241, 4567 and 4577.

Chance, B., and Williams, G. R. (1956). *Advan. Enzymol.* 17, 65.

Fernández-Morán, H. (1962). *Circulation* 26, 1039.

Gaebler, O. H., ed. (1956). "Enzymes: Units of Biological Structure and Function." Academic Press, New York.

Green, D. E. (1959). *Advan. Enzymol.* 21, 73.

Green, D. E. (1962). *Comp. Biochem. Physiol.* 4, 81.

Greenberg, D. M., ed. (1954). *Chemical Pathways of Metabolism*, Vol. I, Academic Press, New York. (Krebs, H. A., The Tricarboxylic Acid Cycle, p. 109; Pardee, A. B., Free Energy and Metabolism, p. 1).

Harper, H. A. (1965). "Review of Physiological Chemistry," 10th ed. Lange Med. Publ., Los Altos, California.

Klotz, I. M. (1957). "Some Principles of Energetics in Biochemical Reactions." Academic Press, New York.

Krebs, H. A. (1954). *Chem. Pathways Metab.* 1, 109.

Lehninger, A. L. (1960). *Sci. Am.* 202, 102.

Lehninger, A. L. (1961). *In* "Biological Structure and Function" (T. W. Goodwin and O. Lindberg, eds.), Vol. 2, p. 45. Academic Press, New York.

Lehninger, A. L. (1964). "The Mitochondrion." Benjamin, New York.

Lehninger, A. L., Wadkins, C. L., Cooper, C., Devlin, T. M. and Gamble, J. L., Jr. (1958). *Science* 128, 450.

Lipmann, F. (1946). "Currents in Biochemical Research," p. 137. Wiley (Interscience), New York.

Novikoff, A. B. (1961). *In* "The Cell" (J. Brachet and A. E. Mirsky, eds.), Vol. 2, p. 299. Academic Press, New York.

Pardee, A. B. (1954). *Chem Pathways Metab.* 1, 1.

Pressman, B. C. (1965). *5th Symp. Mitochondria at Ann. Meeting Am. Soc. Cell Biol., Philadelphia, Pennsylvania* [see Harris, E. J., Cockrell, R., and Pressman, B. C. (1966). *Biochem. J.* 99, 200; Harris, E. J., Catlin, G., and Pressman, B. C. (1967). *Biochemistry* 6, 1360; Höfer, M., and Pressman, B. C. (1966). *Ibid.* 5, 3919.]

Quagliariello, E., Papa, S., Slater, E. C., and Tager, J. M. (1967). "Mitochondrial Structure and Compartmentation." Adriatica Editrice, Bari.

Racker, E. (1968). *Sci. Am.* 218, 32.

Safer, B., and Schwartz, A. (1967). *Circulation Res.* 21, 25.

Schneider, W. C. (1948). *J. Biol. Chem.* 176, 259.

Schneider, W. C. (1959). *Advan. Enzymol.* 21, 1.

Schwartz, A., and Johnson, C. L. (1965). *Life Sci.* 4, 1555.

Schwartz, A., Johnson, C. L., and Starbuck, W. C. (1966). *J. Biol. Chem.* 241, 4505 and 4513.

Tager, J. M., Papa, S., Quagliariello, E., and Slater, E. C. (1966). "Regulation of Metabolic Processes in Mitochondria." Elsevier, Amsterdam, 7.

Umbreit, W. W., Burris, R. H., and Stauffer, J. F. (1964). "Manometric Techniques," 4th ed. Burgess, Minneapolis, Minnesota.

SPECIFIC REFERENCES

ION TRANSPORT

Bartley, W., and Davies, R. E. (1952). *Biochem. J.* **52**, XX.
Bartley, W., and Davies, R. E. (1956). *Biochem. J.* **64**, 754.
Brierley, G. P., Murer, E., and Bachmann, E. (1964). *Arch. Biochem. Biophys.* **105**, 89.
Greenawalt, J. W., Rossi, C. S., and Lehninger, A. L. (1964). *J. Cell Biol.* **23**, 21.
Lardy, H. A., Connelly, J. L., and Johnson, D. (1964). *Biochemistry* **3**, 1961.
Lehninger, A. L., Carafoli, E., and Rossi, C. S. (1967). *Advan. Enzymol.* **29**, 259.
Mitchell, P., and Moyle, J. (1967). In "Biochemistry of Mitochondria." (E. C. Slater, F. Kaninga, and L. Wojtezak, eds.), p. 53. Academic Press, New York.
Pressman, B. C., (1965a). *Federation Proc.* **24**, 425.
Pressman, B. C. (1965b). *Proc. Natl. Acad. Sci. U.S.* **53**, 1076.
Rasmussen, H., Chance, B., and Ogata, E. (1965). *Proc. Natl. Acad. Sci. U.S.* **53**, 1069.
Rossi, C. S., and Lehninger, A. L. (1964). *J. Biol. Chem.* **239**, 3971.
Slater, E. C., and Cleland, K. W. (1953). *Biochem. J.* **55**, 566.

INHIBITORS

Chance, B. and Hollunger, G. (1963). *J. Biol. Chem.* **238**, 418.
Ernster, L., Dallner, G., and Azzone, G. F. (1963). *J. Biol. Chem.* **238**, 1124.
Estabrook, R. W. (1962). *Biochem. Biophys. Acta* **60**, 249.
Lardy, H. A., Johnson, D., and McMurray, W. C. (1958). *Arch. Biochem. Biophys.* **78**, 587.
Lardy, H. A., Connelly, J. L., and Johnson, D. (1964). *Biochemistry* **3**, 1961.
Potter, V. R., and Reif, A. E. (1952). *J. Biol. Chem.* **194**, 287.

UBIQUINONE AND VITAMIN K

Pumphrey, A. M., and Redfearn, E. R. (1960). *Biochem. J.* **76**, 61.
Redfearn, E. R. (1961). *CIBA Found. Symp. Quinones Electron Transport* p. 346. Churchill, London.
Wosilait, W. D. (1961). *Federation Proc.* **20**, 1005.

COUPLING FACTORS

Conover, T. E., Prairie, R. L., and Racker, E. (1963). *J. Biol. Chem.* **238**, 2831.
Green, D. E., Beyer, R. E., Hansen, M., Smith, A. L., and Webster, G. (1963). *Federation Proc.* **22**, 1460.
Lehninger, A. L. (1962). *J. Biol. Chem.* **237**, 946.
Neubert, D., Rose, T. H., and Lehninger, A. L. (1962). *J. Biol. Chem.* **237**, 2025.
Penefsky, H. S., Pullman, M. E., Datta, A., and Racker, E. (1960). *J. Biol. Chem.* **235**, 3322 and 3330.
Pullman, M. E., and Monroy, G. C. (1963). *J. Biol. Chem.* **238**, 3762.
Racker, E. (1967). *Federation Proc.* **26**, 1335.
Racker, E., and Conover, T. E. (1963). *Federation Proc.* **22**, 1088.

Racker, E., Tyler, D. D., Estabrook, R. W., Conover, T. E., Parsons, D. F., and Chance, B. (1965). *In* "Oxidases and Related Redox Systems" (T. E. King, H. S. Mason, and M. Morrison, eds.), Vol. 2, p. 1077, Wiley, New York.

PHOSHORYLATED INTERMEDIATES ("HIGH ENERGY")

Ahmed, K., Judah, J. D., and Gallagher, C. H. (1961). *Nature* **191**, 1309.
Peter, J. B., Hultquist, D. E., DeLuca, M., Kreil, G., and Boyer, P. D. (1963). *J. Biol. Chem.* **238**, 1182.
Slater, E. C. (1956). *Proc. 3rd Intern. Congr. Biochem., Brussels, 1955* p. 264. Academic Press, New York.
Slater, E. C., and Kemp, A., Jr. (1964). *Nature* **204**, 1268.
Slater, E. C., Kemp, A., Jr., and Tager, J. M. (1964). *Nature* **201**, 781.

CONTRACTILE-SWELLING CYCLE

Gebicki, J. M., and Hunter, F. E., Jr. (1964). *J. Biol. Chem.* **239**, 631.
Hackenbrock, C. R. (1966). *Federation Proc.* **25**, 414; (1966). *J. Cell Biol.* **30**, 269.
Lehninger, A. L. (1959). *J. Biol. Chem.* **234**, 2465.
Ohnishi, T., Kawamura, H., Takeo, K., and Watanabe, S. (1964). *J. Biochem. (Tokyo)* **56**, 273.
Packer, L. (1963). *J. Cell Biol.* **18**, 487 and 495.
Raaflaub, J. (1953). *Helv. Physiol. Pharmacol. Acta* **11**, 142.

HISTONES AND MITOCHONDRIA

Schwartz, A. (1965). *J. Biol. Chem.* **240**, 939 and 944.
Johnson, C. L., Mauritzen, C. M., Starbuck, W. C. and Schwartz, A. (1967). *Biochemistry* **6**, 1121.

DNA AND RNA IN MITOCHONDRIA

Corneo, G., Moore, C., Sanadi, D. R., Grossman, L. I., and Marmur, J. (1966). *Science* **151**, 687.
Gibor, A., and Granick, S. (1964). *Science* **145**, 890.
Kalf, G. F. (1964). *Biochemistry* **3**, 1702.
Luck, D. J. L., and Reich, E. (1964). *Proc. Natl. Acad. Sci. U.S.* **52**, 931.
Marshak, A., and Calvet, F. (1949). *J. Cellular Comp. Physiol.* **34**, 451.
Meves, F. (1908). *Arch. Mikroskop. Anat, Entwicklungsmech.* **72**, 816.
Nass, S. (1967). *Biochim. Biophys. Acta* **145**, 60.
Nass, M. M. K. (1966). *Proc. Natl. Acad. Sci. U.S.* **56**, 1215.
Nass, M. M. K., and Nass, S. (1963). *J. Cell Biol.* **19**, 593 and 613.
Nass, S., Nass, M. M. K., and Hennix, U. (1965). *Biochim. Biophys. Acta* **95**, 426.
Neubert, D., and Helge, H. (1965). *Biochem. Biophys. Res. Commun.* **18**, 600.
Philipson, L., and Zetterqvist, Ö. (1964). *Biochim. Biophys. Acta* **91**, 171.
Pinchot, G. B., and Hormanski, M. (1962). *Proc. Natl. Acad. Sci. U.S.* **48**, 1970.
Schmidt, G., and Thannhauser, S. J. (1945). *J. Biol. Chem.* **161**, 83.
Siekevitz, P. (1952). *J. Biol. Chem.* **195**, 549.
Sinclair, J. H., Stevens, B. J., Gross, N., and Rabinowitz, M. (1967). *Biochim. Biophys. Acta* **145**, 528.
Swift, H., Kislev, N., and Bogorad, L. (1964). *J. Cell Biol.* **23**, 91A.

MITOCHONDRIAL MORPHOGENESIS AND MEMBRANES

Chance, B., and Parsons, D. F. (1963). *Science* **142**, 1176.
Chance, B., Parsons, D. F., and Williams, G. R. (1964). *Science* **143**, 136.
Fernández-Morán, H. (1962). *Circulation* **26**, 1039.
Fernández-Morán, H., Oda, T., Blair, P. V., and Green, D. E. (1964). *J. Cell Biol.* **22**, 63.
Robertson, J. D. (1959). *Biochem. Soc. Symp (Cambridge, Engl.)* **16**, 3.
Schjeide, O. A., McCandless, R. G., and Munn, R. J. (1964). *Nature* **203**, 158.
Schnaitman, C. A., Erwin, V. G., and Greenawalt, J. W. (1967). *J. Cell Biol.* **32**, 719.

TRANSHYDROGENASE REACTIONS

Lee, C.-P., and Ernster, L. (1966). *In* "Regulation of Metabolic Processes in Mitochondria" (J. M. Tager *et al.*, eds) Vol. 7, p. 218, Elsevier, Amsterdam.

Chapter IV

THE NUCLEUS AND ITS INCLUSIONS

I. INTRODUCTION

One of the most fascinating of subcellular constituents is the nucleus, an organelle, described in 1833 by the botanist Robert Brown, in the cellular juices from orchids. Its function in cellular activity has been gradually elucidated, and, today, there is at least one complete branch of science devoted to the study of these functions, the science of genetics. However, recognition of the importance of this organelle was not immediately afforded but was severely hampered by the lack of suitable methods for the isolation of relatively pure nuclei in good yield. With the development of adequate methodology, the elucidation of nuclear activities progressed rapidly.

II. ISOLATION OF NUCLEI

A number of methods for the isolation of nuclei exist, each of which possesses certain merits. The cell membrane is ruptured by grinding the tissue in a vessel with a glass pestle, e.g., homogenization. Recently, Teflon and glass-coated Teflon have replaced glass in the construction of the pestle, resulting in a diminished production of heat and in the absence of small glass particles which may prove inhibitory to certain enzymes. The temperature at which these procedures are conducted is extremely critical and must approach 4°C.

One of the earliest recorded procedures for the isolation of relatively pure nuclei was introduced by Behrens in 1932 and involved the utilization of nonaqueous techniques. The tissue suspension was frozen and lyophilized, i.e., freeze-dried, and then worked up in organic solvents such as mixtures of carbon tetrachloride with benzene or cyclohexane. The nuclei were subsequently concentrated by centrifugation. The non-aqueous techniques do not leach out any water-soluble macromolecules and in this regard, offer some advantage.

The aqueous techniques employ a variety of materials as media for homogenization. The substance most commonly employed in a homogenizing medium is sucrose, and methods employing the latter were introduced by W. C. Schneider. Sucrose is used in concentrations ranging from 0.20 to 0.88 M. Many investigators have included calcium chloride (0.002 − 0.005 M) in the sucrose solutions which may function by preventing the loss of macromolecules from the nucleus.

Citric acid solutions (2.5 − 5.0%) were introduced by Stoneberg and Haven in 1940 and exploited by Marshak and co-workers as media for the isolation of nuclei. The use of these solutions of low pH, i.e., pH 3 − 4, facilitates the isolation of intact nuclei by preventing the loss of water-soluble organic molecules. Pepsin has also been employed as an adjuvant in this treatment to further eliminate cytoplasmic contamination. The citric acid method does, however, produce a "hardening" of the nuclear envelope, although the method has been improved by incorporating isotonic sucrose solutions into the homogenizing medium.

Most of the methods do not yield a sufficiently pure preparation of nuclei in good yield. Presently, the method of choice involves the sedimentation of nuclei from very dense sucrose solutions, i.e., 2.2 M, containing 3mM calcium chloride and further purification in 1 M sucrose (Chauveau procedure). An example of the quality of the nuclei isolated in this fashion is presented in Fig. 4.1.

Unfortunately, no universal protocol is available for the isolation of purified nuclei from a variety of tissues. The techniques are dependent upon the particular tissue, the interests of the investigator, and the relative degree of purity which is required. All the procedures must be monitored by phase or light microscopy of unstained preparations, or stained with azure C (Fig. 4.2).

III. MORPHOLOGY OF THE NUCLEUS

Reference will be made to the electron micrograph of a nucleus illustrated in Fig. 4.3. The nucleus possesses, in general, a spherical shape although in some cells (e.g., leukocytes, spermatozoa) the shape may be irregular or nonspherical. Its size is variable but appears to be related to the amount of cytoplasm. The diameter of the smallest nucleus in some fungi is 1 μ while the largest nucleus, found in some oocytes, may be as large as several hundred microns. The average volume of animal and plant nuclei ranges from 8 to 2000 cubic microns.

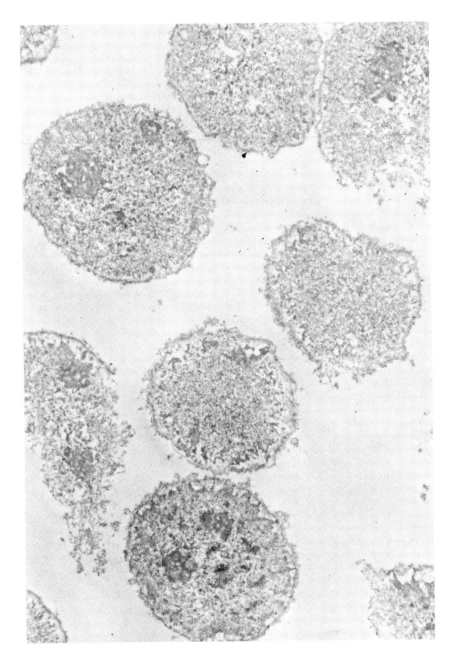

Fig. 4.1. Isolated nuclei. Nuclei were isolated from rat liver by the Chauveau method and processed for electron microscopy, i.e., fixed in glutaraldehyde, stained in OsO₄ and post-stained with uranyl acetate. × 10,000. Courtesy of H. Busch and K. Shanker.

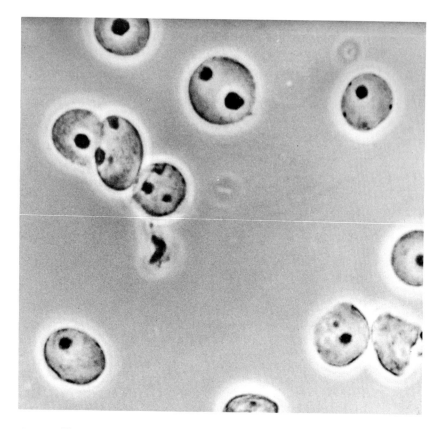

Fig. 4.2. Phase microscopy of isolated nuclei. Nuclei, isolated from the livers of "weanling" rats by the Chauveau method, were viewed under phase microscopy. × 1800. Note the pronounced nucleoli.

The nucleus contains (1) a *nuclear membrane* which bounds the nucleus separating the latter from the cytoplasm (NM); (2) *nuclear sap* which fills the nucleus and in which the nuclear inclusions are embedded; (3) interlaced filaments called *chromatin* (CH), the basophilic nuclear material composed of deoxyribonucleoproteins; (4) interchromatinic particles (I) made up largely of ribonucleoprotein and whose function is not completely understood; (5) *nucleoli*, very dense intensely staining structures (Nu).

1. NUCLEAR MEMBRANE

Since no method for the mass isolation of this structure exists, much of our knowledge of the nuclear membrane stems from data which has been obtained from electron microscopy. Electron micrographs reveal the nuclear envelope as two concentric membranes (Figs. 4.4 and 4.5). The outside layer (closest to the cytoplasm), which is approximately 90 Å in width, may have attached dense bodies of ribonucleoprotein called *ribosomes*. The structure and function of the ribosomes will be discussed in Chapter V. The innermost layer (nuclear side), which is also approximately 90 Å in width, is separated from the outer by a variable distance of approximately 140 Å. Each of these two parts of the nuclear membrane, i.e., the nuclear and cytoplasmic, consists of a trilamellar unit membrane resembling other membranes found within the cell. It has

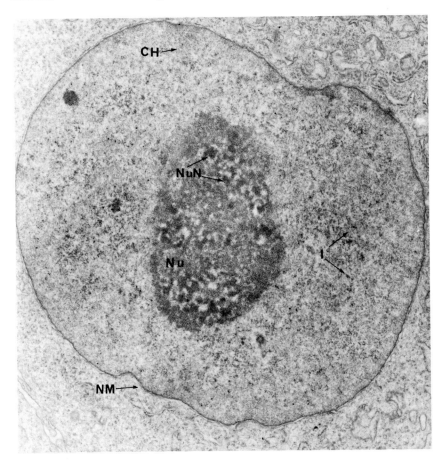

Fig. 4.3. Electron micrograph of the nucleus (Walker carcinosarcoma) *in situ*. NM, nuclear membrane; Nu, nucleolus; NuN, nucleolonema; CH, chromatin; I, interchromatinic particles. The light areas within the nucleolus are the internucleolonema regions. The Walker tumor cell had been fixed with glutaraldehyde, stained in OsO_4 and poststained with uranyl acetate. × 19,000. Courtesy of H. Busch and K. Shankar.

been suggested that all unit membranes have a common origin. The nature of this progenitor, however, is obscure although certain evidence has implicated the endoplasmic reticulum. The inner and outer layers appear to fuse to a fine fibrillar structure (in perpendicular sections) and form a thin monolayer for a short distance, the annulus or nuclear pore (NP) (see Figs. 4.4 and 4.5).

The existence of the nuclear pores has suggested a role for them in the transport of materials between the nucleus and the cytoplasm. The number of nuclear pores on a membrane bears some relation to the biosynthetic activity of the cell; the germinal vesicles of oocytes possess

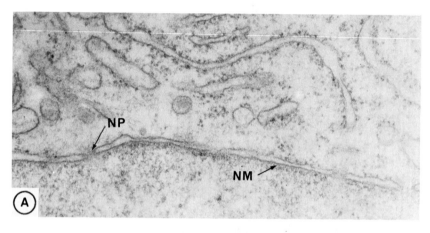

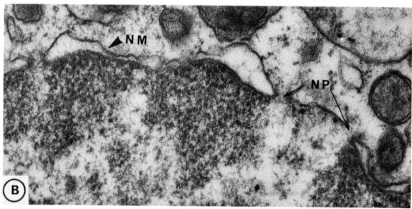

Fig. 4.4. Nuclear membrane and pores. (A) Nuclear membrane (NM) of a Walker tumor cell *in situ.* The cell was fixed in glutaraldehyde, stained in OsO_4 and poststained with uranyl acetate. × 58,000. Note the nuclear pores (NP). Courtesy of H. Busch and K. Shankar. (B) A portion of the nucleus of an undifferentiated cell from human aorta atherma, stained with uranyl acetate. × 44,600. Courtesy of R. M. O'Neal and G. Adams.

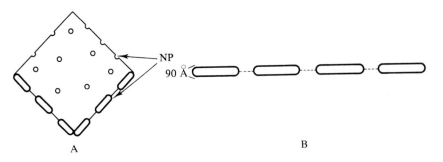

Fig. 4.5. Diagrammatic representation of the (A) nuclear membrane and (B) nuclear pores.

nuclei with a great number of pores. The nucleus of the oocyte from the growing frog possesses 40% more pores per unit surface area than the mature oocyte nucleus. In the nucleus from the primary spermatocyte, those regions of the membrane in which are centered the nucleoli and chromosomes (i.e., the hyper-active centers) have many pores.

Another view of the nuclear membrane depicting the nuclear pores and trilamellar membranes is presented in Fig. 4.5B. The diameter of the nuclear pores may vary from 300 to 1000 Å, but is large enough to permit the passage of macromolecules.

2. CHROMOSOMES, NUCLEOLI

The morphology and biochemical functions of these structures will be discussed later.

IV. NUCLEAR FUNCTIONS

The nucleus occupies a central role in the direction of cellular activity. Some concept of the extent of the nuclear involvement may be inferred from the consequences of the removal of this organelle from the cell. The latter has been accomplished by (1) removal of the nucleus with a hooked needle or micropipette; (2) centrifugation of the cell to the point of rupture into two, one of which contains a nucleus, and the other is enucleated; (3) cleft of the cell by means of a glass needle into two portions, again, one nucleated, and the other enucleated, a surgical procedure which is termed *merotomy*. Needless to say, these operations have only been performed on rather large cells, e.g., the protozoan, *Amoeba proteus*. The resultant enucleated half of the cell assumes a more spherical shape, exhibits a loss in motility and activity, and a complete unresponsiveness to the environment although some chemical activity does persist. Although the digestion of nutriment that had been accepted prior to enucleation proceeds, no further ingestion occurs. A profound reduction in the ribonucleic acid (RNA) content of the enucleated portion may be observed while the RNA in the nucleated half remains essentially constant. The nucleated half, in contrast, reacts normally to all environmental stimuli, indicating that the manipulative procedures per se were not responsible for the above effects. Furthermore, reimplantation of the nucleus into the enucleated cell will restore the original activity if the latter procedure is performed within a reasonable length of time.

Some interesting observations stem from investigations with the single-celled algae, *Acetabularia*. The algae possesses a foot, stalk, and a cap which is characteristic of the species (see Fig. 4.6). When the cap is removed from the alga, a process of regeneration occurs, culminating in the re-formation of a cap identical in all respects with the amputated structure. If a nucleus is transferred from an alga of one species to another species (possessing a different cap), from which the cap had been removed, regeneration again takes place. The resultant cap may be either intermediate in character between the two species or may be identical to the cap of the donor nucleus. The determining factor is the length of the stalk remaining after decapitation. If only the cap has been removed, the regenerated cap will be intermediate in nature. If, however, some of the stalk as well as the cap has been extirpated, the regenerated cap will be of the species of the donor nucleus. If now, the newly regenerated cap is amputated (whether intermediate or not) differentiation will ensue, resulting in the formation of a cap typical of the donor of the originally transplanted nucleus. The existence of a *morphogenetic substance elaborated by the nucleus* and stored in the cytoplasm can account for these observations. Amputation of the nucleus and a portion of the stalk will remove the determinant structures and the implanted nucleus now is the sole morphogenetic factor.

Other evidence for the role of deoxyribonucleic acid (DNA) in morphogenesis is derived from studies with amphibian eggs. In the frog egg, as in most organisms, a single cell, the fertilized egg or zygote, is the parent of all organs and tissues. The cell actively divides, quantitatively increasing in number, and produces the embryonic structure, the blastula, although exact definition of the organs and tissues is not yet achieved at this stage of development. The cells of the blastula then undergo rearrangements during which most of the future organs are made (gastrulation). The further development of organs and tissues rapidly ensues. Enucleation of the unfertilized frog egg has little effect upon the cleavage of the cell but inhibits further morphogenesis into the blastula stage. In a series of experiments, Briggs and King were able to show that transplantation of nuclei from cells in blastula or early gastrula into enucleated unfertilized eggs reinstated the process of morphogenesis.

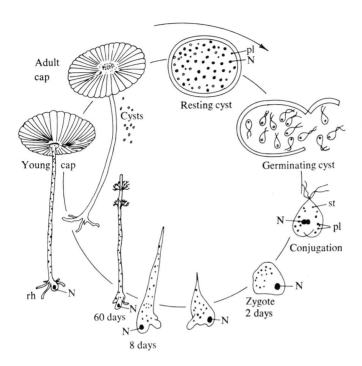

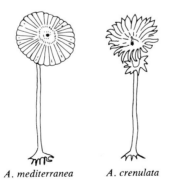

A. mediterranea   *A. crenulata*

Fig. 4.6. Life cycle of *Acetabularia*. Upper: Life cycle of *A. mediterranea*. N, nucleus; rh, rhizoid; st, stigma on isogametes; pl, plastids. From J. Brachet, "Biochemical Cytology." Academic Press, New York, 1957. Lower: Two species of *Acetabularia*. The caps are about 1 inch diameter; the stalks, 2-6 cm in length. From J. Hammerling, *Intern. Rev. Cytol.* **2**, 475 (1953).

That the nucleus does play an important role in RNA metabolism may be inferred from the interesting experiment of Goldstein and Plaut who cultivated the protozoan, *Tetrahymena*, in the presence of phosphate-[32]P. The radioactively labeled organisms were then fed to ameba, the ^{32}P which was liberated into the cytoplasm of the ameba was absorbed into the nucleus of this organism and there was assembled into RNA. After digestion of the *Tetrahymena* by the ameba, the nucleus was removed and implanted into the enucleated ameba. Within 12 hours after implantation, ^{32}P-labeled RNA was observed in the cytoplasm. If the labeled nucleus was implanted into an intact ameba, ^{32}P-labeled RNA was also noted in the cytoplasm. In the latter, however, no label could be found in the original nucleus of the ameba. These experiments suggested *the transmittance of RNA or a precursor thereof from the nucleus to the cytoplasm and furthermore,* that *the passage was unidirectional.* The nucleus then appears vital not only for the normal operation of the cytoplasm but for the differentiation of the cytoplasm, and for maintenance of the intracellular RNA content.

V. CHEMICAL COMPOSITION

The chemical composition of the nucleus has been studied by histochemical and cytochemical techniques, autoradiography, as well as by biochemical analysis of isolated nuclei. Cytochemistry has contributed the specific and sensitive stain for DNA, the Feulgen reaction, which is based upon the reaction of an aldehyde group with leucofuchsin (see Chapter I). The characteristic purplish-blue color has been demonstrated almost exclusively within the nucleus of the cell and its intensity has been measured for the quantitative estimation of nuclear DNA.

A number of additional staining processes have been employed in the histochemical analysis of the nuclei and have revealed the presence of both RNA and DNA. Among these may be included acridine orange which stains both DNA and RNA although yielding different colors; methyl green, a characteristic stain for DNA; azure C and pyronine which intensely stain RNA.

These cytochemical methods may be improved and the specificity enhanced by the use of either ribonuclease, first crystallized by Kunitz in 1940, or deoxyribonuclease, obtained by Fischer and his colleagues in 1941. Treatment of the cell with either of these enzymes will specifically effect hydrolysis of RNA or DNA, respectively. A comparison of the staining properties of an enzyme-treated and control cell preparation will reveal the location of either nucleic acid. With these techniques, RNA has been demonstrated within the nucleolus and DNA has been demonstrated in the chromatin material and in a region around the nucleolus, the nucleolus-associated chromatin (see p. 196).

The localization of the nucleic acids within the nucleus has been corroborated by autoradiographic studies in which relatively specific precursors have been employed. Thymidine and either uridine or cytidine labeled with ^{3}H or ^{14}C may be utilized as precursors for DNA and RNA, respectively. The autoradiographic studies are also controlled with the use of ribonuclease and deoxyribonuclease to show that the grain counts do indeed represent polynucleotide material.

An additional method which has indicated the presence of nucleic acids within the nucleus is ultraviolet photomicrography. The density of photomicrographs at 260 mμ, i.e., the absorption peak of nucleic acids, will establish semiquantitatively the amount of the nucleic acids present in the nuclear preparation.

Recently, the electron microscope has performed the lion's share of the work in the elucidation of the intranuclear location of the nucleic acids. The scope of the analysis has been expanded by the use of thin sections of tissues which have been treated with either proteolytic enzymes, i.e., pepsin, or the nucleases.

Modifications of some of these techniques have revealed the presence in the nucleus of lipids, basic proteins, other proteins including enzymes, inorganic salts, and various phosphate-containing compounds. The ensuing sections will be concerned with a more detailed discussion of some of these nuclear constituents.

A. DNA

DNA is a polynucleotide with a molecular weight in mammalian cells of approximately 10^9. It is composed of the two pyrimidines (Fig. 4.7), thymine (I) and cytosine (II); two purines, adenine (III) and guanine (IV); a sugar, 2'-deoxyribose; and phosphorus. The mammalian diploid cell contains approximately 6 pg of DNA, over 95% of which is present in the nucleus.

The pyrimidines are linked to the 2'-deoxyribose and phosphorus to form the deoxyribonucleotides (Fig. 4.8): deoxythymidine 5'-monophosphate (V), deoxycytidine 5'-monophosphate (VI), deoxyadenosine 5'-monophosphate (VII), and deoxyguanosine 5'-monophosphate (VIII).

Two chains of deoxyribonucleotides linked by C_3'-C_5' phosphate ester bonds (Fig. 4.9) are arranged in a double helical structure as originally suggested by Watson and Crick (see Figs. 4.10 and 4.11). The double helix of DNA is held together by hydrogen bonds between the adenine of one chain and thymine of the other, and the guanine of one chain and

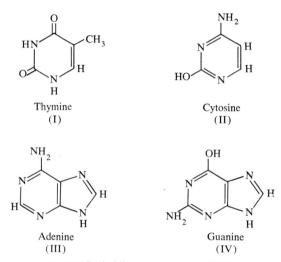

Thymine
(I)

Cytosine
(II)

Adenine
(III)

Guanine
(IV)

Fig. 4.7. Pyrimidines and purines of DNA.

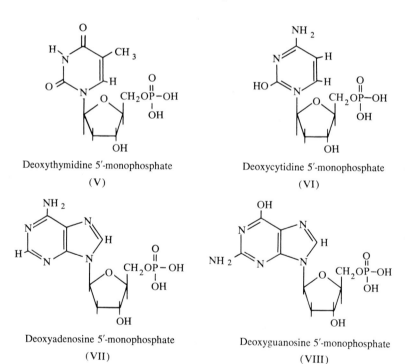

Deoxythymidine 5′-monophosphate
(V)

Deoxycytidine 5′-monophosphate
(VI)

Deoxyadenosine 5′-monophosphate
(VII)

Deoxyguanosine 5′-monophosphate
(VIII)

Fig. 4.8. Deoxyribonucleotides of DNA.

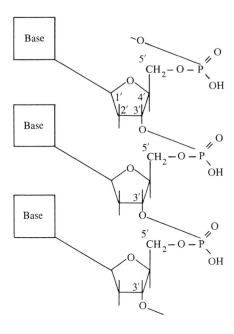

Fig. 4.9. C_3'-C_5' phosphodiester bonds.

the cytosine of the other (see Fig. 4.12). In addition to hydrogen bonding, several other forces are involved in the stabilization of the secondary structure including π-π electron interactions between the stacked purines and pyrimidines, and hydrophobic bonding. That the DNA molecule must possess tertiary structure may be inferred from the following calculation. Assuming a molecular weight (MW) of a nucleotide as 300, the number of nucleotide pairs present in the diploid cell nucleus must approximate the

$$\frac{\text{No. of grams DNA/nucleus}}{\text{MW of nucleotide}} \times \text{Avogadro's number}$$

or

$$\frac{10 \times 10^{-12}}{300} \times 6 \times 10^{23} \quad \text{or} \quad 20 \times 10^{9}$$

If the distance between the members of the nucleotide pair is 3.4 Å, then the total length of a DNA helix will be 3×10^{10} Å or 2 *meters*! Without tertiary structure, the DNA molecule could not be confined within the nucleus and indeed within the cell.

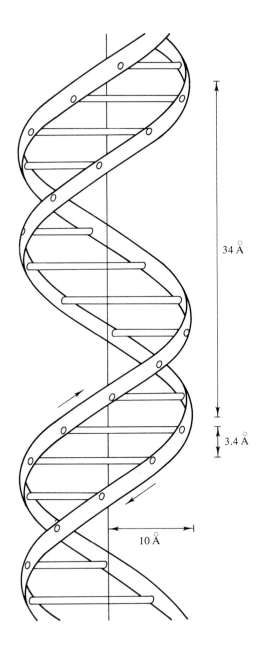

Fig. 4.10. Double helical form of DNA.

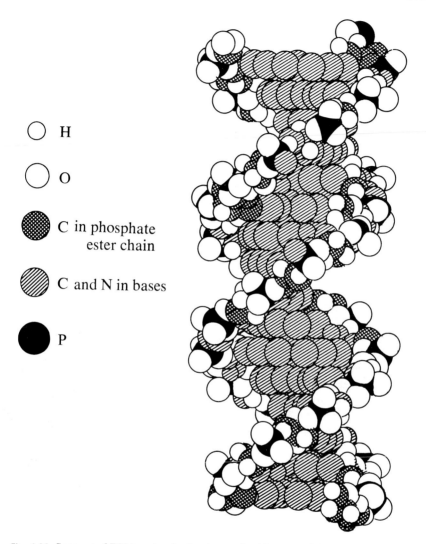

○ H

○ O

⬤ C in phosphate
     ester chain

⬤ C and N in bases

⚫ P

Fig. 4.11. Segment of DNA molecule showing nucleotide core and deoxyribosephosphate spirals on the surface of the DNA. From L. D. Hamilton, *Bull. Cancer Progr.* 5, 159 (1955).

1. SOME PROPERTIES OF DNA

a. HYPERCHROMICITY. The ultraviolet light absorption, i.e., at 260 mμ, of a polydeoxyribonucleotide is considerably less than the sum of a mixture of deoxyribonucleotides of the same composition. The degradation of the DNA to its component deoxyribonucleotides is attended by

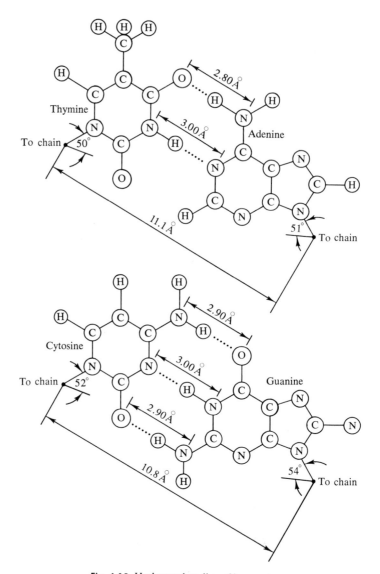

Fig. 4.12. Hydrogen bonding of base pairs.

an increase in the absorption at 260 mμ. This effect is referred to as *hyperchromicity;* it is usually expressed as a percentage increase. Hyperchromicity as high as 60–70% has been observed upon total degradation of certain DNA samples.

Two components contribute to the hyperchromic effect. (1) the separation of the two strands of the double helix and (2) the hydrolysis of the individual internucleotide bonds. The hyperchromic effect is based upon the structural arrangement of the purine and pyrimidine bases in stacks. To some extent, an interaction occurs between the π electrons of the adjacent rings and a redistribution of electrons and subsequent stabilization of the structure take place. Thus, a new ultraviolet absorption results which is a property of the entire molecule.

b. HEAT DENATURATION. As DNA is heated, the double-stranded nature of the molecule is disrupted, resulting in the appearance of localized unbonded base pairs, i.e., open loops. Extensive opening occurs at the ends of the DNA and at sections of the helix rich in adenine-thymine pairs. At some critical temperature, which is a property of the specific DNA molecule, a massive breakdown of the double-stranded structure occurs. This temperature is referred to as the transition temperature, denaturation temperature, or simply, as the T_m. The denaturation process is accompanied by a significant hyperchromic effect, an increase in the optical rotation, and a decrease in the viscosity. Each of these properties can be used to characterize the DNA (see Fig. 4.13). An excellent correlation is apparent between the percentage GMP + CMP in the DNA and its T_m; the higher the former, the greater is the T_m.

The denatured strands do not separate completely but remain partially united. Upon cooling, some re-formation of the double-stranded structure can occur. If the rate of cooling is conducted slowly, there is a great tendency for the DNA chains to re-form the original helix, with restoration of the biological activity. This process is called *annealing*. If the DNA is rapidly cooled, there is little tendency to re-form the original double-stranded structure.

c. ACTION OF POLYAMINES. In the mammalian cell, DNA is not present as such, but combined with several varieties of protein, including basic proteins, e.g., histones (see p. 208). The combination with histones stabilizes the DNA structure in general, causes an increase in the T_m. In addition to the histones and other basic proteins, several polybasic amines have widespread occurrence. These include spermine, spermidine, putrescine, and cadaverine. These substances also produce an increase in the T_m when combined with DNA.

d. FACTORS AFFECTING DENATURATION OR STRAND SEPARATION. The extent of denaturation as measured by the T_m is markedly affected by pH, and by ionic strength. The T_m was sharply diminished when the pH was lowered below 4 or raised above 9. This effect is based upon the loss of the ability of the bases to form hydrogen bonds in the pH regions. The T_m is also increased by increasing the ionic strength of the medium.

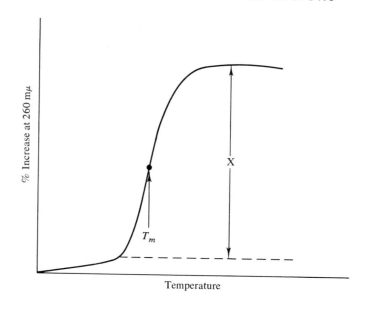

Fig. 4.13. Determination of T_m (temperature at which $X/2$ occurs).

2. RING STRUCTURE OF DNA

In nature, not all DNA is of the linear double helical type. Fiers and Sinsheimer (1962) noted the absence of a free 3'-hydroxyl *and* a 5'-hydroxyl in the infective form of the DNA bacteriophage, ϕX-174. Under proper conditions of velocity sedimentation, two distinct components, (A) and (B), could be found. One of these (A) could be readily converted into the other (B) by mild treatment with deoxyribonuclease. When viewed in the electron microscope, A appeared as a ring structure while B represented an opened and twisted form of ϕX-174 DNA. The ring structure was covalently linked, double stranded, and possessed a molecular weight of 3.2×10^6. The "uninfective" form of the virus, in contrast, occurs as single-stranded, with a molecular weight of 1.6×10^6. Only upon infection of a cell does the virus assume a double-stranded, ringlike replicative form. The ring confers a certain degree of resistance to exonucleases, i.e., lytic enzymes attacking at the ends of the nucleic acid, and to phosphodiesterases.

The ring structure as a replicative form apparently is common to many, if not all, viruses. Recently, such a structure has been found in mitochondria from mammalian cells and as such, differs markedly from nuclear DNA.

3. DNA AS THE GENETIC MATERIAL

The question arises — what is the function of the DNA whose structure we have just discussed? Since the isolation of DNA by Miescher in 1869 from the nuclei of leukocytes obtained from pus, a concerted effort has been expended in an attempt to elucidate the function of this polynucleotide. These efforts have culminated in the following hypothesis — DNA is the gene material. Perhaps before proceeding further, we should define exactly what we mean by the term, the gene. *The gene is a factor intimately involved in the transmission of hereditary characteristics to the progeny.* What then is the evidence for the postulation of DNA as the genetic substance?

In 1928, Griffith inoculated mice with cultures of pneumococci, some of which were attenuated and appeared morphologically to be without capsules (called the "rough" type) and some of which were virulent and encapsulated (the "smooth" type). He observed that only after infection with the "smooth" forms, i.e., the virulent, encapsulated pneumococci, did pneumonia result. Furthermore, heat-killed "smooth" forms were no longer infective. However, when the latter were mixed with the non-infectious "rough" form, the resultant mixture appeared highly infectious. The infectivity was accompanied by a transformation from the "rough" to the "smooth" type and this newly acquired characteristic was inheritable. Griffith concluded that the heat-stable factor, present in the heat-killed "smooth" type, was responsible for this conversion.

Alloway (1932) confirmed and extended Griffith's findings with the partial purification of the component which possessed this "transforming" activity. He was able to definitely establish the noncellular identity of this factor with the use of Berkfeld-filtered extracts. It was not until 1944, however, that the identity of the "transforming" factor was recognized by Avery *et al.*, as DNA.

With these efforts was born the concept of a biological material, DNA, which is able to transmit inheritable characteristics. Since that time, a host of examples of "transforming" factors have appeared and in each case, it has been DNA that was proved responsible.

A considerable body of evidence confirming the DNA-gene thesis also stems from studies with viruses, and we are indebted to A. D. Hershey for the initial observations. The bacteriophage particle, depicted in Fig. 4.14, is composed of a head (usually hexagonal) containing the coat protein and a tail around which is wrapped a tail fiber; the DNA is contained within the head. Upon infection of the host bacteria, *Escherichia coli*, with a T_2 bacteriophage, the tail of the phage attaches itself to the bacterium, and the phage releases a small amount of the enzyme, lysozyme, from presumably the distal tail region, which aids in the puncture of the bacterial cell wall. The phage DNA is then injected through the puncture into the host bacterial cell and the synthesis of viral proteins *by the bacteria* immediately ensues. Inheritable characteristics have been transmitted by viral DNA.

In addition to the more direct evidence cited above, many other observations suggest the relationship of the gene material to DNA. The greatest sensitivity of certain bacterial species to mutation, i.e., alteration in an inheritable characteristic, occurs at wavelengths in the ultraviolet region — regions of the greatest absorption of DNA. At wavelengths in the far ultraviolet where there is less absorption of DNA, a smaller number of mutations may be induced.

Cytochemical studies have established the intracellular location of the gene and the chromosome, i.e., the linear arrangement of genes, within

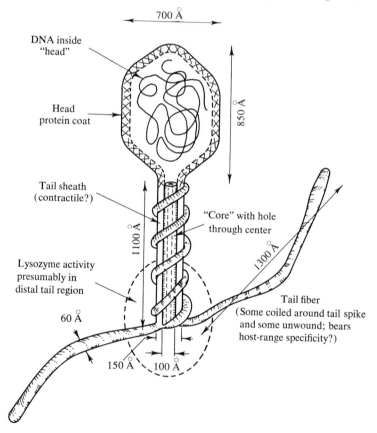

Fig. 4.14. Bacteriophage particle.

the nucleus, and in addition, a number of investigators, for example, Feulgen, Brachet, Caspersson, Boivin, Mirsky and Ris, have localized DNA exclusively within the nucleus. Boivin, in 1947, determined the amount of DNA per set of chromosomes in the nuclei of various tissues and observed a remarkable constancy in cells of a given species. The amount of DNA per haploid set of chromosomes, as in the nuclei of sperm cells, is one half of the amount found in the diploid nuclei. Much of the earlier work, however, was based upon the counting of the number of nuclei per unit volume, and the determination of the amount of DNA per unit volume from which the quantity of DNA per nucleus could be calculated. With the introduction of optical methods, i.e., microspectro-photometry, by Leuchtenberger and others, measurements could be made upon individual nuclei. The quantity of DNA per nucleus was found to approximate ratios of 1:2:4. The constancy of DNA per nucleus appeared directly related to the number of chromosome sets, i.e., to the ploidy. Chromosomal replication in cells may occur without a concomitant cell division, resulting in diploid, tetraploid, or octoploid sets of chromosomes in the nucleus. The constancy of DNA per nucleus of various cells as related to ploidy is presented in Table 4.1.

TABLE 4.1

Average Amount of DNA in Somatic Tissues of Mice in Optical Measurements from Feulgen Slides Expressed in Arbitrary Units[a]

Cell type	Subgroup	DNA (units)	Ploidy
Liver	1	3.34	2
	2	6.77	4
	3	13.20	8
Pancreas	1	3.10	2
	2	6.36	4
	3	12.40	8
Thymus	1	3.28	2
	2	6.17	4
Lymphocytes	1	3.20	2
	2	6.00	4
Sertoli cells	1	3.00	2
	2	6.40	4
Kidney tubule		3.14	2
Intestinal epithelium		2.97	2
Spleen		3.12	2
Ganglion cells		3.14	2
Testes interstituem		3.05	2
Spermatid		1.68	1

[a]Adapted from the data of H. Swift, *Physiol. Zool.* **23**, 169 (1950).

The DNA within the nucleus is stable; metabolic studies with isotopes have established the low rate of turnover of DNA. One would indeed expect to find such unique stability as a property of the genetic material.

Finally, immediately prior to cell division (see Chapter VIII), the amount of DNA within the nucleus doubles, and subsequently, equal partition of the DNA into each of the daughter cells takes place. This corresponds to the acquisition by the daughter cells of the parental characteristics.

In light of this overwhelming body of evidence, the only conclusion that may be drawn is that DNA is the carrier of genetic information as the gene constituent of the chromosome.

4. SYNTHESIS OF DNA

Our present state of knowledge of the intranuclear synthesis of DNA is the result of the brilliant work of Kornberg and his colleagues who employed extracts obtained from exponentially growing *Escherichia coli*. They demonstrated the existence of an enzyme, DNA polymerase (DNA nucleotidyltransferase), which catalyzes the synthesis of DNA from the four deoxyribonucleoside triphosphates, Mg^{2+}, and a template DNA according to the following:

$$\begin{matrix} \text{d-GTP} \\ \text{d-CTP} \\ \text{d-ATP} \\ \text{d-TTP} \end{matrix} \quad + \quad \text{template} - \text{DNA} \xrightarrow{Mg^{2+}} \text{DNA} + \text{pyrophosphate}$$

The template appears to be a vital component of the reaction mixture since in its absence, the reaction proceeds only very slowly, if at all. Heat-denatured DNA (presumably single stranded) and the DNA from the bacteriophage ϕX-174 (also single stranded) are more effective templates than naturally occurring double-stranded DNA. These findings have led to the assumption that the deoxyribonucleotides are formed into the chains of DNA on the single-stranded DNA template by hydrogen bonding to the complementary base (Fig. 4.15). The DNA synthesized by this process is then dependent upon the template that is employed. That this is indeed so, may be inferred from the data abstracted from the manuscript of Lehman *et al.* (1958) and presented in Table 4.2. The enzyme source, i.e., an *E. coli* extract, was employed to synthesize DNA of totally different composition, under conditions where only the template was varied. A remarkable agreement in the chemical composition of the template DNA and the synthesized product may be noted. It is evident, then, that the composition of the newly formed DNA is dependent upon that of the template and that specificity is not conferred by the enzyme.

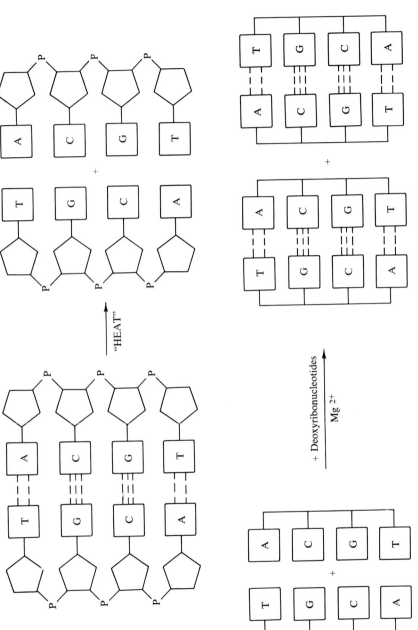

Fig. 4.15. Replication of DNA. Role of primer.

TABLE 4.2

Composition of Enzymically Synthesized DNA[a]

DNA (source)	A[b]	T	G	C
Micrococcus phlei				
Primer	0.65	0.66	1.35	1.34
Product	0.66	0.80	1.17	1.34
T$_2$ Bacteriophage				
Primer	1.31	1.32	0.67	0.70
Product	1.33	1.29	0.69	0.70
Calf thymus				
Primer	1.14	1.05	0.90	0.85
Product	1.19	1.19	0.81	0.83

[a]Abstracted from the data of I. R. Lehman *et al., Proc. Natl. Acad. Sci. U.S.* 44, 1191 (1958).

[b]Molar proportions; A, adenine; T, thymine; G, guanine; C, cytosine.

The template DNA has not been observed *in situ* (with the exception of DNA of ϕX-174 and the question of its formation *in vivo* remains a mystery.

Additional information relating to the identity of the "product"-DNA with the template DNA has been obtained from nearest neighbor frequency analysis, an ingenious technique devised by Josse *et al.* (1961). The enzyme, DNA polymerase, is incubated in four independent reaction mixtures with Mg^{2+}, template DNA; i.e., heated DNA, and the four deoxyribonucleoside triphosphates, one of which is labeled with ^{32}P in the α-phosphate. A different radioactive nucleotide is used in each reaction mixture. After a suitable incubation, the polydeoxyribonucleotide product is isolated, and degraded with DNase II and a 3'-phosphodiesterase. The ^{32}P originally attached to a 5'-hydroxyl of a particular nucleotide is released in ester linkage to the 3'-hydroxyl of the adjacent nucleotide. The dinucleotide frequencies of the product and template DNA may be compared: the results of such an experiment are presented in Table 4.3. From this experiment, several conclusions may be drawn: (1) the product of the enzymic reaction is formed by copying both strands of the DNA template. The overall dinucleotide frequency of the product and template DNA was identical, i.e., ApA = TpT; GpG = CpC. (2) The two strands of the product are of opposite polarities since the dinucleotide frequencies of complementary pairs are identical, e.g., ApG = 0.045 = CpT and $\neq$ TpC; GpA = TpC $\neq$ CpT.

Additional evidence has established the catalytic role of the DNA template in the reaction, the initiation of the reaction from the free

3'-hydroxyl end of the DNA strand, and further, the apparent identities of the product and template DNA. The latter is based upon the comparison of the physical properties of the native DNA with the product polydeoxyribonucleotide. This question is of such vital importance in establishing the function of DNA polymerase *in vivo* that the task has occupied the activities of a number of investigators for at least 10 years.

a. OTHER FUNCTIONS FOR DNA POLYMERASE. In addition to having the responsibility for the synthesis of DNA, the polymerase also serves as part of a "repair" mechanism. If an incorrect base has been inserted into a DNA strand or a deletion has occurred by some means, it is the function of the polymerase to introduce the correct base in juxtaposition to its complementary base. The initial stage of this repair takes place with the "ejection" of the incorrect base by a nuclease enzyme which monitors the gene for such abnormalities. The repair is schematically presented in a simplified version in Fig. 4.16.

b. INTRACELLULAR LOCALIZATION OF POLYMERASE. In mammalian systems, the DNA is located in a discrete region within the cell, the nucleus, whereas in bacteria, this structure is more diffuse in the cytoplasm. We have already noted the autoradiographic evidence implicating the nucleus as the site of DNA synthesis. Early attempts, however, to obtain DNA polymerase from nuclear preparations from mammalian sources proved unsuccessful. The enzyme appeared to be present in the soluble portion of the cell, the cytoplasm. The "soluble" DNA polymerase has been found in preparations from Ehrlich ascites cells (mouse), HeLa cells (tissue culture), regenerating liver (rat), and the Novikoff hepatoma (rat), and appears to be ubiquitous in nature.

TABLE 4.3

Dinucleotide Frequencies in the Product of DNA Polymerase
Action Using DNA from *Mycobacterium phlei* as Primer[a]

ApA	0.024	CpA	0.063
ApG[b]	(0.045)	CpG	0.139
ApC	0.064	CpC	0.090
ApT	0.031	CpT	(0.045)
GpA	0.056	TpA	0.012
GpG	0.090	TpG	0.063
GpC	0.122	TpC	0.061
GpT	0.060	TpT	0.026

[a]M. J. Bessman, *in* "Molecular Genetics" (J. H. Taylor, ed.), Part I, p. 1. Academic Press, New York, 1963.
[b]Phosphate ester bridge between 3'-hydroxyl of dAMP to 5'-hydroxyl of dGMP.

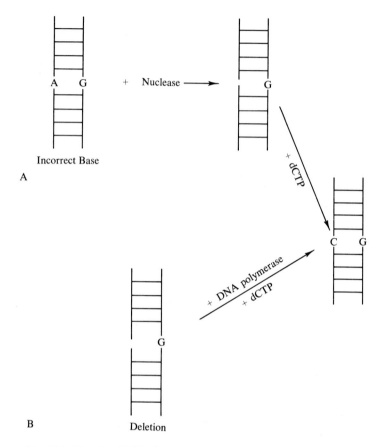

Fig. 4.16. "Repair of DNA by a nuclease (A) and DNA polymerase (B).

The question naturally arises of where the DNA synthesis is accomplished within the cell. Is DNA synthesized in the cytoplasm and then transported to the nucleus where it then resides? If this is true, how is the autoradiographic data reconcilable with this hypothesis? An alternative hypothesis exists which would account for all observations. DNA polymerase of the nucleus may be rendered soluble under extremely mild conditions, e.g., homogenization, so in the process of isolation of the nuclei, the enzyme is lost. A practical resolution to this important controversy has recently been supplied by Keir, Smellie, and Siebert (1963), who prepared nuclei by nonaqueous techniques from regenerating rat liver. The specific activity of DNA polymerase of these nuclei was significantly higher than that of either the cytoplasm or the whole homogenate, while nuclei isolated by the conventional aqueous techniques were devoid of activity. DNA polymerase may then be a representative member of a class of "leaky" enzymes, i.e., easily removed from their usual location.

5. FUNCTIONS OF DNA

DNA fulfills two major functions to which a more intensive consideration will be given in later chapters.

1. *Replication.* The individual strands of the double-helical structure of a DNA molecule form complementary pairs with the deoxyribonu-cleotides which then are united, via phosphodiester linkages into another DNA strand. In this manner, DNA is able to perpetuate itself. It is this mechanism which is vital prior to mitosis.

2. *Transcription.* The strands of the DNA molecule form complementary pairs with ribonucleotides with the subsequent formation of a polyribonucleotide which acts as a transmitter of genetic information from the gene, DNA, in the process of protein synthesis. The polyribonucleotide so produced is known by the operational term, "messenger" RNA (see Chapter V).

6. MITOMYCIN AND PORFIROMYCIN

The mitomycins represent a series of bacteriocidal, cytotoxic, and mutagenic antibiotics possessing the basic formula depicted in Fig. 4.17. The basic nucleus of the mitomycins and the porfiromycins is the mitosane ring structure. The potent activity of the antibiotics may be attributed to the presence of the very reactive aziridine ring structure attached

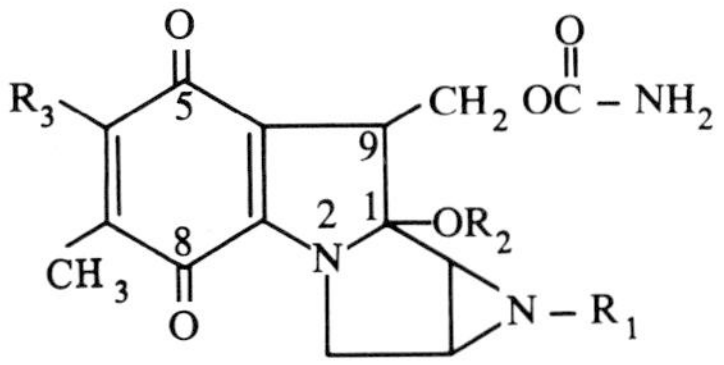

Mitomycin A: R_1 = H; R_2 = CH_3 ; R_3 = CH_3O –

Mitomycin B: R_1 = CH_3 ; R_2 = H; R_3 = CH_3O –

Mitomycin C: R_1 = H; R_2 = CH_3; R_3 = CH_3O –

7-Hydroxyporfiromycin: R_1 = CH_3; R_2 = CH_3 ; R_3 = OH

N-Methylmitomycin A: R_1 = CH_3; R_2 = CH_3 ; R_3 = CH_3O –

Porfiromycin: R_1 = CH_3 ; R_2 = CH_3; R_3 = NH_2

Fig. 4.17. The structure of the mitomycins and porfiromycins.

at positions 1 and 2, the methylurethan group at position 9 and the quinone at positions 5 and 8.

In addition to the above-named properties, the mitomycins have been reported to fragment mammalian chromosomes, an observation which suggests a direct effect of the antibiotic upon DNA. Additional evidence favoring this site of action is based upon the selective inhibition of DNA synthesis in mammalian cells in tissue culture; no effects upon either RNA or protein synthesis were observed. A depolymerization of DNA with a concomitant accumulation of DNA precursors has been noted in cells which were treated with the antibiotic.

Exposure of bacterial cells to the antibiotic for brief periods *in vivo* results in the cross-linking of the DNA chains which is directly proportional to the concentration of the antibiotic, the length of time the DNA is in contact with the agent, and the temperature at which the exposure was conducted. The resultant cross-linkages are stable and may be correlated with the mortality rate of the bacterial cell. These studies suggest that mitomycin may inhibit bacteria by virtue of an "alkylation" of the DNA molecule.

Attempts to alkylate the DNA molecule *in vitro* with the mitomycins were, at first, disappointing until it was realized that a prior activation of the mitomycin ring was necessary. The metabolic activation was demonstrated to be an NADPH-dependent enzymic reduction to a hydroquinone derivative. The resultant activated molecule is very labile and unless the reaction is conducted in the presence of DNA, no evidence for the existence of the lethal product may be obtained. The structural possibilities of this molecule have been considered and are indicated in Fig. 4.18. The reactive sites are indicated as A, B, and C.

The activated mitomycin molecule then resembles a bifunctional alkylating agent with four carbon chains linking the reactive regions. The alkylation is believed to occur at a guanine and/or cytosine moiety of DNA since DNA with high guanine-cytosine content is more heavily cross-linked.

Mitomycin C, an antibiotic isolated from *Streptomyces caespitosus*, has been employed as an agent for dissociating viral from the host cell DNA biosynthesis. At low concentrations, a profound drop in the DNA content of the HeLa cell has been demonstrated, with little or no effect upon RNA or protein (Magee and Miller, 1962). Viral DNA synthesis and indeed, the production of infectious virus, is still observed in HeLa cells which have been infected with the DNA virus, vaccinia, although the HeLa cells per se are no longer able to maintain their own DNA. The antibiotic then offers a convenient means for ascertaining the relative contributions of the cell and of the virus to the overall DNA biosynthesis in an infected cell.

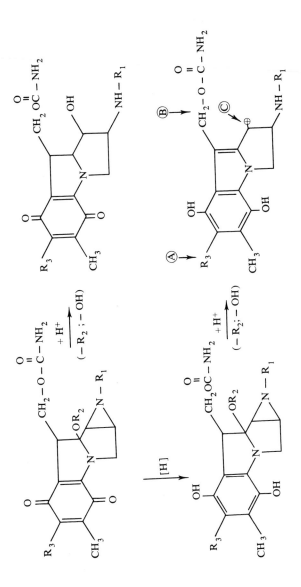

Fig. 4.18. Activation of mitomycin: three active sites.

7. PHLEOMYCIN

Phleomycin has been isolated from *Streptomyces verticullatis* and found to selectively inhibit DNA synthesis. The antibiotic inhibits the DNA polymerase reaction *in vitro*. The inhibition is dependent upon the primer DNA and not the enzyme itself. The exact nature of the inhibition has not been defined and indeed, the structure of the antibiotic is not known. The binding of the antibiotic to the DNA appears to take place at the regions containing adenine-thymine pairs and accordingly the T_m of the DNA is altered.

B. RNA

Of the total amount of RNA existing within the cell, $10-30\%$ is located within the nucleus and of this approximately 50% is associated with the intranuclear inclusion, the nucleolus. Perhaps before continuing our discussion of the second major substance found in the nucleus, RNA, a word about the nucleolus would be in order.

1. NUCLEOLUS

The nucleolus is a membraneless intranuclear body including: a system of vacuolelike bodies; the internucleolonema regions; ribonucleoprotein inclusions which may consist of particles and filaments (see Figs. 4.19–4.21 for light and electron micrographs). This organelle was described by Fontana in 1781, observed by Schleiden in 1838, and later reported by Ogata *et al.* in 1883, who referred to this structure as a "plasmosome." These intranuclear components are the densest structures within the cell, consisting of up to 90% protein and RNA and are prominent in cells that are actively synthesizing protein. The shape of the nucleoli may vary from rounded or spherical to angular or oblong (see Fig. 4.19 for shapes of isolated nucleoli), while the number within the nucleus also may vary from one to several thousand, the latter present in some oocytes. Within the rat liver nucleus, 1–6 nucleoli are observed; the average value is 2. In slow-growing mammalian embryos, the greatest number of nucleoli appear during the earlier stages of growth; in fast developing mammalian embryos, there exists a paucity of nucleoli although the nucleoli are larger. Electron micrographs sometimes depict an intimate relationship between the nucleolus and the cytoplasm, as inferred from the proximity of the nucleolus to the nuclear membrane. Upon careful scrutiny of these electron micrographs, a passage from the nucleolus to the nuclear membrane may be seen, an observation in support of the nucleolus-cytoplasm relationship.

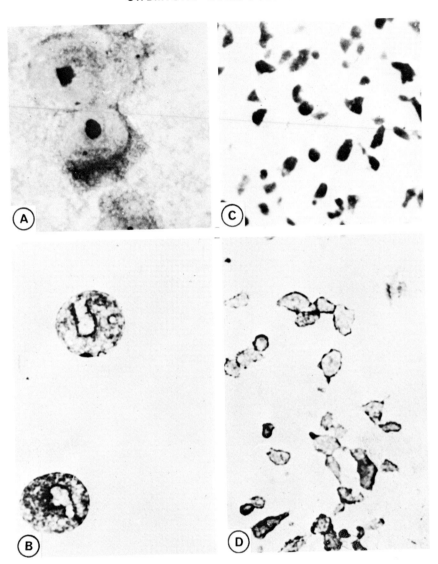

Fig. 4.19. Nucleoli. A, Smear of isolated nuclei from Walker tumor (stained with toluidine blue). × 1800. Note the intense staining of the nucleoli and cytoplasmic basophilic structures. B, Smear of isolated nuclei from Walker tumor stained with toluidine blue after hydrolysis with 1 N HCl. The nucleolus no longer stains; the nucleolus-associated chromatin is stained. × 1800. C, Isolated nucleoli from Walker tumor stained with toluidine blue. × 1800. D, Isolated nucleoi from Walker tumor stained with toluidine blue after hydrolysis with 1 N HCl. The nucleolus-associated chromatin is stained while the rest of the nucleolus is not. × 1800. Courtesy of H. Busch.

Nucleoli may be isolated by methods involving disruption of the nuclear structures by either sonication or by decompression (see Busch *et al.*, 1963). The isolated nucleoli stain negatively for DNA (Fig. 4.19) although a region emanating from and surrounding the nucleolus proper is apparent which does contain DNA, the nucleolus-associated chromatin (Fig. 4.22). The nucleolus-associated chromatin may play an important role in the direction of nucleolar activities. The chemical composition of the nucleolar preparations includes 1 − 14% RNA, 10 − 18% DNA, and up to 70% by dry weight of protein.

The availability of methods for obtaining sufficient quantities of nucleoli for biochemical studies has made possible the investigation of the enzyme complement of this organelle. These studies have been

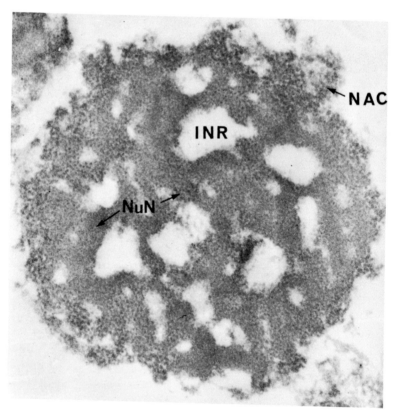

Fig. 4.20. Isolated liver nucleolus. Nucleoli were isolated from rat liver and prepared for electron microscopy by fixing in glutaraldehyde, staining in OsO_4 and poststaining with uranyl acetate. × 46,000. NuN, nucleolonema; NAC, nucleolus-associated chromatin; INR, internucleolonema regions (vacuolelike). Courtesy of H. Busch and K. Shankar.

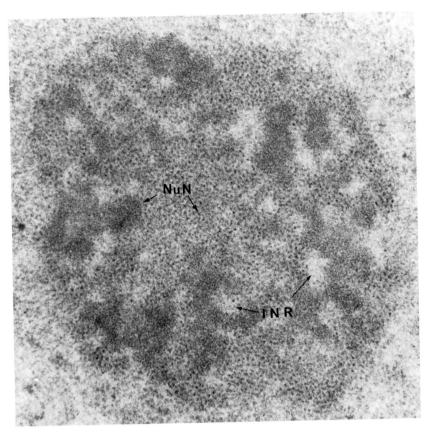

Fig. 4.21. Nucleolus of rat liver *in situ.* Preparation stained with OsO₄. 26,000. NuN, nucleolonema; INR, internucleolonema regions. Note the two types of nucleolonema, particles, and filaments. Courtesy of H. Busch and K. Shankar.

performed by Busch, Siebert, and their colleagues. A summary of their findings is presented in Table 4.4.

Their experiments indicate the preferential localization of RNA polymerase, ribonuclease, and to a lesser extent of adenosinetriphosphatase A in the nucleolus. Polynucleotide phosphorylase activity was also demonstrable in nucleolar preparations.

As we shall see in a later section, the nucleolus is very intimately associated with the synthesis of RNA and indirectly with the manufacture of protein. Let us now return to the major topic of this section, RNA.

2. STRUCTURE OF RNA

RNA is a high molecular weight substance, i.e., 2.5×10^4 to 2×10^6, composed of unbranched chains of ribonucleotides. The major substances included in the structure of RNA are the two pyrimidines, uracil (X) and cytosine (XI); the two purines, adenine (I) and guanine (II); the sugar ribose (XII) and phosphorus (Fig. 4.23).

The chains of RNA are composed of polymers of 5'-ribonucleotides of adenosine (XIII), guanosine (XIV), cytidine (XV), and uridine (XVI) (Fig. 4.24). Certain classes of RNA, e.g., "transfer" RNA (see Chapter V), contain a number of minor components, methylated ribonucleotides, 5,6-dihydrouridine, and pseudouridine monophosphates (ribose phosphate is attached at the C-5 position of the uracil moiety). Additional information about the structures of the various RNA molecules will be presented in the following chapter.

3. SITE OF RNA BIOSYNTHESIS

A variety of elegant techniques have established the site of RNA synthesis as the nucleus. Once manufactured in this organelle, the RNA is transported to other portions of the cell, e.g., the endoplasmic reticulum, for utilization. The foundation for this hypothesis are the studies with

TABLE 4.4
Intranuclear Distribution of Marker Enzymes[a]

Enzyme	Specific activity (mμmoles product/min/mg protein)	
	Nuclei	Nucleoli
Glutamate dehydrogenase	5.4	2.5
Adenylate kinase	0.06	0.02
Catalase	35	18
Acid phosphatase	5.3	2.3
5'-Nucleotidase	6.2	2.5
Glucose-6-phosphatase	33	0
Lactate dehydrogenase	0.22	0.24
Pyruvate kinase	20	40
NAD pyrophosphorylase	7.3	2.2
Adenosine-5'-triphosphatase A	14	31
Acid DNase	0.06[b]	0.12[b]
Alkaline DNase	0.08[b]	0.12[b]
RNase	0.82[b]	5.0[b]
RNA polymerase	140[c]	3300[c]

[a] These values have been abstracted from the data of G. Siebert et al. [J. Biol. Chem. 241, 71 (1966)]. Nucleoli were obtained from the livers of normal rats by the sonication method.

[b] Specific activity is given as the change in absorbance at 260 mμ/30 min/mg protein.
[c] Specific activity is given as counts per minute acid-insoluble material/20 min/mg protein.

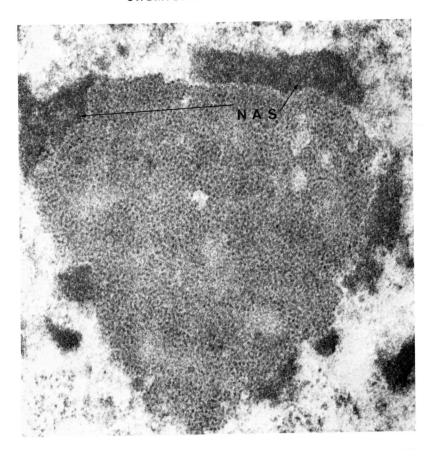

Fig. 4.22. Nucleolus-associated chromatin (NAS). Represents a nucleolus from Walker tumor *in situ* stained with OsO_4. × 79,000. Courtesy of H. Busch and K. Shankar.

isotopic precursors and autoradiography; the investigations by Prescott and his associates (1958) are representative of this approach. He recorded the kinetics of the synthesis of RNA in the protozoan, *Tetrahymena*, in a medium containing cytidine-^{3}H (present in the medium for a brief period, a *pulse*, or continuously present during the growth of the organism). The results are graphically depicted in Fig. 4.25.

The incorporation of the precursor was measured by several of the techniques briefly discussed in Chapter I. The RNA of the nucleus is initially rapidly labeled, reaching a plateau value after a defined period. Cytoplasmic RNA, on the other hand, is labeled rapidly only after the synthesis of nuclear RNA has plateaued. These kinetics are consistent with the view that RNA synthesized within the nucleus ultimately finds

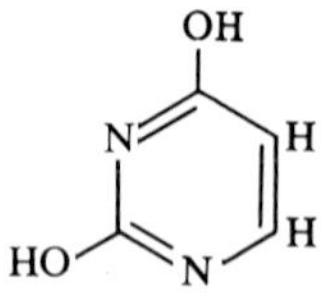

Uracil
(X)

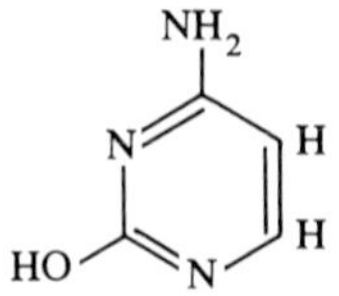

Cytosine
(XI)

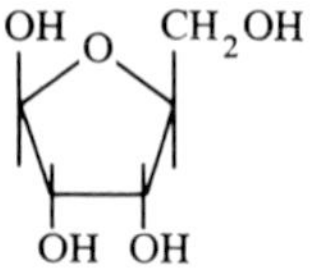

D-Ribose
(XII)

Fig. 4.23. Constituents of RNA.

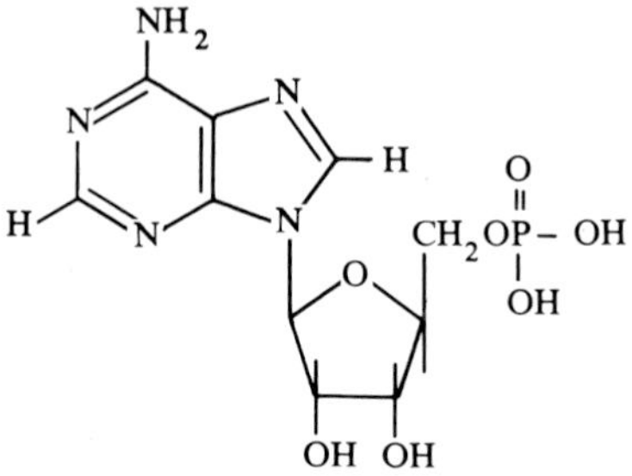

(XIII)

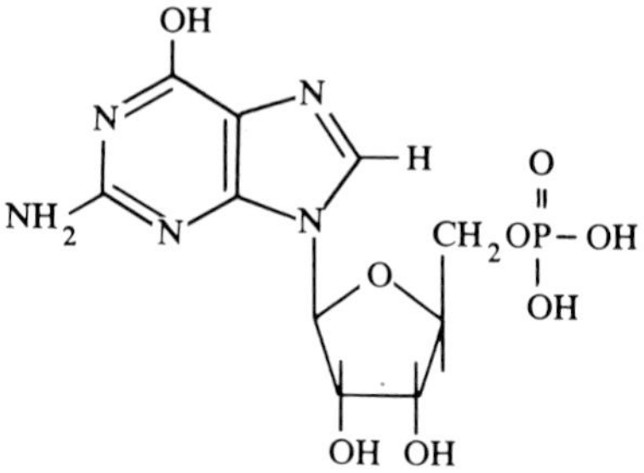

(XIV)

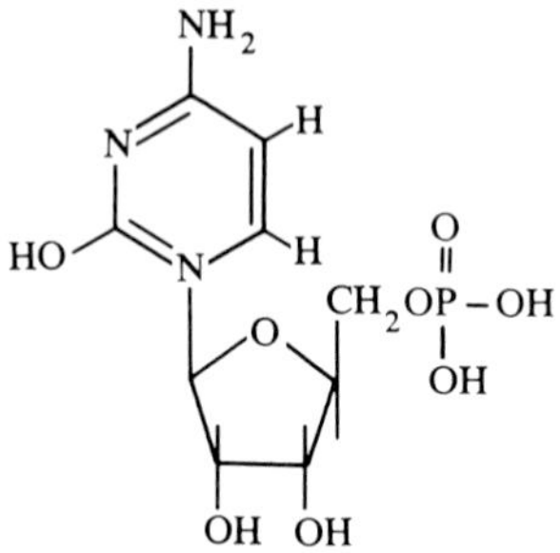

(XV)

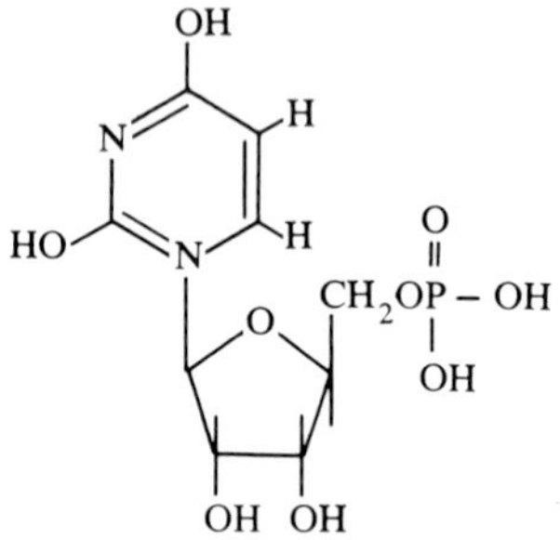

(XVI)

Fig. 4.24. Ribonucleotides of RNA.

its way to the cytoplasm. The inability of enucleated cells to incorporate cytidine-^{3}H or uridine-^{3}H into RNA is in accordance with this hypothesis.

The following interesting experiment has also augmented our knowledge of the events of intracellular RNA synthesis. The nucleoli of HeLa cells maintained in tissue culture had been selectively irradiated by means of an ultraviolet beam microirradiator, a procedure which "kills" this organelle. The treated cells were unable to incorporate the radioactive precursors into RNA. One may conclude from these studies that the

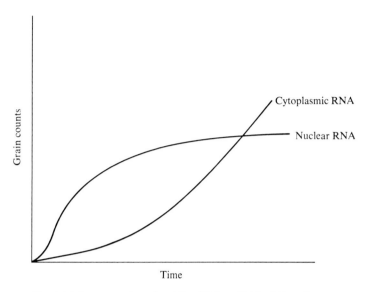

Fig. **4.25.** Incorporation of cytidine-³H into RNA of *Tetrahymena*.

nucleolus is a major intranuclear site of RNA synthesis within, at least, the HeLa cell.

The RNA which is synthesized either within the nucleus or the nucleolus is then transferred through the nuclear pores to positions where RNA is required. The exact mechanism, however, by which this transport is accomplished is still not understood with any certainty.

4. MECHANISM OF RNA SYNTHESIS

In 1955, Grunberg-Manago and Ochoa reported the isolation of an enzyme, polynucleotide phosphorylase, from extracts of the bacteria, *Azotobacter vinelandii*, that catalyzed the formation of polynucleotide material from ribonucleoside diphosphates. The enzyme could act upon single nucleoside diphosphates or mixtures thereof, forming a polymer composed of either one base or a random mixture of several bases (Fig. 4.26). It was this enzyme which was presumed responsible for RNA synthesis within the cell. Certain inconsistencies, however, were immediately apparent. The enzyme was located in the soluble portion of bacterial or mammalian extracts, whereas autoradiographic studies suggested the nucleolus (or nucleus) as the site of RNA synthesis. It will be recalled that a similar objection could be levied against the synthesis of DNA by the "apparently" soluble enzyme, DNA polymerase. The enzyme also exhibited no apparent specificity for substrate and required no co-factor which would confer any degree of specificity. How then

$$n \text{ XDP} \xrightleftharpoons{+ \text{Mg}^{2+}} \left[\text{XP} \right]_n + n \text{P}_i$$
$$(\text{RNA})$$

Fig. 4.26. Polynucleotide phosphorylase.

could a particular RNA species be synthesized? That the enzyme utilized nucleoside diphosphates was yet another peculiarity. The nucleoside triphosphates are present in greater concentration in mammalian tissues and ideally then, these nucleotides would represent better substrates for such an enzyme. These apparent inconsistencies with existing data and thought cast considerable doubt as to the role of this enzyme in the synthesis of RNA within the cell.

This healthy skepticism was rewarded by the direct observation by Weiss in 1960 (see Weiss, 1962), of RNA synthesis which *was* catalyzed by an enzyme present within the mammalian cell nucleus (rat liver, calf thymus) and which required nucleoside triphosphates as substrates (Fig. 4.27). The enzyme called RNA polymerase (nucleotidyltransferase) is ubiquitously found in the nuclei of mammalian cells, and requires DNA, as judged by the marked sensitivity to DNase, and the lack of any effect of RNase. It has subsequently been found that DNA acts as a template for RNA synthesis in a fashion analogous to its role in the reaction catalyzed by DNA polymerase.

The nature of the primer DNA has been studied. It has been found that native DNA is a better primer than denatured DNA and no major differences in priming activity may be found with DNA's from a variety of sources. Treatment of DNA with nitrous acid, which deaminates mainly the cytosine moiety, or removal of the purines from DNA by reaction with acid, i.e., apurinic acid, destroyed priming activity. Irradiation of DNA with either X-ray or ultraviolet rays is accompanied by a loss in activity in proportion to the dose of radiation. Deoxyribonucleoprotein can prime the reaction, although with much less efficiency.

These data would be consistent with the hypothesis of an assembly of the complementary ribonucleoside monophosphates upon a DNA template to form a DNA-RNA hybrid, with the subsequent formation of

$$\left.\begin{array}{l} \text{ATP} \\ \text{CTP} \\ \text{GTP} \\ \text{UTP} \end{array}\right\} + \text{DNA} \xrightarrow{\text{Mg}^{2+}} \underline{\text{RNA}} + \text{pyrophosphate} + \text{DNA}$$

Fig. 4.27. RNA polymerase.

polyribonucleotide (RNA). The mechanism contains a built-in specificity since the composition of the polyribonucleotide should be dependent upon the DNA template. The experimental evidence is in agreement with this supposition and is presented in Table 4.5.

We may conclude from these data that the RNA produced in this enzyme-catalyzed reaction is dependent upon the base ratio of the DNA template. This hypothesis would presuppose the existence of naturally occurring DNA-RNA hybrids and indeed, such structures have been observed in *E. coli* by Spiegelman and colleagues. Additional information about this mechanism of biosynthesis will be presented in the next chapter.

A number of viruses are found in nature whose sole components are RNA and protein, e.g., the Mengo virus, polio-virus. It is obvious that the synthesis of these viral RNA's cannot be DNA directed and may in fact, be RNA mediated. This hypothesis has borne the test of experimentation with the isolation of an RNA synthetase which catalyzes the formation of polyribonnucleotide on an RNA template.

In mammalian cells, the bulk of RNA biosynthesis, if not all, is DNA dependent. Indeed, RNA polymerase is found within the nucleus of the mammalian cell tightly bound to the chromatin, i.e., to the deoxyribonucleoprotein. Only in a few cases has the enzyme been brought into solution from this particulate matter without loss in activity.

If we assume that native DNA is the normal primer in the RNA polymerase reaction, several questions remain to be answered. Is one or both strands copied during the polymerase reaction? What is the starting point for enzyme attachment? What is the direction of copying?

Although the *in vitro* evidence is available suggesting that both strands of the DNA *can* serve as templates in the RNA polymerase reaction, the

TABLE 4.5

Influence of DNA in RNA Synthesis[a]

Source of DNA	Template DNA		RNA Product	
	$\frac{A+T}{G+C}$	$\frac{A+G}{T+C}$	$\frac{A+U}{G+C}$	$\frac{A+G}{U+C}$
Micrococcus lysodeikticus	0.40	1.0	0.48	1.01
Escherichia coli	1.0	1.0	0.93	0.98
Thymus (calf)	1.35	1.0	1.52	1.93
T_2 Bacteriophage	1.86	1.0	1.85	0.96

[a]Data extracted from J. Furth *et al., Biochem. Biophys. Res. Commun.* **4**, 362 (1961).

results from a variety of bacterial systems have indicated that *in vivo* the information from only one strand is transcribed. An experiment from Spiegelman's laboratory which demonstrates this point employs the circular double-stranded replicative DNA of φX-174 bacteriophage as the biological tool. With the latter as a template in the RNA polymerase reaction, an asymmetric transcription (i.e., only one strand was copied) resulted. If the replicative form was disrupted by sonic oscillation, both strands could be copied.

RNA polymerase is found as an enzyme-chromatin complex within the mammalian nucleus. A model for this interaction has been proposed by Watson (1965) and is offered in Fig. 4.28. The attachment of RNA polymerase to the DNA leads to the "unwinding" of a region of the DNA helix at which complimentary bases polymerize into an RNA molecule. The other strand, i.e., not active in the transcription process, may be the site of DNA replication and hence, serves as the site of attachment of DNA polymerase.

The RNA products transcribed from a DNA strand are much smaller than the DNA template. Therefore, one could speculate on the number of enzyme molecules that take part in the enzyme-chromatin complex. Such experiments and calculations have been performed with bacteriophage systems, in which the DNA is smaller than in mammalian organisms. With the former, it is supposed that at least 50 enzyme molecules can bind to the genome, presumably at specific starting points. The nature of these starting points is unknown but could represent breaks in the phosphodiester linkages of one of the DNA chains or a unique sequence of bases.

The problem of the direction of RNA synthesis has been approached by several laboratories. These concur in establishing the growth of the RNA chain from its 5′ end.

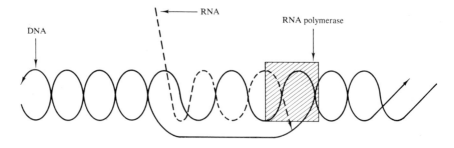

Fig. 4.28. Transcription on one DNA strand.

5. ACTINOMYCIN D (MERACTINOMYCIN)

One of the most useful tools in nucleic acid biochemistry is an antibiotic, actinomycin D, isolated from a strain of *Streptomyces* by S. Waksman in 1954. The agent possesses significant inhibitory efficacy against many bacteria and animal tumors, but because of its toxicity has been employed only with extreme care. Recently, the actinomycins have shown some efficacy as an immunosuppressive agent in the heterotransplantation of organs.

The structure of the antibiotic has been elucidated and shown to consist of a chromophoric group, actinocin, and two peptide-lactone groups of equal structure. A family of actinomycins have been isolated which contain an identical actinocin portion and varying polypeptides. The structure of actinomycin D (meractinomycin) is presented in Fig. 4.29. The polypeptide of actinomycin D contains the D-amino acid, D-valine, and the two methylated amino acids, methylglycine or sarcosine and *N*-methylvaline.

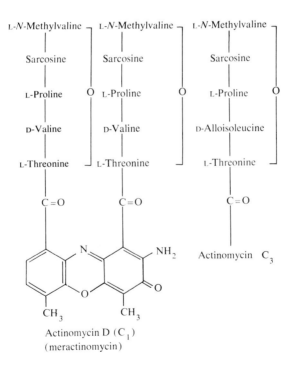

Fig. 4.29. The structure of the actinomycins.

The mechanism of action of the polypeptide antibiotic is closely linked to the synthesis of RNA. In bacterial cells and in cultured mammalian cells, RNA but not DNA biosynthesis is inhibited by actinomycin D. Reich and co-workers (1961) have observed the dissolution of nucleoli in mouse fibroblasts, cultured in a medium containing the antibiotic. Further, the cytoplasm of these cells, which is normally rich in RNA, was almost devoid of the nucleic acid after a 72-hour exposure to this agent. Confirmation of this profound effect upon RNA has stemmed from a number of laboratories and the site of action of the antibiotic has been defined as DNA-directed synthesis, catalyzed by RNA polymerase.

It is clear from the results from several laboratories that actinomycin D complexes specifically with DNA and not with RNA by a noncovalent bond. The consequence of this event is a marked change in the absorption spectrum of the antibiotic, its peak extinctions in the visible range being reduced and shifted to longer wavelengths. Similar changes in spectrum may be induced by apyrimidinic DNA (pyrimidines have been removed), synthetic deoxyguanylic-deoxycytidylic polymers, crab deoxyadenylic-thymidylic polymers (which contain less than 2% guanine), and indeed, by deoxyguanosine itself. Native DNA however, is $10-20$ times more active in binding actinomycin than is deoxyguanosine. The amount of the DNA-actinomycin complexation is related but not directly proportional to the guanine content of the DNA. The exact nature of and requirements for the binding are still not clear.

The inhibition of RNA synthesis by actinomycin D is not mediated by a direct effect upon RNA polymerase, or by a competition for substrates or cofactors, but is entirely dependent upon the formation of a DNA-actinomycin D complex. A model for the structure of the DNA-actinomycin complex has been constructed in which the antibiotic is viewed as interculating in the minor groove of the DNA helix. Such a model would fit the existing evidence on structure-activity relationships. The lactones stabilize the peptide chains in a structure which would allow the formation of additional hydrogen bonding between the peptide-amino groups and the phosphodiester oxygens of the DNA in juxtaposition to the guanine-actinocin interaction.

It has further been proposed that the region of complex formation between DNA and actinomycin, i.e., the minor groove, is the site at which RNA is produced, and hence during the interaction, the RNA polymerase is displaced from this groove. Accordingly, DNA synthesis as catalyzed by DNA polymerase may take place in the major groove.

Interestingly, Reich and co-workers (1962) have found that actinomycin D inhibited the DNA-containing virus, vaccinia, but was without effect upon the RNA-containing Mengo virus. The explanation lies in the lack of any effect of the antibiotic upon RNA-primed RNA polymerase. One may, at least theoretically, separate the contributions of DNA-directed and RNA-directed RNA polymerases to the overall production of RNA by the cell with this agent.

6. EFFECT OF OTHER ANTIBIOTICS UPON RNA POLYMERASE

A number of antibiotics exert an inhibition upon the RNA polymerase reaction. These include chromomycin A_3, mithromycin, and olivomycin. The mechanism of action appears similar to actinomycin D, in that guanine residues in the DNA are required. In addition, bivalent metal ions, e.g., Mg^{2+} or Mn^{2+}, are needed for assisting in the complexation between the antibiotic and the DNA.

Duanomycin and echinomycin also inhibit the template activity of polydeoxyribonucleotides in the RNA polymerase reaction. Their action differs slightly, however, from that of actinomycin D and is not entirely known.

C. Nuclear Proteins

The third major constituent of the nucleus is a group of nuclear proteins, including the basic proteins, histones and protamines, acidic proteins (residual protein), lipoproteins, and nuclear enzymes. The basic proteins are closely associated with DNA and, in fact, the amount of the former within the nucleus is directly related to the quantity of DNA. The acidic nuclear proteins, on the other hand, are more closely associated with RNA, e.g., the structural component of the nuclear ribosomes. The nuclear membrane is composed of lipoproteins, as are the other membranes of the cell.

1. BASIC NUCLEAR PROTEINS

In 1872, Miescher turned his attention from pus cells as the source of his "nuclein" to salmon spermatozoa. The fish of the upper Rhine river proved a fertile source for these cells and Miescher, who was accustomed to the difficulties of obtaining sufficient experimental material from purulent bandages, was suddenly faced with an almost inexhaustible supply. He isolated two substances from the spermatozoa, DNA and a nitrogen-rich base with a relatively simple structural formula which he called *protamine*. It was Kossel who established the proteinaceous nature of this nitrogenous material of salmon spermatozoa. Miescher had been unable to isolate protamine from unripe salmon testis but was successful in finding another basic protein to which Kossel subsequently gave the name *histone*.

In protamines, arginine may comprise more than 80% of the total amino acid residues. The molecule is a small one by general protein standards, with a range in molecular weight from 4000 to 12,000. The function of the protamines was recognized very early and indeed, Miescher suggested that these basic entities may merely neutralize the phosphoric acid groups of the DNA. Presumably, the protamines would function by virtue of the binding with DNA and in so doing, would maintain the DNA structure compact and stable within the sperm.

The histones are more complex molecules with molecular weights between 5000 − 30,000, although they too possess a large number of basic amino acids, i.e., 23 − 30% of the total amino acid residues. The histones have been loosely classified into the following operational categories: very lysine rich, slightly lysine rich, and arginine rich. A number of individual proteins exist within each of these categories. The histones are remarkably stable to both high and low pH's, an observation used in the preparation of these basic proteins from tissues.

It was Kossel who observed that the histones always occurred in combination with DNA, a complex referred to as *nucleoprotein* or *nucleohistone*. Nucleohistones occur in the chromatin of the nucleus, in which the histone moiety may represent some 60 − 70% of the dry weight of the chromatin. Additional information on the relationship of histones to chromosomal structure will be presented in a later section.

The synthesis of the histones *in vivo* occurs most probably within the nucleus of the cell, a finding which suggests that not all protein synthesis takes place in the endoplasmic reticulum (see Chapter V). In many mammalian cells, nuclear ribosomes have been described, and perhaps these structures are functional in the synthesis of the histones. Cytoplasmic synthesis of the histones has also been reported.

Although the most active rates of histone synthesis have been observed in rapidly proliferating tissues, appreciable synthetic activity is still observed in the resting tissues, e.g., liver. DNA replication, on the other hand, is virtually nonexistent in resting tissues; DNA is replicated perhaps once a year in the liver parenchymal cell. It is evident that the rates of synthesis of histones and DNA in the nucleohistone complex must occur independently of one another.

The functions of the histones still remain uncertain although a number of hypotheses have been advanced during the past 20 years. These include:

1. Protection of the DNA molecule from the action of destructive enzymes, e.g., DNase, hence fulfilling a structural role.

2. Regulation of genetic function by binding to sites on the DNA molecule not involved in the transmission of genetic information.

3. Regulation of protein synthesis by binding with RNA.

4. Possession of enzymic function, e.g., RNase activity, which may result in regulation or stabilization of RNA molecules.

Recently, evidence has been presented, principally by Allfrey *et al.* (1963) and by Bonner *et al.* (1963, 1964), indicating that the histones may inhibit or repress a number of biosynthetic activities, including nuclear protein and nuclear ATP syntheses. The evidence accumulated by these investigators is also strongly suggestive of the involvement of histones in genetic regulation. This evidence includes:

1. The addition of histones to isolated thymus nuclei caused a profound inhibition of RNA synthesis from labeled precursors. The most potent of the histones in this regard was the arginine-rich group; the lysine-rich histones were the least inhibitory.

2. The addition of small amounts of the enzyme, trypsin, a proteolytic enzyme which attacks the peptide bond at an arginine or lysine linkage, to an isolated thymus nuclear preparation, significantly *increased* the incorporation of labeled precursors into RNA. Under these conditions, 70% of the nuclear histones have been degraded within 30 minutes. After the inactivation of the trypsin, the reintroduction of histones to the preparation reduced the level of synthesis of RNA.

3. RNA polymerase activity in thymus nuclei is markedly inhibited in the presence of the histones. The most effective inhibitors were the arginine-rich histones. The most potent inhibitors of RNA polymerase isolated from pea seedlings, however, were the lysine-rich histones.

4. The base composition of the product RNA formed in the system to which the histones had been added differed significantly from that present in the uninhibited reaction. These results suggest the transcription of different sequences of the DNA strand.

These provocative findings suggest different roles for the different histones, but are certainly in accord with the general hypothesis, namely, that histones may be regulators of RNA synthesis and hence of protein synthesis. The definitive experiments have not been performed. Accordingly, recognition of their function must await further study.

2. HORMONES AS GENETIC REGULATORS

A body of evidence has accumulated suggesting the RNA polymerase-chromatin complex as the site of action of a number of hormones. An increase in the activity of this aggregate enzyme in the target tissue can be shown after administration of thyroid hormone, estrogens, corticosteroids, androgens, growth hormone, or insulin. In many of these cases, the enhanced activity is due not to an elaboration of more *enzyme* but to an *activation* of the chromatin, presumably by removal of the histone regulators. The manner in which the latter is accomplished is not understood.

The increased activity of the RNA polymerase-chromatin complex is accompanied first by an increase in the rapidly labeled nuclear RNA and later by an elaboration of specific proteins.

3. CHROMOSOMES

What is generally envisioned as a chromosome is the structure which is observed during *metaphase* (see Chapter VIII) while the interphase chromosome exists as a diffuse and elongated thread throughout the nucleus. The DNA of the metaphase chromosome occurs as an apparently continuous thread (see Fig. 4.30A). a nonhistone protein perhaps assists in maintaining the continuity of the mammalian chromosome. These proteins have been referred to as "linkers" (see Chapter VIII) and may be the acidic nuclear proteins.

One of the concepts of chromosomal structure proposed by Mirsky and Ris (1950) pictures the chromosome as a cylinder of DNA with a protein core (see Fig. 4.30B). The purines and pyrimidines of the DNA would lie on the exposed surface of the chromosome. Zubay and Doty (1959), however, suggested that the chromosome consisted of a core of DNA with the deoxyribose moiety exposed to the surface; the histones would be situated on the outside surface of the DNA molecule. The arrangement of the histones on the exterior of the chromosomal structure would allow for the easy removal and addition of these proteins without gross distortion of the chromosome, in accordance with the hypothesis depicting the histones as regulators of genetic function by blockade of the transcription process. The function of the residual proteins as components of these chromosomal models is not known. It is clear that our knowledge of chromosomal structure is far from complete.

4. ACIDIC NUCLEAR PROTEINS

Not all of the nuclear proteins are soluble in dilute acid, i.e., are basic proteins, a finding recognized as early as 1893 by Lilienfeld. An appreciable amount of the total nuclear protein is soluble only in dilute alkali. This alkali-soluble material has been referred to as *"chromosomin"* by Stedman and Stedman in 1944 who also suggested a role for these substances in genetic transmission.

Another function for the acid-insoluble, alkali-soluble proteins has been suggested by Mirsky and Pollister who in 1942 isolated a substance which comprised 10% of the total deoxyribonucleoprotein mass. The alkali-soluble protein, the tryptophan protein, was believed to function as a thread around which the chromosomal structure was formed.

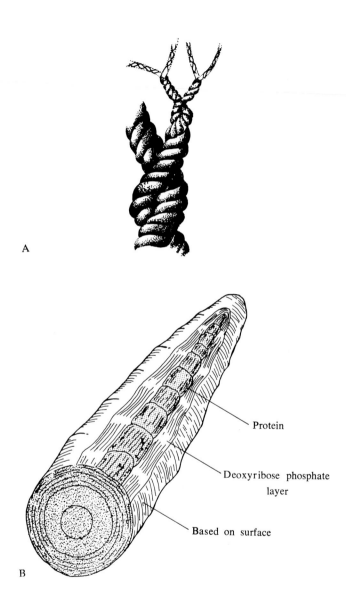

A

B

Protein

Deoxyribose phosphate
layer

Based on surface

Fig. 4.30. Chromosome structure. (A) Steffensen's concept of chromosome (1959). (B) Mirsky and Ris concept (1950).

Little definitive evidence, however, for either role has been produced. The further elucidation of the mechanisms by which these proteins operate is hampered by their relative insolubility in the common laboratory reagents.

5. NUCLEAR ENZYMES

The enzymology of the nucleus has recently been the subject of a review by Siebert and Humphrey. The renewed interest in the nuclear enzymes may be attributed to the refinement in the techniques for the isolation of pure nuclei in good yield. In particular, the application of improved nonaqueous techniques has clarified some glaring inconsistencies between cytochemical and biochemical experimentation.

The intracellular distribution of DNA polymerase has already been considered (see p. 189), and only the conclusion from these studies will be presented here. DNA polymerase is found in the soluble portion of the cytoplasm if the nuclei are prepared by conventional aqueous techniques. If sufficient care is observed in the utilization of nonaqueous techniques, a greatly enhanced enzymic activity may be noted in the nuclei. DNA polymerase represents, then, an extremely soluble nuclear enzyme which may be leached from the nuclei into the aqueous isolation media.

Generally, the nuclear enzymes fall into three classes based upon their solubility. The first group is comprised of those enzymes which may be extracted from nuclei prepared by the nonaqueous technique with 0.14 M sodium chloride containing 2 mM ethylenediaminetetraacetate. Examples of these enzymes are DNA polymerase, and the glycolytic enzymes, e.g., malic and isocitrate dehydrogenases.

Group two may be characterized by their solubility in sodium chloride solutions with a concentration higher than 1 M. In this category falls NAD pyrophosphorylase, an enzyme to which considerable attention will be given later in this section. These enzymes exist tightly bound to the nucleoproteins of the nucleus. Additional members of this category include RNA polymerase, polyadenylate-synthesizing enzyme, and a DNA-degrading enzyme.

The third class of nuclear enzymes may be represented by those proteins which are insoluble in saline solutions but may be obtained in soluble form with the use of detergents, e.g., 1% digitonin or deoxycholate.

a. PRODUCTION OF ADENOSINE TRIPHOSPHATE (ATP). Until a few years ago, the mitochondria were believed to be the sole site for the intracellular productions of ATP (see Chapter III). Today, there is little doubt that ATP may be produced within the nucleus. The principal mechanism for this production would involve not oxidative phosphorylation but glycolysis, a process measurable in isolated mammalian nuclear preparations. What is the function of the ATP which is of nuclear origin? Perhaps the nucleus may supply its own means for the production of free energy for the vital nuclear functions, DNA and RNA syntheses, and may not depend upon the mitochondrial supply of ATP.

b. NUCLEAR PROTEIN SYNTHESIS. The results from a number of autoradiographic studies indicate that protein synthesis occurs within the nucleus and, in particular, in the formation of the histones. An additional consideration of this topic will be presented in Chapter V.

c. NAD PYROPHOSPHORYLASE. Nicotinamide-adenine dinucleotide (NAD) (Fig. 4.31) is an important biological substance involved as a coenzyme in many dehydrogenation reactions. These reactions include the citric acid cycle, the electron transport system, glycolysis, the synthesis of purines and pyrimidines, etc. Hence, the synthesis of NAD is most vital in the maintenance of mammalian cell function.

Leder and Handler first recognized the ability of human erythrocytes to effect the conversion of nicotinamide (see above) to NAD in 1951. They succeeded in isolating two major products from the enzymic reaction mixture, one of which was NAD, and the other, nicotinamide mononucleotide (Fig. 4.32). Although these investigators sought the

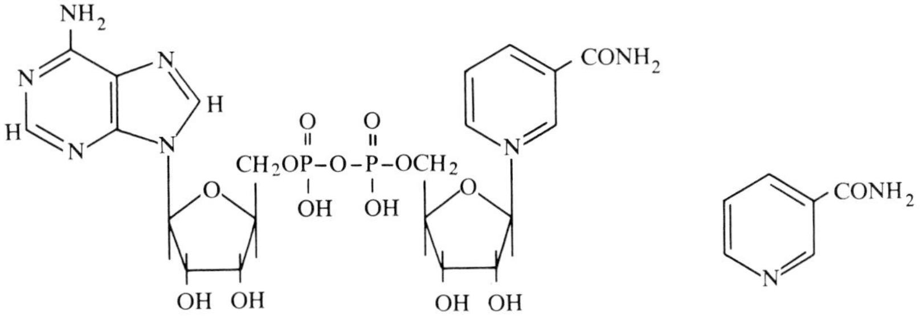

Fig. 4.31. Structures of coenzyme I (NAD) and nicotinamide.

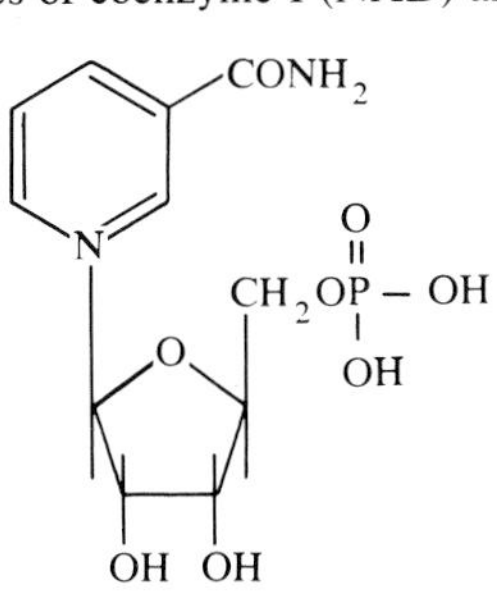

Fig. 4.32. Nicotinamide mononucleotide.

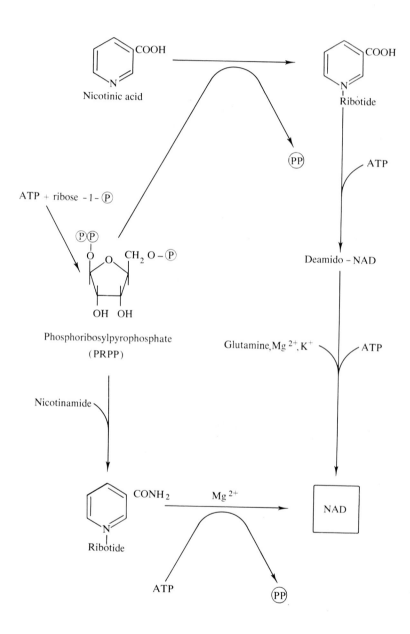

Fig. 4.33. Biosynthesis of NAD.

presence of the riboside in the incubation mixture, none could be found. Further, maximal enzymic activity was observed only in the presence of an energy source, ATP, or an ATP-generating system. These results suggested the direct conversion of nicotinamide to the ribotide without prior formation of the riboside. Subsequently, Handler and his associates were able to establish the biosynthetic pathway for NAD formation as presented below (Fig. 4.33).

Most of the enzymes involved in this pathway are located in the cytoplasm, i.e., soluble enzymes. However, NAD pyrophosphorylase, the enzyme catalyzing the formation of NAD from nicotinamide mononucleotide and ATP, is located exclusively within the nucleus. The nucleus, by affecting the synthesis of this rate-determining enzyme, may possibly exert a regulatory control over cytoplasmic events which utilize NAD (e.g., glycolysis).

VI. CONCLUSION

We have very rapidly perused some of the activities of the nucleus and its inclusions, and have indicated the role played by the nucleus in the direction of cellular functions. The DNA found exclusively within the nucleus is the basic unit of the gene, the structure responsible for inheritable characteristics, and directly controls its own replication. The DNA is also capable of transcription, and in this manner exerts some control of protein synthesis. RNA, an important constituent involved in protein synthesis, at least in part, is manufactured within the nucleus, in the nucleolus. Finally, the rate-determining enzyme in the biosynthetic pathway for NAD, the coenzyme for many dehydrogenations, is located within the nucleus.

GENERAL REFERENCES

Bourne, G. H., ed. (1964). "Cytology and Cell Physiology." Academic Press, New York.
Brachet, J. (1957). "Biochemical Cytology." Academic Press, New York.
Busch, H. (1962). "An Introduction to the Biochemistry of the Cancer Cell." Academic Press, New York.
Busch, H., ed. (1963). "The Nucleus of the Cancer Cell." Academic Press, New York.
Caspersson, T. (1950). "Cell Growth and Cell Function." Norton, New York.
Chargaff, E., and Davidson, J. N., eds. (1955a). "The Nucleic Acids," Vol. 1. Academic Press, New York.
Chargaff, E., and Davidson, J. N., eds. (1955b). "The Nucleic Acids," Vol. 2. Academic Press, New York.
Chargaff, E., and Davidson, J. N., eds. (1960). "The Nucleic Acids," Vol. 3. Academic Press, New York.

McElroy, W. D., and Glass, B., eds. (1957). "The Chemical Basis of Heredity." Johns Hopkins Press, Baltimore, Maryland.
Morton, R. K. (1961). *Australian J. Sci.* **24**, 260-278.
Steiner, R. F., and Beers, R. F., Jr. (1961). "Polynucleotides," Elsevier, Amsterdam.
Taylor, J. H., ed. (1963). "Molecular Genetics," Part I. Academic Press, New York.
Watson, J. D. (1965). "Molecular Biology of the Gene." Benjamin, New York.

SPECIFIC REFERENCES

ISOLATION OF NUCLEI

Allfrey, V. (1959). *In* "The Cell" (J. Brachet and A. E. Mirsky, eds.), pp. 193-290. Academic Press, New York.
Behrens, M. (1932). *Z. Physiol. Chem.* **209**, 59-74.
Busch, H., Starbuck, W. C., and Davis, J. R. (1959). *Cancer Res.* **19**, 684-687.
Chauveau, J., Mouley, J., and Rouiller, C. H. (1956). *Exptl. Cell Res.* **11**, 317-321.
Maggio, R., Siekevitz, P., and Palade, G. E. (1963). *J. Cell Biol.* **18**, 267-291.
Marshak, A. (1941). *J. Gen. Physiol.* **25**, 275-291.
Rees, K. A., and Rowland, G. F. (1961). *Biochem. J.* **78**, 89-95.
Schneider, W. C. (1945). *J. Biol. Chem.* **161**, 293-303.

NUCLEAR MEMBRANE

Baud, C. A. (1959). *In* "Problemes d'ultrastructure et de functions nucleaires" (J. A. Thomas, ed.). Paris.
Schnitzer, S. W. (1960). *Intern. Rev. Cytol.* **10**, 137-163.
Watson, M. L. (1955). *J. Biophys. Biochem. Cytol.* **1**, 257-270.
Watson, M. L. (1959). *J. Biophys. Biochem. Cytol.* **6**, 147-162.

NUCLEAR ENZYMES

Siebert, G., Villalobos, J., Jr., Ro, T. S., Steele, W. J., Lindenmayer, G., Adams, H., and Busch, H. (1966). *J. Biol. Chem.* **241**, 71-78.

NUCLEOLUS

Busch, H., Byvoet, P., and Smetana, K. (1963). *Cancer Res.* **23**, 313-339.
Caspersson, T., and Schultz, J. (1940). *Proc. Natl. Acad. Sci. U.S.* **26**, 507-515.
Maggio, R., Siekevitz, P., and Palade, G. E. (1963). *J. Cell Biol.* **18**, 293-312.
Monty, K. J., Litt, M., Kay, E. R. M., and Dounce, A. L. (1956). *J. Biochem. Biophys. Cytol.* **2**, 127-145.
Sirlin, J. L. (1962). *Progr. Biophys. Biophys. Chem.* **12**, 27-66.
Swift, H. (1959). *Symp. Mol. Biol.* pp. 266-303. Univ. of Chicago Press, Chicago, Illinois.
Vincent, W. S. (1955). *Intern. Rev. Cytol.* **5**, 269-299.

NUCLEAR AND CHROMOSOMAL CONSTITUENTS

Allfrey, V. G., Mirsky, A. E., and Stern, H. (1955). *Advan. Enzymol.* **16**, 411-500.
Busch, H., and Davis, J. R. (1958). *Cancer Res.* **18**, 1241-1256.
Feulgen, R. (1913). *Z. Physiol. Chem.* **84**, 309-328.
Keir, H. M., Smellie, R. M., and Siebert, G. (1962). *Nature* **196**, 752-754.
Phillips, D. M. P. (1962). *Progr. Biophys. Biophys. Chem.* **12**, 211-281.

Steffensen, D. A. (1959). *Brookhaven Symp. Biol.* **12**, 103-118.
Swift, H. (1950). *Physiol. Zool.* **23**, 169-198.
Zubay, G., and Doty, P. (1959). *J. Mol. Biol.* **1**, 1-20.

DNA AS THE GENE MATERIAL

Alloway, J. L. (1932). *J. Exptl. Med.* **55**, 91-99.
Alloway, J. L. (1933). *J. Exptl.Med.* **57**, 265-278.
Avery, O. T., MacLeod, C. M., and McCarty, M. (1944). *J.Exptl. Med.* **79**, 137-157.
Bessman, M. J. (1963). *In* "Molecular Genetics" (J. H. Taylor, ed.) Part I, pp. 1-64. Academic Press, New York.
Fiers, W., and Sinsheimer, R. L. (1962). *J. Mol. Biol.* **5**, 408-424.
Furth, J., Hurwitz, J., and Goldman, A. (1961). *Biochem. Biophys. Res. Commun.* **4**, 362-364.
Griffith, F. (1928). *J. Hyg.* **27**, 113-159.
Hamilton, L. D. (1955). *Bull. Cancer Progr.* **5**, 159.
Hammerling, J. (1953). *Intern. Rev. Cytol.* **2**, 475-498.
Hayashi, M., Hayashi, M. N., and Spiegelman, S. (1964). *Proc. Natl. Acad. Sci. U. S.* **51**, 351-359.
Hershey, A. D., and Chase, M. (1956). *J. Gen. Physiol.* **36**, 39-56.
Hotchkiss, R. D. (1952) *in* "Phosphorus Metabolism" (W. D. McElroy and B. Glass, eds.), Vol. 2. Johns Hopkins Press, Baltimore, Maryland.
Josse, J., Kaiser, A. D., and Kornberg, A. (1962). *J. Biol. Chem.* **236**, 864-875.
Kornberg, A. (1960). *Science* **131**, 1503-1508.
Lehman, I. R., Zimmerman, S. B., Adler, J., Bessman, M. J., Simms, E. S., and Kornberg, A. (1958). *Proc. Natl. Acad. Sci. U. S.* **44**, 1191-1196.
Perry, R. P., Hell, A., and Errera, M. (1961). *Biochim. Biophys. Acta* **49**, 47-57.
Watson, J. D., and Crick, F. H. C. (1953). *Nature* **171**, 737-738 and 964-967.
Weiss, S. B. (1962). *Federation Proc.* **21**, 120-126.
Weiss, S. B., and Nakamoto, T. (1962). *Proc. Natl. Acad. Sci. U. S.* **48**, 880-887.
Zalokar, M. (1960). *Exptl. Cell Res.* **19**, 559-576.

ACTINOMYCIN D

Karnofsky, D. A., and Clarkson, B. D. (1963). *Ann. Rev. Pharmacol.* **3**, 357-428.
Reich, E. (1963). *Cancer Res.* **23**, 1428-1441.
Reich, E., Franklin, R. M., Shatkin, A. J., and Tatum, E. L. (1961). *Science* **134**, 556-557.
Reich, E., Franklin, R. M., Shatkin, A. J., and Tatum, E. L. (1962). *Proc. Natl. Acad. Sci. U. S.* **48**, 238-245.
Vining, L. C., and Waksman, S. A. (1954). *Science* **120**, 389-390.

NICOTINAMIDE-ADENINE DINUCLEOTIDE

Hogeboom, G. H., and Schneider, W. C. (1952). *J. Biol. Chem.* **197**, 611-620.
Kornberg, A. (1957). *Advan. Enzymol.* **18**, 191-240.
Preiss, J., and Handler, P. (1958). *J. Biol. Chem.* **233**, 488-492 and 492-496.

NUCLEAR ENZYMES

Roodyn, D. B. (1959). *Intern. Rev. Cytol.* **8**, 279-344.
Siebert, G., and Humphrey, G. B. (1965). *Advan. Enzymol.* **27**, 329-388.

BASIC NUCLEAR PROTEINS

Allfrey, V. G., Littau, V. C., and Mirsky, A. E. (1963). *Proc. Natl. Acad. Sci. U. S.* 49, 414-421.

Bonner, J., and Huang, R. C. (1963). *J. Mol. Biol.* 6, 169-174.

Bonner, J., and Ts'o, P., eds. (1964). "The Nucleohistones." Holden-Day (1965), San Francisco, California.

Busch, H. (1965). "Histones and Other Nuclear Proteins." Academic Press, New York.

MITOMYCIN C

Magee, W. E., and Miller, O. V. (1962). *Biochim. Biophys. Acta* 55, 818-826.

Schwartz, H. S. (1963). *Science* 142, 1181-1187.

Szybalski, W., and Iyer, K. N. (1964). *Federation Proc.* 23, 946-957.

Chapter V

THE ENDOPLASMIC
RETICULUM

I. INTRODUCTION

In 1897, Garnier, Prenant, Bouin, and several other investigators reported the presence within cells of the pancreas and of the salivary gland, of a specialized region in the cytoplasm which stained intensely with basic dyes such as safranine, gentian violet, or toluidine blue. Furthermore, the intensity of the basophilic staining of this fibrillar material appeared dependent upon the stage of secretory activity within the cell, i.e., the basophilia was pronounced prior to the onset of a new cycle of secretory activity.

Garnier also noted the lateral displacement of the fibrillar material about the nucleus and frequently the total envelopment of the latter structure. Despite the paucity of techniques available, he exhibited a remarkable astuteness in the interpretation of his data. He proposed that the relationship of the nucleus and the basophilic filaments, which he named the *"ergastoplasm,"* was not only topographical, but that both structures may actively participate in the secretory activity of the glandular cell.

Although Garnier was first to offer some interpretation of the role of the ergastoplasm, the cellular organelle had been recognized some 30 years previously, in 1869, by Pflüger and also by Langerhans. The ergastoplasm, as most of the other subcellular organelles, passed through a period where its very existence as a natural entity was doubted and indeed, many investigators looked at this structure as a "mitochondrial body." Interest in the ergastoplasm waned while investigators became more enraptured with mitochondrial structure and function.

In 1940, Rees, employing the new technique of polarization microscopy, confirmed the existence of the ergastoplasm as a distinct structure. Complete acceptance was accorded with the advent of electron microscopy. In the very first electron micrographs, Porter and colleagues observed a permeation of the cytoplasm of the chick microphage by a network which resembled a "lacelike reticulum." They referred to this structure as "endoplasmic reticulum." With the development of techniques in electron microscopy, Garnier's ergastoplasm was soon established as a portion of the endoplasmic reticulum (Fig. 5.1).

II. "ROUGH" AND "SMOOTH" ENDOPLASMIC RETICULUM

Porter and co-workers observed the presence of three structures which constituted the endoplasmic reticulum.

1. Intracytoplasmic membranes, *"smooth" or agranular endoplasmic reticulum,* consisting of filaments or lamellae, 50 Å in thickness and which were generally disposed in parallel array. The morphological appearance of these membranes is in accordance with the unit membrane concept of Robertson (see Chapter II). Their lipid content is approximately 30 — 50% by dry weight, of which over 50% is in the form of phospholipid, i.e., lecithins and cephalins.

2. A space enclosed within the membranes presenting a tubular or canalicular appearance.

3. Small dense opaque granules, 130 – 150 Å in diameter (Fig. 5.1), which consist largely of ribonucleic acid and protein, ribonucleoprotein, and hence have been referred to as ribonucleoprotein bodies or *ribosomes* by H. M. Dintzis. The ribosomes may occur free either individually or in clusters in the cytoplasm or attached to the endoplasmic reticulum. The latter structure when studded with these dense granules represents the *"rough"* or *granular endoplasmic reticulum.*

Sjostrand has adopted another terminology in describing the various types of endoplasmic reticulum. The "rough" and "smooth" endoplasmic reticulum were identical to the α-cytomembranes and β-cytomembranes. The latter nomenclature, however, will not be employed in this treatise.

The endoplasmic reticulum is found ubiquitously in mammalian cells with the exception of the red blood cell and some embryonic cells and is particularly abundant in exocrine pancreatic cells, salivary gland cells, plasma cells, and in certain nerve cells; the tubular cells of the kidney and the muscle cells are only sparsely infiltrated with these structures.

The dimensions of the membranes of the endoplasmic reticulum are very similar to those of the plasma membrane, an observation which has suggested the origin of the former from the latter. The endoplasmic reticulum may represent a series of complex invaginations of the plasma membrane which enfold cavities, thus providing a direct route from the interior to the exterior of the cell. The merit of this theory of the origin of the endoplasmic reticulum lies in the postulation of a transport system into the cell, a system which would provide for the circulation of nutrient material obtained from the cellular environment. The theory, however, has never been fully substantiated, as revealed in the conclusion of Porter in 1959.

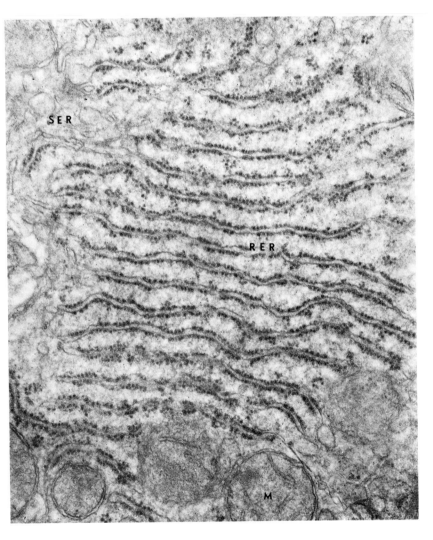

Fig. 5.1. Endoplasmic reticulum in monkey liver. Section of monkey liver stained with $OsO_4 \times 75,000$. RER, rough endoplasmic reticulum; SER, smooth type. Courtesy of J. J. Ghidone.

That the membrane of the endoplasmic reticulum may closely approach the internal surface of the plasma membrane has been repeatedly observed, but any direct patency of the endoplasmic reticulum cavity with the outside must be of extremely short duration, for it has never been clearly demonstrated.

Although many of the inferences on the functions of the endoplasmic reticulum stem from electron microscopic evidence, the definitive proof has come from the studies with a structure which does not occur naturally within the cell but is produced during the course of the tissue preparation. This structure is the *microsome.*

A. Microsomes

In 1943, Claude described in a fraction of a rat liver or chicken tumor homogenate from which the nuclei and mitochondria had been removed, a series of minute, submicroscopic bodies. The bodies could only be sedimented with very high centrifugal forces. Claude called these bodies *microsomes.* That the microsomes were not naturally occurring subcellular organelles but were artifacts of the isolation procedure has been convincingly shown by Palade and Siekevitz (see Siekevitz, 1963). In the electron microscope, the microsomes appeared as membranous structures $500-3000$ Å in length and chemical analysis has revealed the presence of large amounts of RNA. What is the relationship of these structures to the naturally occurring organelles of the cell? The microsomal fraction actually consists of vesicular and tubular fragments of the endoplasmic reticulum, most of which are derived from the "rough" type. The fragmentation occurs readily during homogenization of the tissue.

The ribosomes of the "rough" endoplasmic reticulum may be separated from the membranous portions by treatment with the surface-acting agent, sodium deoxycholate (DOC) at 0.26%, a procedure devised by Palade and Siekevitz. The detergent solubilizes the lipoprotein membrane, releasing the granules, which may sediment after centrifugation at high speeds, i.e., $100,000$ g for 2 hours. Another fraction, in addition to the solubilized portions of the endoplasmic reticulum and the sedimented granules, is also obtained, a loose fluffy layer situated atop the latter fraction; the latter consists of the smooth-surfaced membranes.

The ribosomes contain the bulk of the microsomal RNA consisting of $40-60\%$ RNA and $60-40\%$ protein. The interrelationships of rough-, smooth-surfaced endoplasmic reticulum, microsomes, and ribosomes are indicated in Fig. 5.2. Ernster *et al.* (1962) have achieved a partial chemical characterization of these fractions, and their data are presented in Table 5.1.

MICROSOMES

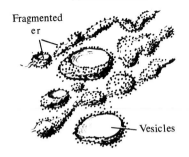

RNP Particles

Membranes

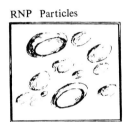
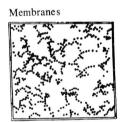

ERGASTOPLASM

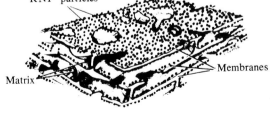

Fig. 5.2. Diagram of endoplasmic reticulum of a cell and relationship to microsomes. From De Robertis *et al.,* "General Cytology," 3rd ed. Saunders, Philadelphia, Pennsylvania, 1965.

TABLE 5.1

Distribution of Protein, RNA, and Phospholipid
in Microsomal Subfractions obtained by DOC
Treatment[a]

Fraction	Protein (mg/gm liver)	RNA (mg/gm liver)	Phospholipid (mg/gm liver)
Microsomes	20.6	4.23	12.5
Subfractions after DOC Treatment: Clear			
supernatant (4.3 ml/gm liver)	7.1	0.12	3.6
M fraction[b] (0.7 ml/gm liver)	5.1	0.69	6.6
P fraction[c]	6.2	3.36	2.9

[a]From L. Ernster, P. Siekevitz, and G. E. Palade, *J. Cell Biol.* 15, 541 (1962)
[b]Loose packed fluffy layer ("smooth-surfaced" membranes).
[c]Tightly packed pellet (ribosomes).

The data of Table 5.1 demonstrate the effectiveness of DOC in rendering the phospholipid and protein components of the microsomal fraction soluble, i.e., 30% appears in the clear supernatant preparation. The bulk of the RNA remained in the pellet under these conditions. The rough- and smooth-surfaced endoplasmic reticulum cannot be separated, however, by deoxycholate although procedures involving differential centrifugation have been developed to accomplish this task (see Chapter I).

B. Ribosomes

Ribosomes may be prepared by treatment of the microsomal preparations with either sodium deoxycholate, Lubrol W, or sodium perfluorooctonoate, rendering the lipoprotein components soluble; the pelleted material after high speed centrifugation consists principally of ribosomes. These bodies are composed of RNA and protein as mentioned previously, magnesium and/or polyamines, e.g., spermine or spermidine. The ribosomal proteins, after treatment with urea, have been separated by electrophoresis into at least 20 fractions and fulfill principally a structural or nonenzymic function. The range in molecular weights of these proteins varies from 12,000 to 25,000. Although latent RNase and DNase activities were initially observed in the ribosomes from *Escherichia coli*, subsequent studies have shown that absorption of these enzymes had occurred during the isolation procedure.

The structure of the ribosomes is markedly affected by the concentration of the magnesium in the isolation medium. At concentrations less than 10^{-4} M, the ribosome exists as two basic units with sedimentation constants of 30 S and 50 S. In a medium containing magnesium at a concentration greater than 10^{-3} M, these units aggregate to form 70 S and 100 S structures. These interrelationships have been depicted in Fig. 5.3. The properties of the various ribosomal structures of $E.$ $coli$ are presented in Table 5.2.

The RNA may be released from the ribosomal structure by treatment with sodium dodecylsulfate-phenol mixture. The RNA isolated from a 30 S or 50 S ribosome has a sedimentation constant of approximately 16 or 28 S, respectively, and a molecular weight of 0.55 or 1.2 × 10⁶, respectively. The ribosomal RNA is very stable in $vivo$, exhibiting a slow turnover.

Ribosomes isolated from mammalian cells sediment at approximately 80 S and are also composed of 2 subunits, with sedimentation constants of approximately 33 S and 54 S (Petermann and Pavlovec, 1966). The molecular weights of the 33 S and 54 S ribosomal subunits have been calculated to be 1.4 and 2.8 × 10⁶, respectively. Associated with these ribosomal subunits are RNA molecules with molecular weights of 0.6 and 1.6 × 10⁶. Increasing the concentration of Mg^{2+} of the medium also results in aggregation of the ribosomal subunits leading to the production of particles with sedimentation constants > 80 S.

Warner (1966) has extracted the proteins from HeLa cell ribosomes and could distinguish three classes of these substances. (1) Class A proteins, representing 60% of the total ribosomal proteins, enter the ribosomes along with the newly formed ribosomal RNA. It is believed that these substances fulfill a true structural role. (2) Class B proteins are firmly bound to ribosomes after extraction of the latter from the cell although the Class B proteins exchange with soluble substances in $vivo$.

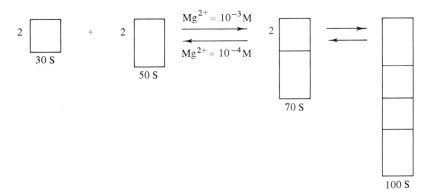

Fig. 5.3. Ribosomal interrelationships.

TABLE 5.2

Properties of *E. coli* Ribosomes

	Dimensions of particle	Molecular weight	RNA associated with ribosome	Molecular weight of RNA
30 S	95×170 Å	0.85×10^6	16 S	0.55×10^6
50 S	140×170 Å	1.8×10^6	28 S	1.2×10^6
70 S	200×170 Å	2.8×10^6	16 and 28 S	–
100 S	–	5.9×10^6	16 and 28 S	–

The Class B proteins represent $15-20\%$ of the total ribosomal protein and perhaps have some enzymatic functions. (3) Class C proteins, representing 20% of the total, are less firmly bound to the ribosome.

Synthesis of Ribosomal RNA. As we have seen, the nuclear RNA becomes labeled more rapidly than cytoplasmic RNA upon incubation of mammalian cells with radioactive precursors, e.g., uridine-^{14}C (see Chapter IV). Since ribosomal RNA comprises approximately 85% of the total RNA mass of cell, it is provocative to suppose the nucleus as the site of synthesis of ribosomal RNA within the cell.

In a series of chemical as well as autoradiographic studies, Perry (1965), as well as others, has confirmed this hypothesis and indeed has established the intranuclear site as the nucleolus. The synthesis is catalyzed by RNA polymerase and is dependent upon DNA.

The complete mechanism of synthesis and indeed the *function* of ribosomal RNA are not clearly understood. That a small number of methylated bases are present in ribosomal RNA is known. The synthesis of the latter apparently is conducted in two steps. (1) Under the direction of a segment of the gene and RNA polymerase, a ribosomal RNA precursor is manufactured. (2) The latter polyribonucleotide is methylated (see Chapter IV) to yield ribosomal RNA which then may be packaged into the ribosomal structure. The function of the methylated bases in ribosomal RNA is unknown. Their presence is associated with a decrease in secondary structure of the RNA and theoretically, the capacity for hydrogen bonding should be altered. However, attempts to demonstrate the diminished efficiency have not been fruitful (McConkey and Dubin, 1965). The mechanism of methylation will be discussed in a later section.

The existence of high molecular weight precursors of ribosomal RNA within the nuclei of mammalian cells certainly indicates the complexity of the problem. In this regard, sucrose density gradient ultracentrifugation techniques have proved most useful (see Chapter I). With this technique, the ribosomal precursors have been shown to possess sedimentation constants of 35 S and 45 S and may be produced within the nucleolus at the nucleolus-associated chromatin. The results of a sedimentation analysis by Steele and Busch (1967) are depicted in Fig. 5.4. In these studies, actinomycin D was administered to rats to block further synthesis of RNA in the liver, a labeled precursor was injected, and the nucleolar RNA was extracted at periodic intervals. It is evident from the data that much of rapidly labeled RNA of liver nucleolar preparations is present in the 28 S, 35 S, and 45 S and greater regions of the gradient. Periodically after administration of the antibiotic, a gradual disappearance of both the radioactivity and the ultraviolet absorption may be observed in these regions. The most rapidly affected RNA was in the 55 S region, closely followed by the 45 S RNA. These kinetic studies are indicative of a parent-progeny relationship and are in accordance with the derivation of the 28 S ribosomal RNA component from the 35 S, 45 S, and 55 S nucleolar components. Recent studies have also indicated that the methylation process occurs at the stage of 45 S RNA in the nucleolus.

Darnell and colleagues (Vaughan *et al.*, 1967) have studied the synthesis of ribosomal constituents in HeLa cells. In these cells, ribosomal precursor RNA is synthesized within the nucleolus as a 45 S molecule. The latter is converted to 28 S and 16 S RNA; the 16 S RNA rapidly passes out of the nucleus into the cytoplasm. Both the 16 and 28 S ribosomal RNA combine within the cytoplasm to form eventually the ribosome. The stage at which these RNA molecules associate with ribosomal proteins is not known at present.

Recently, the existence of an additional RNA molecule which forms part of the ribosomal structure has been established in mammalian cells. This RNA has a sedimentation of 5 S and is believed to encompass part of the 50 S ribosomal subunit.

The relationship of the 16 S and 28 S RNA components of the ribosome has also come under consideration. The initial evidence bearing upon this point emanated from Spiegelman's laboratory (1963) with bacterial systems. He employed a technique based upon complementarity of certain regions of the genome with the ribosomal RNA components. In this technique, use is made of hybrid formation between the genome and isotopically labeled ribosomal RNA. The resultant hybrid is subjected to RNase treatment to eliminate unpaired RNA (the hybrid is resistant to this treatment under these conditions), allowing the detection of regions in the DNA which are complementary to the 16 S and 23 S ribosomal RNA's. These studies with bacteria have established the existence of multiple independent sites on the genome for each of the ribosomal constituents. Furthermore, the proportion of the total bacterial

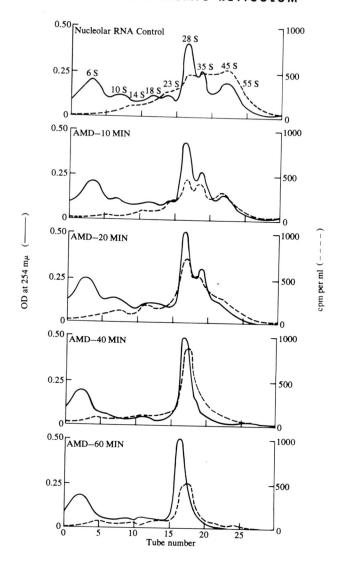

Fig. 5.4. Sedimentation profile of isolated nucleolar RNA after actinomycin. Nucleoli were isolated from the livers of control and actinomycin D(AMD)-treated rats (150 μg/kg body wt) after 10, 20, 40, and 60 minutes. The rats were given orotic acid-6-^{14}C as a RNA precursor at 40 minutes prior to death. The arrow indicates the direction of sedimentation. The actinomycin stops any further synthesis of RNA, allowing only the shift of the 35 S and 45 S RNA to the 28 S position. Courtesy of W. J. Steele and H. Busch.

genome devoted to transcription of ribosomal RNA was in the order of 0.3%.

Similar studies have been performed with the ribosomal RNA of *Drosophila melanogaster* by Spiegelman's group (1965) and of HeLa and other human systems by Attardi *et al.* (1965). The former workers have shown that the DNA complementary to ribosomal RNA is located in a specialized region of the nucleus of *Drosophila*, the *nucleolar organizer* (see Chapter VIII) and the genome contains approximately 200 sites per diploid set for each of the ribosomal RNA's, comprising about 0.27% of the genome. Attardi and co-workers have calculated the fraction of the human genome complementary to 18 S and 28 S ribosomal RNA to be approximately 2×10^{-5} and 5×10^{-5}, respectively. These values would correspond to $400-600$ stretches of the length of the 18 S molecule and $200-400$ stretches of the 28 S RNA in the HeLa cell genome.

Subsequently, a number of investigators with mammalian systems have shown the base compositions of the 16 S and 28 S RNA to differ and accordingly are in agreement with the above proposals that the two ribosomal RNA moieties are derived from unique segments of the genome.

Ribosomal RNA and protein have a relatively long half-life. The rates of ribosomal synthesis and catabolism have been determined in rat liver by Hirsch and Hiatt (1967) and by Loeb *et al.* (1965). Both groups have calculated the identical figure for the half-life, 5 days. It is apparent that the replacement rate of ribosomes greatly exceeds the turnover time of the hepatic parenchymal cell, i.e., greater than 200 days.

III. FUNCTIONS OF ENDOPLASMIC RETICULUM

A variety of biochemical activities encompassing the entire gamut of metabolic events take place within the endoplasmic reticulum, many of which may be related to its canalicular structure. These activities include:

1. *Carbohydrate Metabolism*
   a. Biosynthetic systems responsible for glycogen synthesis, although not contained within, are intimately associated with the microsomes.
   b. Degradative enzymes, i.e., glucose-6-phosphatase, are located exclusively in the membranous portions.
2. *Protein Metabolism*
   a. The synthetic mechanisms by which the protein structure is elaborated are located in the "rough" endoplasmic reticulum.
   b. Degradative enzymes, e.g., peptidases and amino acid oxidases, are contained within the "smooth" endoplasmic reticulum.

3. *Lipid Metabolism*

   a. Fatty acid, steroid, and phospholipid biosynthesis are functions of the "smooth" endoplasmic reticulum.

   b. The metabolism of steroids, including oxidation and reduction reactions, also occurs within the "smooth" endoplasmic reticulum.

   c. Esterases, e.g., cholinesterase, have been reported in the "smooth" endoplasmic reticulum.

4. *Electron Transport and Related Processes*

   a. Certain elements of the electron transport system, e.g., NADH and NADPH-cytochrome b_5 reductases are present within the "smooth" endoplasmic reticulum.

   b. The microsomal membranes possess an active ATPase complex which may be related to ion transport in this organelle.

A discussion of these biochemical activities as related to cellular function will be taken up in the rest of this chapter.

A. "Rough" Endoplasmic Reticulum and Protein Synthesis

Protein synthesis is recognized as the principal activity of the "rough" endoplasmic reticulum although progress in this area has been gratifyingly rapid only within the past decade. The mechanisms underlying protein synthesis are so interwoven with those of nucleic acid chemistry that only after considerable headway had been achieved in the latter area could an understanding of protein synthesis evolve. The earliest theories of protein synthesis attributed the process to the proteolytic enzymes, the cathepsins, those enzymes which hydrolytically cleave proteins to peptides. The synthesis of a new protein was achieved after the breakdown products of an expendable protein had been reassembled into the required polypeptide by both reactions, i.e., catabolism and anabolism, catalyzed by the proteolytic enzymes. However, evidence was soon forthcoming which was not in support of protein synthesis from peptide fragments but favored the *de novo* synthesis from amino acids.

a. NUCLEIC ACIDS AND PROTEIN SYNTHESIS. In 1941, Caspersson, using the newly devised technique of ultraviolet microspectrophotometry, and Brachet, with the more classical cytochemical techniques, independently observed a profound relationship between the capacity of a tissue for protein synthesis and its content of RNA. RNA was abundant in rapidly proliferating cells and cells having secretory functions, e.g., exocrine cells of the pancreas or pepsin-producing cells of the gastrointestinal tract. Other cells devoid of large amounts of protein, although physiologically active, possessed only small amounts of RNA. The more quantitative biochemical techniques of Davidson and his colleagues with mammalian systems, and of Gale and Folkes in bacteria have confirmed these observations.

Additional studies established further the dependence of protein synthesis *not* upon DNA but upon RNA *synthesis*. One of the more interesting of these studies embodies the phenomenon of "thymine-less" death in a thymine-requiring bacterium. In the absence of thymine, and hence, of DNA biosynthesis, these organisms increase in size, the RNA content may even double, although they do not form colonies. Since no synthesis of DNA can occur, the cells eventually die. Thus, in these thymine-requiring organisms, the processes of protein and RNA syntheses are divorced from that of DNA.

Ultraviolet irradiation of bacteria produces little effect upon either RNA or protein synthesis under appropriate conditions, but effects a profound reduction in the capacity of the cell to synthesize DNA. Other work with pyrimidine-requiring mutants of *E. coli* has also shown the dependence of enzyme synthesis, i.e., protein synthesis, upon exogenously supplied uracil or other pyrimidines required for the manufacture of RNA.

The most striking evidence for the protein-RNA relationship is derived from a study with enucleated amebas and algae, and from the naturally occurring enucleated mammalian cell, the reticulocyte. In amebas, after the removal of the nucleus, a reduction may be observed in the capacity to incorporate ^{14}C-labeled amino acids into protein only after a considerable decrease in the amount of RNA had taken place. The same general conclusions have been drawn from studies with enucleated algae. The mature reticulocyte, although capable of synthesizing hemoglobin and producing RNA, does not contain any nucleus, and hence has no DNA. It is clear from these observations that protein synthesis can take place in the absence of the nucleus and that RNA may in some manner be involved in this process.

b. ROLE OF THE ENDOPLASMIC RETICULUM IN PROTEIN SYNTHESIS. The importance of the microsome in protein synthesis was established principally through the efforts of Borsook (1950), Hultin (1950), Siekevitz (see 1963), and Zamecnik and his colleagues (1954), who observed within a short interval after the administration of a variety of ^{14}C-labeled amino acids *in vivo* that the microsomal protein of rat liver became rapidly labeled. This conclusion is clearly demonstrated in some experiments from Zamecnik's group that are depicted in Fig. 5.5.

The microsomal fraction, isolated from rats given a tracer dose of leucine-^{14}C (Fig. 5.5A) was treated with sodium deoxycholate which dissolved the lipid components and the bulk of the microsomal protein (lipoprotein), leaving a residue consisting mostly of ribosomes. A rapid rate of incorporation into ribosomal protein ensued for 5 minutes, after

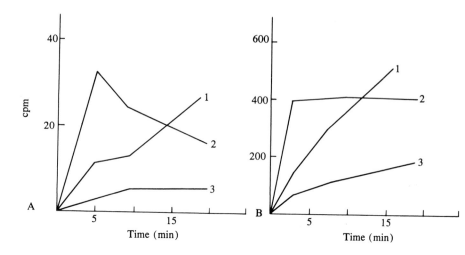

Fig. 5.5. Protein synthesis in rat liver *in vivo*. (A) Microsomal lipoprotein. (B) Microsomal ribonucleoprotein. (C) Soluble protein. Adapted from J. W. Littlefield *et al., J. Biol. Chem.* 217, 111 (1955).

which time the incorporation fell precipitously. Similar results were obtained after the administration to rats of a larger dose of leucine-[14]C (Fig. 5.5B). The ribosomal protein was again labeled more rapidly and possessed a greater specific activity than the other constituents of the microsome. After several minutes, the other proteins of the microsome and indeed of the cell, became labeled. These well-designed experiments leave little doubt as to the role of the ribonucleoprotein particle in protein synthesis, and further, the specific activities of the ribosomal and soluble proteins suggest a possible precursor-product relationship. The role these ribosomes played in protein synthesis had to await the design of more suitable *in vitro* systems.

c. IN VITRO STUDIES ON PROTEIN SYNTHESIS. Eventually, successful incorporation of ^{14}C-labeled amino acids into proteins of cell-free mammalian systems, i.e., homogenates, was reported by Borsook, Siekevitz, and Zamecnik and their associates. The incorporation was dependent upon the presence of actively phosphorylating mitochondria, was enhanced in the presence of Krebs cycle intermediates, e.g., α-ketoglutarate, and was suppressed by dinitrophenol, cyanide, azide, and other inhibitors or uncouplers of oxidative phosphorylation. An active glycolytic system or ATP would adequately substitute for the mitochondria in these *in vitro* experiments. The purpose of the mitochondrial or the glycolytic systems was an indirect but vital one, the production of ATP, the ultimate biological energy source. Prior to these demonstrations, Lipmann, in 1949, had suggested an ATP-mediated activation of amino acids by a mechanism analogous to the activation of fatty acids, a hypothesis which now looked particularly attractive.

In addition to a pool of amino acids and a source of energy for activation, microsomes with intact ribosomes, a cofactor, guanosine 5′ triphosphate, GTP, and a nondialyzable soluble fraction, i.e., a fraction which would not sediment after centrifugation at 100,000 g for 60 minutes, were required for protein synthesis *in vitro*. The requirement for intake microsomes was reminiscent of the *in vivo* experiments with labeled amino acids—the most rapidly labeled organelle in the intact animal was the microsome. In later experiments, microsomal preparations could be replaced by the ribosomes themselves.

The nondialyzable soluble fraction required for protein biosynthesis *in vitro* contained certain enzymes which activated amino acids in the presence of ATP. These enzymes were concentrated by precipitation at pH 5.0 and the reactions they catalyze were formulated as follows:

$$\text{enzyme} + \text{ATP} \longrightarrow \text{enzyme} - \text{AMP} + \text{pyrophosphate}$$
$$\text{enzyme} - \text{AMP} + \text{amino acid} \longrightarrow \text{enzyme} - \text{AMP} - \text{amino acid}$$

Both processes are accomplished by a single enzyme with at least two reactive sites.

The activation process consists of the formation of an enzyme-amino acyl-AMP complex. The AMP was linked to the amino acid by an ester bond between the amino acyl carboxyl and the 3′-hydroxyl of the AMP (Fig. 5.6).

For each amino acid, there exists a specific amino acid activating enzyme and a number of the latter have been isolated and partially

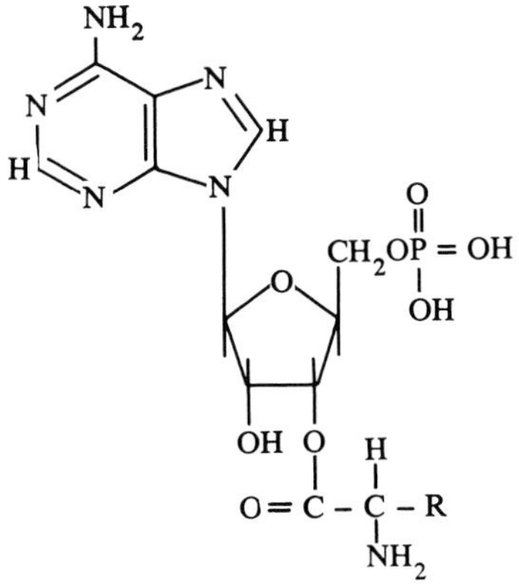

Fig. 5.6. Amino acyl-AMP.

purified. Indeed, in some cases, more than one enzyme has been isolated that activates a single amino acid, e.g., methionine and phenylalanine activating enzymes.

The necessity in protein synthesis for another RNA molecule distinct from the RNA of the ribosome was indicated from the effects of the RNase upon cellular systems. Ribonuclease is rapidly taken up by amebas, and by frog or starfish oocytes by a process of pinocytosis (see Chapter VII), and within a short time, the amount of basophilia is markedly diminished. Concomitant with the reduction in basophilic areas, a profound inhibition in the rate of the incorporation of amino acids into protein may be observed, an effect that is not mediated through a reduction in energy production. In certain systems, e.g., plants, a partial reactivation of protein synthesis may be achieved upon addition of RNA. These studies intimated the existence of a ribonuclease-sensitive substance that was vital in protein synthetic processes.

B. Transfer RNA

A clarification of the role of RNA as an intermediate in the amino acid activation process and in the assembly of these subunits into a polypeptide structure is largely the effort of Zamecnik and his colleagues (see Zamecnik, 1960). The RNA was present in the nondialyzable soluble fraction that had previously been shown to contain the amino acid activating enzymes; this RNA was also precipitated at pH 5.0 ("pH 5.0 fraction"). This RNA moiety was designated "soluble" RNA or transfer RNA (tRNA) in view of its function as a vehicle for the transfer of activated amino acids to the ribosomes:

$$\text{amino acyl} - \text{AMP} + \text{tRNA} \longrightarrow \text{amino acyl} - \text{tRNA} + \text{AMP}$$

The existence of multiple "transfer" RNA molecules may be inferred from the following: the lack of competition between various amino acids for attachment to tRNA; the specificity of the amino acid-activating enzymes catalyzing the activation of the individual amino acids to tRNA; the separation of tRNA into fractions which possessed enhanced reactivities with particular amino acids. A number of these tRNA molecules have since been purified and a specific tRNA does indeed exist for each natural amino acid. Indeed, in some instances, several tRNA molecules specific for a single amino acid have been observed, e.g., two types of methionine, leucine, and threonine tRNA. It is interesting that although the actual isolation of tRNA by Zamecnik's group came as a surprise to the scientific community, Crick, several years prior, had suggested the existence of an adaptor molecule specific for each amino acid as an intermediate in protein synthesis.

Transfer RNA possesses a number of interesting and unique properties. The molecule is quite small compared to the ribosomal RNA, with a molecular weight of 25,000 − 40,000 and a sedimentation constant of 4 − 6 S. Of the 100 or so nucleotides which it contains, three terminal nucleotides are common to all types of tRNA and occur at the amino acid-acceptor region.

RNA-p-C-p-C-p-A[1]

The amino acids are bound in ester linkage to the 3′-hydroxyl of the ribosyl group of the terminal adenyl moiety of tRNA. The pCpCpA residue can be enzymically cleaved from tRNA, resulting in the total loss in amino acid-acceptor activity. The acceptor activity of this altered RNA can be restored enzymically by the sequence of reactions depicted below:

altered RNA + CTP ⟶ altered RNA − pCp + pyrophosphate
altered RNA − pCp + CTP ⟶ altered RNA − pCpCp + pyrophosphate
altered RNA − pCpCp + ATP ⟶ tRNA + pyrophosphate

The base composition of tRNA from a variety of sources is presented in Table 5.3.

A most interesting feature of these molecules is the presence of several unusual nucleotides in the primary structure. These "odd" bases include:

1. *Pseudouridine* or 5-ribosyluracil, which may account for 25% of the uridine moieties of tRNA, e.g., five residues per molecule (Fig. 5.7).

2. *Methylated purines and pyrimidines* present to the extent of three residues per molecule (Fig. 5.8). The types and amounts of the methylated bases occurring in rat liver tRNA are tabulated in Table 5.4.

3. *Methylhypoxanthine and 5,6-dihydrouracil* (Fig. 5.9). The latter is naturally occurring but has previously been considered only as an intermediate in the *catabolism* of uracil.

The tertiary structure of tRNA is also quite unique. Ribosomal RNA consists of single chains of polyribonucleotides while tRNA has a number of double helical regions. Alanine tRNA has been obtained in a homogeneous form by Holley and co-workers (1965) who have determined the nucleotide sequence of this molecule. The structural possibilities have

[1] In accordance with the abbreviations of the *Journal of Biological Chemistry*, p-C indicates CMP-3′, C-p, CMP-5′, ApC-phosphate ester from 5′-hydroxyl of AMP to the 3′-hydroxyl of CMP.

TABLE 5.3

Base Composition (Mole %) of Some Transfer Ribonucleic Acids[a]

Source	Adenine	Uracil	Guanine	Cytosine	Pseudouridine[b]	Reference
E. coli	18.4	16.0	31.2	28.4	5.9	Berg (1959)
	21.0	19.3	31.5	27.2	1.0	Osawa (1960)
Yeast	21.8	22.0	27.2	25.9	3.0	Osawa (1960)
	18.3	19.8	29.0	28.0	3.8	Moniev et al. (1960)
Rabbit liver	18	17	22	29	3.5	Singer and Cantoni (1960)
Rat liver	19.7	18.2	29.2	28.8	4.0	Osawa (1960)

[a] From K. McQuillan, Progr. Biophys. Biophys. Chem. 12, 67 (1962).
[b] May include other unusual bases.

been explored by these investigators and are presented in Fig. 5.10. The internal double helix held together by hydrogen bonds is apparent in each of the three possibilities while regions of uncoiling may also be noted in these models. The unusual bases do not seemingly occur at any special sites, i.e., at regions that might define the loose coiling. However, the existence of a sequence of three bases, defined by two dihydrouracil moieties, may be noted. This triplet which is present in a loosely coiled area may comprise the *"anti-codon"* region of tRNA, i.e., the amino acid specifying region.

The structures of two serine-transfer RNA molecules isolated from brewer's yeast have recently been determined by Zachau and his colleagues (1966). The "clover-leaf" model for the structure of serine-transfer RNA II is presented in Fig. 5.11. Serine-transfer RNA I differs from serine-transfer RNA II in only three nucleotides. The positions and the nature of these nucleotides are indicated by the arrows in Fig. 5.11.

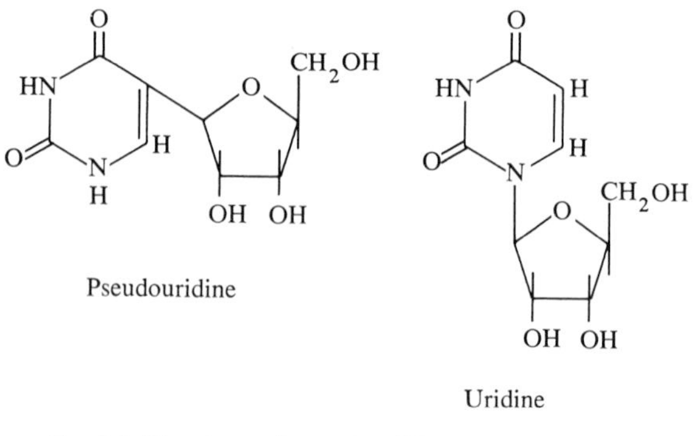

Fig. 5.7. Structure of pseudouridine and uridine.

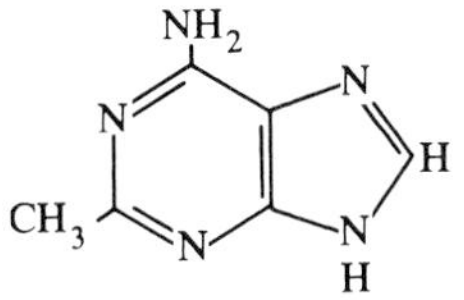

2-Methyladenine 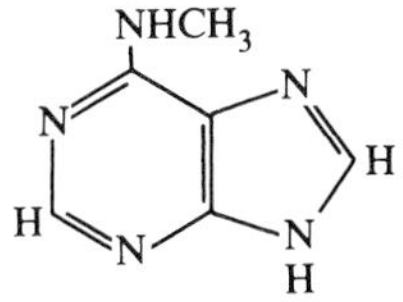   6-*N*-Methyladenine 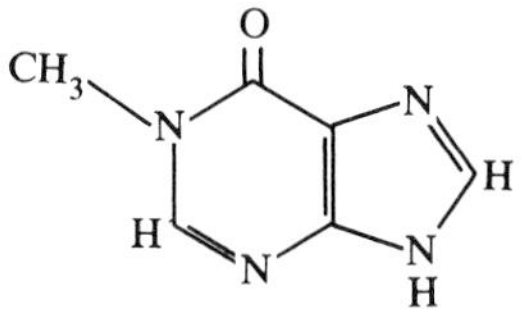   1-Methylhypoxanthine

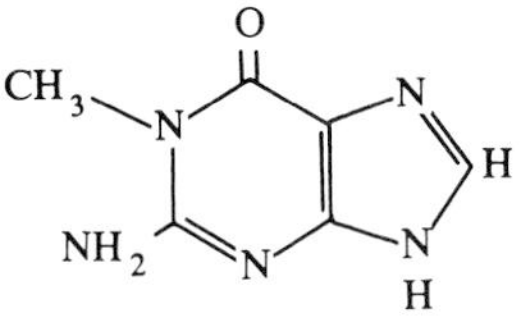

1-Methylguanine 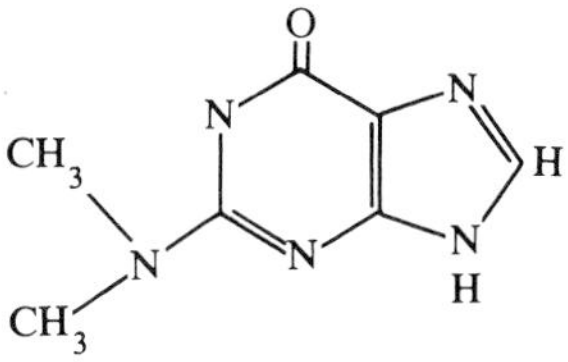   *N,N,*-Dimethylguanine 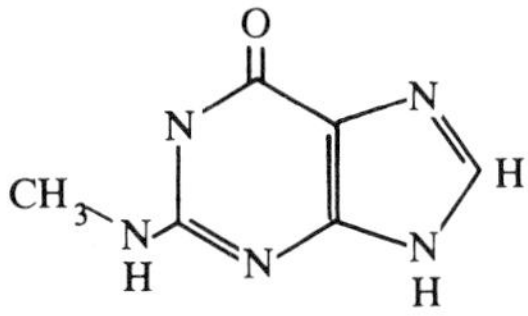   *N*-Methylguanine

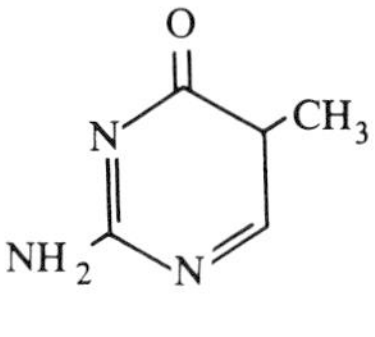

5-Methylcytosine

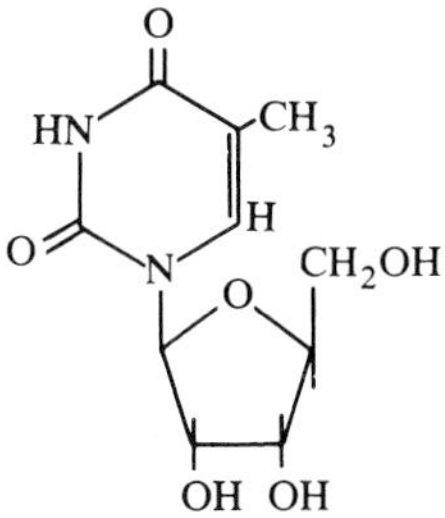

Thymine riboside

Fig. 5.8. Methylated bases in transfer RNA.

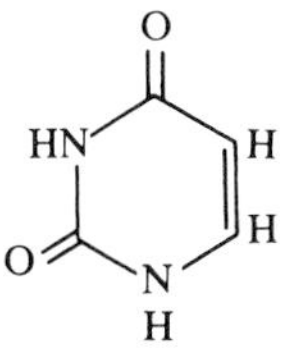

Uracil 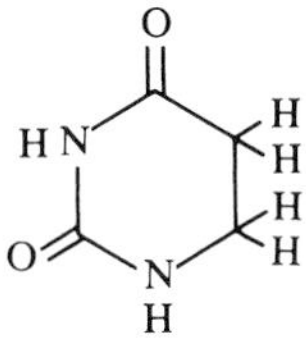   5,6-Dihydrouracil

Fig. 5.9. Structure of uracil and dihydrouracil.

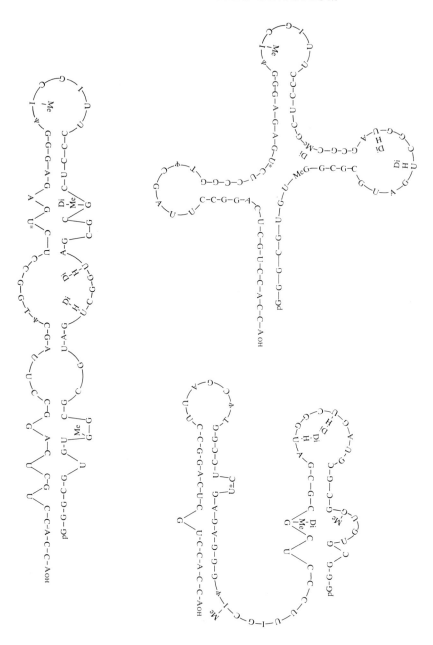

Fig. 5.10. Schematic representation of three conformations of alanyl-transfer RNA. ψ pseudo UMP; Me-I, 1-methyl IMP; diHU, 5, 6-dihydro UMP; DiMe-G, N-dimethyl GMP; Me-G, l-methyl GMP; T, ribo TMP. (From Holley *et al.*, 1965.)

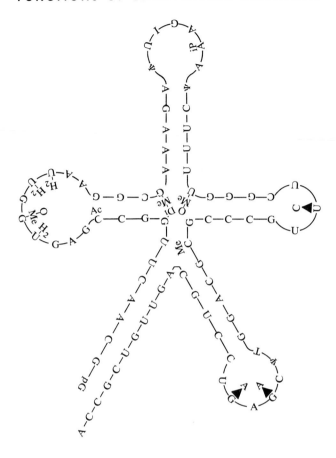

Fig. 5.11. Structure of serine-transfer RNA II.

a. SYNTHESIS OF TRANSFER RNA. The synthesis of an RNA that resembles tRNA appears to take place within the nucleus of the cell. This conclusion is founded upon the following evidence:

1. Only the nucleus and therein, the nuclear RNA, are labeled after a brief incubation of HeLa cells with cytidine-^{3}H. Zonal ultracentrifugation analysis (sucrose density gradient studies—see Chapter I) has established the similarity of this labeled material with tRNA.

2. Isolated nuclei from a variety of sources (when incubated with a labeled precursor) can synthesize a small molecular weight RNA that resembles tRNA. The synthesis of this RNA requires the presence of DNA. Further, the newly synthesized RNA can esterify amino acids.

3. Mammalian cell nuclei can incorporate pseudouridine triphosphate

TABLE 5.4

Methylated Purines and Pyrimidines of Rat Liver tRNA[a]

	Moles/100 μmoles of uridine
Pseudouridine	25
5-Methylcytosine	10
6-Methylaminopurine	0.1
6-Dimethylaminopurine	0.1
1-Methylguanine	3.3
2-Methylamino-6-hydroxypurine	2.3
2-Dimethylamino-6-hydroxypurine	3.0

[a] From D. B. Dunn, *Biochim. Biophys. Acta* 34, 287 (1959).

into a small molecular weight RNA. Pseudouridine is an important constituent of tRNA. This synthesis also is dependent upon DNA. Evidence has also recently accumulated localizing the synthesis of tRNA to the nucleolus of at least the pea seedling.

The synthesis of tRNA is apparently a two-step process. In the first step, an *un*methylated polyribonucleotide arises as the product of a DNA-mediated RNA polymerase reaction. The second stage results in the methylation of the methyl-poor RNA to yield tRNA.

The breakthrough in our understanding of these two stages in the synthesis of tRNA occurred in 1954. E. Borek was studying the growth requirements of a methionine-requiring mutant of *E. coli* $K_{12}W_6$. He noticed that during methionine starvation of this mutant organism, an RNA

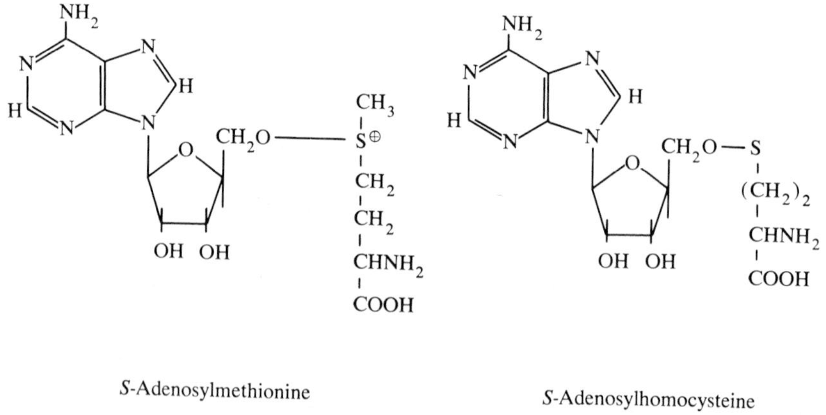

S-Adenosylmethionine   S-Adenosylhomocysteine

Fig. 5.12. Structure of S-adenosylmethionine and S-adenosylhomocysteine.

moiety accumulated within the cell that possessed a low molecular weight, was low in methylated bases, and could accept amino acids. Groups of highly specific enzymes have been isolated from the *E. coli* and subsequently from a number of other species that can methylate the "methyl-poor" tRNA in the presence of the biological methylating agent, S-adenosylmethionine (see Fig. 5.12)—the RNA methylases.

"methyl-poor" tRNA + S-adenosylmethionine ⟶ tRNA + S-adenosylhomocysteine

The *E. coli* RNA methylase will *not* methylate *E. coli* tRNA unless the latter has been isolated from organisms subjected to methionine starvation. However, the enzyme *will* methylate tRNA's from other sources. The tRNA molecule from a particular source, although fully methylated with respect to the methylating enzymes of this source, can accept additional methyl groups in the presence of a heterologous enzyme. Therefore, the methylase will catalyze the methylation of the transfer RNA precursor at particular sites. When these sites are occupied, further methylation can be accomplished only by heterologous enzymes. The methylation reaction as observed by several investigators may be accomplished within the nucleolus.

In summary, tRNA is synthesized in two steps, the first of which is the formation of a polyribonucleotide precursor by RNA polymerase and second, the methylation of the latter.

b. FUNCTIONS OF THE METHYL GROUP. The function of the methylated bases in transfer RNA has been the subject of much conjecture. Initially, it was proposed that the methylated bases either conferred upon the transfer RNA a marked resistance to the action of catabolic enzymes, e.g., ribonuclease, or were required for the esterification of amino acids. These hypotheses have not borne the test of experimentation. The most attractive hypothesis has been derived from the studies of Revel and Littauer (1965)—the methylated bases are required in maintaining the rigorous specificity of amino acid acceptance.

c. PRESENCE OF THIOL GROUPS IN TRANSFER RNA. Recently, several investigators have described the occurrence in a transfer RNA from both bacterial and mammalian sources of sulfur-containing nucleotides (Lipsett, 1965; Carbon *et al.*, 1965). These nucleotides include 4-thiouridine monophosphate and 2-thiopyrimidine ribonucleotide (Fig. 5.13). The addition of the sulfur to the transfer RNA takes place at the polynucleotide stage in a manner analogous to the methylation process. The immediate sulfur donor is cysteine (Fig. 5.13). The function of the thiol groups is unclear.

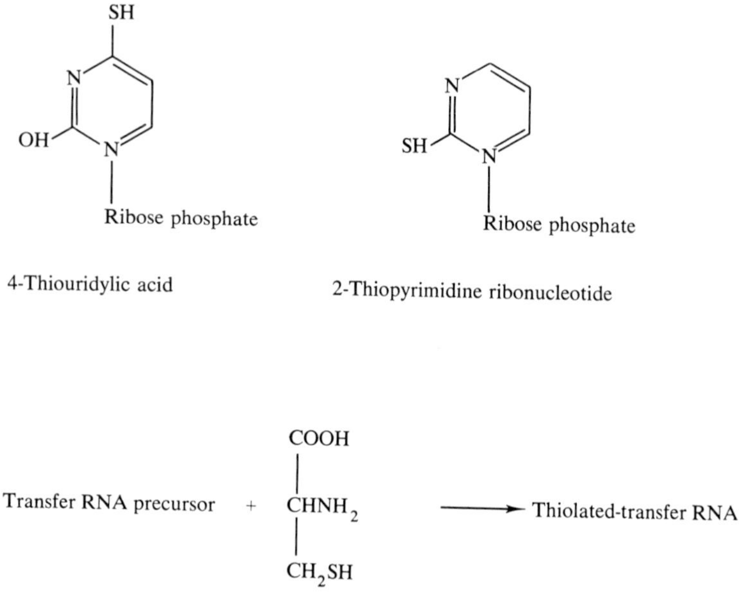

SH

OH

Ribose phosphate

4-Thiouridylic acid

SH

Ribose phosphate

2-Thiopyrimidine ribonucleotide

COOH
|
Transfer RNA precursor　+　CHNH$_2$　⟶　Thiolated-transfer RNA
|
CH$_2$SH

Fig. 5.13. Sulfur-containing nucleotides in transfer RNA.

d. GENETIC LOCI CORRESPONDING TO TRANSFER RNA. Molecular hybridization (Chapter IV) has been utilized to ascertain the percentage of the genome devoted to the transcription of transfer RNA. The studies from Spiegelman's laboratory (Ritossa et al., 1966) with the genome from *Drosophila melanogaster* suggest the presence of 12.5 templates per haploid set for the transcription of each of the 60 different kinds of transfer RNA.

C. Messenger RNA

The mechanisms proposed for protein synthesis in 1955 did not account for the existing facts, and some glaring inconsistencies in the schema were apparent. Should not the gene material, DNA, play a greater role in protein synthesis than had been ascribed? If our phenotypic events, i.e., physical characteristics, are indeed hereditary, the gene through the mediation of the chromosome must be involved in some manner. In addition to this hypothetical consideration, several laboratories had observed effects of DNase upon the synthesis of proteins *in vitro*, thus implying a function for DNA.

The solution to these vexing problems came from the laboratories of Hershey (1953) and of Volkin and Astrachan (see Volkin, 1962) based upon an initial observation of Cohen in 1948 that a small fraction of the RNA of *E. coli* after infection with T_2 bacteriophage turned over rapidly in the absence of net synthesis of RNA. Volkin and Astrachan noted that during the infection of *E. coli* by T2 bacteriophage, a species of RNA was synthesized whose composition closely resembled the base ratio *not* of the host DNA, but of that of the bacteriophage. With the use of brief isotope assimilation periods, pulse-labeling, the newly synthesized RNA was shown to undergo a rapid breakdown and renewal during the period in which biochemical transformations were occurring in the bacteria, i.e., the synthesis of new proteins associated with bacteriophage infection. The feeling that this new RNA species was *only* associated with phage infection was quickly disspelled by several laboratories. Various types of bacteria were pulse-labeled with ^{32}P and the distribution of the label within the RNA was determined. The results adapted from Volkin (1962) are reported in Table 5.5 and clearly point out the resemblance of the newly synthesized RNA, not to the bulk RNA of the cell, ribosomal RNA, but to DNA of the bacteria.

How is this DNA-like RNA synthesized and what is its function? In 1961, Hall and Spiegelman slow-cooled mixtures of partially purified RNA obtained from T2-infected *E. coli* with a single-stranded T2 DNA that had been produced by heating to 100°C. They observed the formation of a hybrid molecule. The requirements for the formation of the hybrid molecule were rigorous, i.e., *E. coli* DNA would *not* hybridize with this RNA. The hybrid formation presumed complementarity between the nucleotide sequences of the RNA and T2 DNA and suggested that the template for the synthesis of this DNA-like RNA was *indeed* single-stranded DNA, in a reaction catalyzed by the enzyme, RNA polymerase.

The existence of this DNA-like RNA provided an answer to the seemingly unsurmountable dilemma presented previously—the transmission of hereditary characteristics. The DNA, located within the nucleus, representing the genetic information, may pass on this information to a RNA molecule, with a complementary structure, a DNA-like RNA, which may then act as an intermediary in protein synthesis. Jacob and Monod (1961) called this RNA species messenger RNA (mRNA) in view of its function as an emissary. Why is it necessary to invoke this *deus ex machina?* Why not allow DNA to act in this capacity per se? Recall that one of the potent arguments in the definition of the genetic material was the remarkable stability and low rate of turnover of DNA. Further, protein synthesis can continue in the absence of DNA in enucleated amebas, algae, and reticulocytes. Additional evidence, originating largely from studies on induced-enzyme formation after infection of bacteria with bacteriophage, has led to the concept of the unstable

TABLE 5.5

Comparison of Relative Radioactivities of RNA Nucleotides with DNA
Compositions in Three Bacterial Species[a]

Determination	Molar proportions				
	A	U (or T)	C	G	$\dfrac{A+U \text{ (or T)}}{G+C}$
Pseudomonas aeruginosa					
RNA composition	1.0	0.89	0.97	1.41	1.26
DNA composition[b]	1.0	0.97	2.02	1.96	0.50
Relative ^{32}P content (RNA)	1.0	1.05	1.59	1.49	0.67
Proteus vulgaris					
RNA composition	1.0	0.83	0.93	1.26	1.20
DNA composition[b]	1.0	0.98	0.69	0.66	1.47
Relative ^{32}P content (RNA)	1.0	1.14	0.70	0.67	1.56
Escherichia coli					
RNA composition	1.0	0.79	0.93	1.32	1.26
DNA composition[b]	1.0	1.0	1.10	1.09	0.92
Relative ^{32}P content (RNA)	1.0	0.87	0.98	1.23	0.85

[a]From E. Volkin, *Federation Proc.* **21**, 112 (1962).
[b]Calculated from Spirin *et al.* (1957).

template, a template whose synthesis is instituted almost immediately after a stimulus and one whose disappearance is also instantaneous. This property of instant synthesis or degradation is in contradistinction to the fundamental aspect of the gene, stability.

The discovery then of mRNA, a species with a base composition closely resembling DNA, whose synthesis occurred rapidly, and whose existence was ephemeral (in bacteria) supplied the answer to this problem. One may also ask if ribosomal RNA may not fulfill this function. The evidence today is in favor of a very stable RNA of the ribosomes, i.e., low turnover. Messenger RNA is then left to act as the messenger in the transmission of genetic information which is responsible for the elaboration of protein, a process referred to as *translation*. This RNA has been defined by Lipmann (1963) ". . . as a sequence-determining template that combines with ribosomes for catalytic functioning in protein synthesis."

Several vexing questions about the localization of messenger RNA synthesis have yet to be satisfactorily answered: (1) What is responsible for the rupture of the hydrogen bonds and subsequent release of the growing messenger polyribonucleotide from the gene? (2) How is the messenger transported from the region of synthesis, the nucleus, to the site of action, the cytoplasm or endoplasmic reticulum?

A tentative solution to these questions has arisen from the laboratories of Nirenberg (Byrne *et al.*, 1964) and Hiatt (Henshaw *et al.*, 1965). The nuclear ribosomes, whose presence has been demonstrated in preparations of isolated nuclei and nucleoli, may play a role in the release of the messenger from the gene. The complex messenger RNA-ribosome, is transferred from the nucleus through the nuclear pores to the cytoplasm where it may function in translation. The finding of a 45 S particle within the liver nucleus and cytoplasm in which is associated a rapidly labeled RNA possessing many of the characteristics of a messenger is certainly consistent with the hypothesis (Henshaw *et al.*, 1965).

D. Polysomes

Unlike the messenger RNA of the bacterial cell, the mammalian counterpart does *not* possess an ephemeral existence and also unlike bacteria, the mammalian cell does not generally possess as short a generation time; bacteria divide in the order of minutes while the average parenchymal cell in the liver may divide less than once a year. The average half-life of total messenger RNA of *Bacillus subtilis* with a generation time of 100 minutes has been estimated as $2\frac{1}{2}$ minutes. The messenger fraction in the cytoplasm of rat liver is stable for at least 40 hours, not appreciably more rapid than ribosomal RNA (Revel and Hiatt, 1964).

Messenger RNA is associated with the bacterial 70 S ribosomes and may be recovered from this complex by decreasing the Mg^{2+} concentration. It is this messenger RNA-70 S ribosome complex which may be considered the basic unit in protein synthesis. The molecular weight of mRNA has been deduced as 2.5×10^5 although it may be much larger. Such a template would contain approximately 600 nucleotides arranged presumably in a single chain of approximately 2000 Å in length. The individual ribosome, on the other hand, with a diameter of 200 Å would have a circumference of 628 Å. The mRNA would have to wrap around the ribosome approximately three times. How would the information which is contained in the inner belt be expressed? This geometrical consideration fostered the search for a larger basic unit. The search was rewarded with the discovery of a structure in preparations of reticulocytes by Warner *et al.* (1963) that sedimented at 170 S and was intimately concerned with the polymerization of amino acids to form hemoglobin *in vitro*. Electron micrographs revealed a unit composed of three to five ribosomes, connected to a tapelike structure; this unit was called a *polyribosome*, or polysome. The tapelike structure that bound the ribosomes was mRNA. In the reticulocyte system, the mRNA is $10-20$ Å in thickness and the ribosomes are spaced along the polymer at 50—100 Å. The mRNA is attached to the 30 S portion of the basic 80 S unit. An electron micrograph depicting polyribosomal formation in isolated preparations from rat liver is presented in Fig. 5.14. The number of ribosomes which partake in the polysomal structure may vary from 5 in the reticulocyte to 30—40 in the HeLa cell and even 50—70 in cells that have been infected with polio virus. An excellent correlation has been obtained between the number of polysomes and the functional capacity of maturing reticulocytes for protein biosynthesis (Marks *et al.*, 1963). The process of maturation in this cell is associated with a decrease not only in the ability of the cell to synthesize proteins, i.e., hemoglobin, but also in the number of polysomes.

The polyribosomal fraction may be concentrated from a tissue homog-

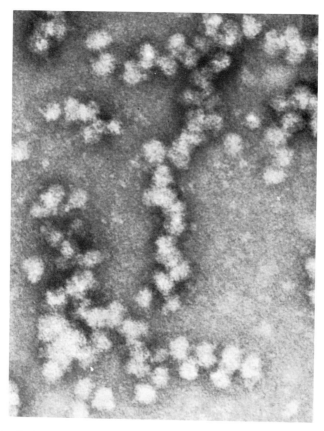

Fig. 5.14. Isolated polyribosomes. A polyribosomal fraction was isolated from rat liver by Wettstein *et al.* (1963) and a sample was processed for electron microscopy. The section was stained with uranyl acetate and a negative contrast was prepared. × 235,000. Courtesy of H. Busch and K. Shankar.

enate by the method of Wettstein *et al.* (1963). The method is dependent upon zonal ultracentrifugation techniques in which a linear sucrose gradient is employed and the centrifugation is carried out only for a short period. The bulk of the monosomes, i.e., only one ribosome attached to the messenger, are removed prior to the sedimentation analysis. The distribution of the ribosomes in the gradient may be ascertained by removing fractions and recording the optical density at 260 mμ, the wavelength of maximum absorption of RNA. The results of polysomal analysis in a homogenate of liver from a weanling rat are depicted in Fig. 5.15. The lighter components in the gradient are represented on the right of the figure. Eight peaks, representing polysomal structures with 1 − 8 attached ribosomes are clearly discernible; a considerable portion of the gradient is composed of polysomes with more than 8 attached ribosomes.

The polyribosome is markedly sensitive to RNase. A rapid dissolution of this structure and the concomitant production of monosomes may be noted in the presence of small amounts of the catabolic enzyme (see Fig. 5.16).

E. Guanosine 5′-Triphosphate

The Zamecnik group reported the necessity for the nucleotide, guanosine 5′-triphosphate (GTP) (Fig. 5.17), in the *in vitro* protein synthesizing system. Its role has befuddled investigators although the nucleo-

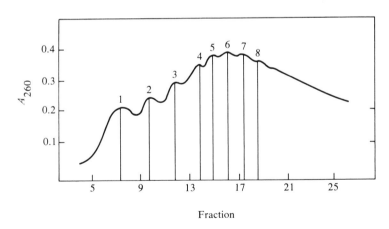

Fraction

Fig. 5.15. Polyribosomal pattern from rat liver. A polyribosomal preparation was isolated from the liver of "weanling" rats by the method of Wettstein *et al.* (1963). The polyribosomes were subjected to a sedimentation analysis; the direction of sedimentation is from left to right. The absorbancy at 260 mμ (A_{260}) is representative of the RNA of the polyribosomes and is used as a measure of the quantity of the latter. Approximately eight peaks are discernible in this pattern.

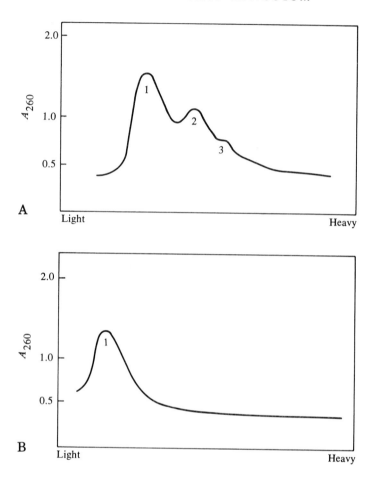

Fig. 5.16. Effect of ribonuclease upon polyribosomes. See legend to Fig. 5.15. Pancreatic ribonuclease, 0.15 (A) and 0.3 μg (B), was added to the polyribosomal preparation. After incubation at 37°C for 5 minutes, the polyribosomes were analyzed as sedimentation techniques. Note the pronounced degradation of the "heavy" polyribosomes.

tide is required in the latter stages of protein synthesis. We will have more to say about GTP in a subsequent section.

F. Coding Problem

The preceding sections have been concerned with the factors required for the elaboration of the structural aspects of a protein molecule. Yet, proteins consist of more than a haphazard array of amino acids, and are compiled in an orderly and exactly defined manner. The determination of

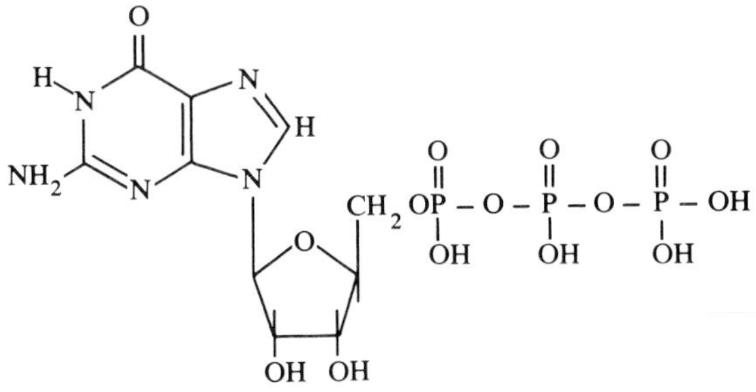

Fig. 5.17. Guanosine 5'-triphosphate.

the mechanism by which this organization is accomplished has evolved only during the past few years.

The number of amino acids constituting most protein is approximately 20, i.e., the "alphabet" consists of 20 amino acids (see Table 5.6). On the other hand, the "alphabet" of the nucleic acids is composed of only

TABLE 5.6

Standard Set of Amino Acids Which
Constitute Proteins

Alanine	Leucine
Arginine	Lysine
Asparagine	Methionine
Aspartic acid	Phenylalanine
Cysteine	Proline
Glutamic acid	Serine
Glutamine	Threonine
Glycine	Tryptophan
Histidine	Tyrosine
Isoleucine	Valine

four bases. How the latter is responsible for incorporation of 20 amino acids into a polypeptide structure comprises the "coding problem." The minimum combination of bases which would be required for the coding of all 20 amino acids is three (two bases would code for 4^2 or 16 amino acids; three bases, 4^3 or 64 amino acids). Obviously, then, more than enough "triplets" of bases, or *codons*, exist to account for peptide bond formation from all amino acids (Table 5.7). A single codon may either not code at all, "nonsense" triplet, or an amino acid may possess several codons, i.e., a degenerate code.

TABLE 5.7

The 64 Possible Three-Letter Codons

AAA	AAG	AAC	AAU
AGA	AGG	AGC	AGU
ACA	ACG	ACC	ACU
AUA	AUG	AUC	AUU
GAA	GAC	GAC	GAU
GGA	GGG	GGC	GGU
GCA	GCG	GCC	GCU
GUA	GUG	GUC	GUU
CAA	CAG	CAC	CAU
CGA	CGG	CGC	CGU
CCA	CCG	CCC	CCU
CUA	CUG	CUC	CUU
UAA	UAG	UAC	UAU
UGA	UGG	UGC	UGU
UCA	UCG	UCC	UCU
UUA	UUG	UUC	UUU

One of the first theories which attempted a solution of the "coding problem" was proposed by G. Gamow in 1954. In his "overlapping" code, the three units of the codon do not follow each other sequentially on the polynucleotide chain but overlap (Fig. 5.18).

According to an overlapping code, a change in a single base should affect the polymerization of at least three sequential amino acids, yet a number of examples exist in which a mutation at the locus of a single amino acid can be seen, e.g., in the hemoglobinopathies. No example has been recorded in which a mutation of two adjacent amino acids has occurred. Crick (1963) has concluded ". . . the weight of evidence certainly suggests that it is a non-overlapping triplet code, heavily degenerated in some semisystematic way, and universal, or nearly so."

The breakthrough in this area of investigation came independently from the laboratories of Nirenberg et al. (1963) and Ochoa (1963). Both investigators employed an in vitro system consisting of "starved" E. coli ribosomes; ATP, amino acids, and the pH 5.0 fraction from E. coli. The ribosomes were preincubated at 37°C to remove the natural messenger RNA, a procedure which then made possible the activation of the ribosomes by exogenous messengers (starved ribosomes). The addition of the "synthetic messenger molecule," polyuridylic acid (poly U), stimulated the incorporation of phenylalanine-^{14}C into a polymeric structure consisting of only this amino acid. The degree of template activity of poly U in initiating the synthesis of polyphenylalanine was dependent upon the number of units composing the former. Chains containing less than 11 residues per poly U molecule were inactive; fifty residues were required for template activity. The optimal activity was recovered with polymers containing more than 250 nucleotides.

The igniting factor in this breakthrough was the discovery and subsequent use by Ochoa of the enzyme polynucleotide phosphorylase (see Chapter IV) in the construction of synthetic polynucleotides of known composition from nucleoside diphosphates. By the use of these synthetic messengers, Ochoa and Nirenberg succeeded in reconstructing a triplet code for the translation of the nucleotides into an amino acid. The initial data suggested the necessity for the inclusion of UMP in *all* codons, a hypothesis which required messenger RNA molecules to possess a large amount of UMP. With the modification of methodology and the availability of many other polymers containing mixed bases, the presence of UMP in all codons was no longer necessary. A new compendium of codewords has been prepared; the data are presented in Table 5.8.

G. Protein Synthesis

Having introduced the major participants, let us now discuss the details of the protein synthetic mechanism. The DNA, with the aid of RNA polymerase, elaborates an RNA molecule whose base composition is complementary to the DNA template and in so doing, the DNA imparts the genetic information to this messenger RNA (see Fig. 5.19). The latter has an ephemeral existence in bacterial and phage-infected systems although evidence has accumulated which favors a longer-lived molecule (e.g., days) in mammalian systems. The messenger RNA activates the pre-existing 80 S ribosomal particles forming the polysome, which may contain from 3 to 60 ribosomes. Concurrently, the amino acids have been activated by individual amino acid-activating enzymes via the intervention of ATP. These amino acyl-AMP-enzyme complexes may be transferred to individual and specific transfer RNA molecules, producing

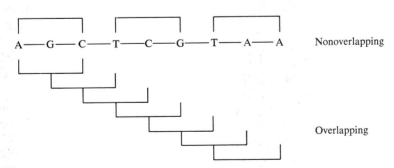

Fig. 5.18. Overlapping and nonoverlapping codes.

TABLE 5.8

Tentative Survey of Codons

Amino acid-^{14}C	Codons[a]
Alanine	GCU, GCC, GCA?, GCG?
Arginine	AGG?, CGG?, AGA, CGU?, CGC?, CGA?
Aspartic acid	GAU, GAC
Asparagine	AAU, AAC
Cysteine	UGU, UGC
Glutamic acid	GAA, GAG
Glutamine	CAA, CAG
Glycine	GGU?, GGC?, GGA?, GGG?
Histidine	CAU, CAC
Isoleucine	AUU, AUC, AUA?
Leucine	CUU, CUC, UUA? UUG
Lysine	AAA, AAG
Methionine	AUG
Phenylalanine	UUU, UUC
Proline	CCC, CCU, CCA, CCG?
Serine	UCA?, UCG?, AGU?, UCU, UCC, AGC?
Threonine	ACA, ACU?, ACC, ACG?
Tryptophan	UGG
Tyrosine	UAU, UAC
Valine	GUU, GUC, GUG, GUA?

[a]First position, representing the 5′ end, is indicated on the left. ?Codons are given on a tentative basis.

amino acyl-tRNA. An interesting experiment of Chapeville *et al.* (1962) has revealed that the recognition of the appropriate site of attachment on the polysome is contained not within the amino acid but within the tRNA, per se. These investigators synthesized cysteinyl-tRNA, reduced the complex chemically to alanyl-tRNA, which was then introduced into a system containing the appropriate synthetic messenger for cysteinyl-tRNA. The results indicated that the alanyl portion of alanyl-tRNA was indeed incorporated into the polypeptide in lieu of cysteine. The tRNA must possess an amino acid specifying site which is therefore distinct from the amino acid attachment site.

How is the information specifying the amino acid contained within the structure of the tRNA molecule? What makes possible the recognition of the site on the polysome? Largely through the efforts of Spencer *et al.* (1962), a partial understanding of these relationships has evolved. Transfer RNA exists in a partial double helical form, as a twisted hairpin, approximately 100 Å in length, with one end containing the point of attachment of the amino acid, the C-C-A trinucleotide (Fig. 5.10). At another portion of the molecule, at the "curl," is located a triplet of nucleotides, which specifies the particular amino acid and determines the site of attachment of the polysome, the amino acid-specifying site. The latter site contains an *anticodon*, i.e., a triplet with a base composition that is complementary to that of the codon of the mRNA in the polysome. The codon then directs the formation of hydrogen bonds between itself and the anticodon, a process which may require NH_4^+ or K^+. (Fig. 5.20).

The nature of the attachment of the amino acyl-tRNA to the polyribosome is a complex one. The messenger molecules are bound to the 30 S subunit of the ribosomes, while the amino acyl-tRNA is in some manner attached to the 50 S subunit. The terminal 5'-phosphate and possibly the entire-pCpCpA end group are involved in this attachment.

The available evidence indicates that proteins grow by stepwise addition of single amino acids at the N-terminal amino ends on the polysome and that the growing polypeptide chain terminates in a single tRNA molecule (Fig. 5.21). We may envision the formation of a polypeptide chain as indicated in Fig. 5.22 taken from Watson (1965): The mRNA picks up a ribosome at one end of its chain and an amino

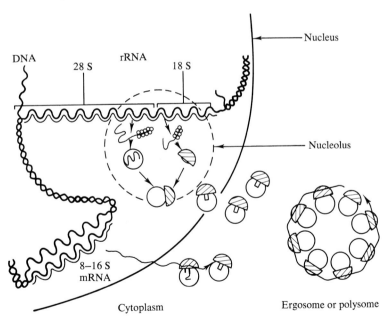

Fig. 5.19. Schematic outline illustrating nuclear synthesis and transfer to cytoplasm of messenger RNA and ribosomes. From H. Noll, *in* "Developmental and Metabolic Control Mechanisms," Williams & Wilkins, Baltimore, Maryland, 1965.

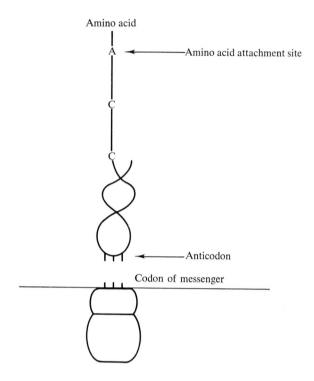

Fig. 5.20. Attachment of amino acyl-tRNA to polysome.

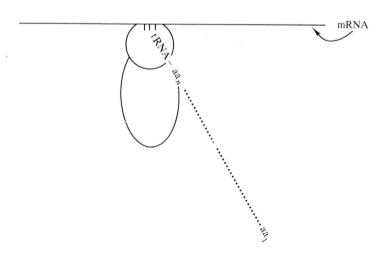

Fig. 5.21. Growing polypeptide chain.

acyl-tRNA corresponding to the N-terminal amino acid locks into position at the codon site on the polysome. At the codon site immediately adjacent to the ribosome attachment, the $n + 1$ region is linked to the incoming amino acyl-tRNA. The latter site may be considered the "ready" position. The peptide bond is formed after a nucleophilic attack of the amino group of the incoming amino acyl-tRNA upon the carboxyl of the peptidyl tRNA, with the subsequent ejection of the latter tRNA. Concomitantly, the messenger RNA is shifted to activate the next codon, the $n + 2$ region, so that the appropriate amino acyl-tRNA may approach.

The formation of the peptide bond requires two enzymes, NH_4^+, and GTP. The binding of the amino acyl-tRNA to its appropriate codon on the polysome is markedly stimulated by NH_4^+. The carboxyl group of the amino acid is already in an activated state by virtue of its involvement in an ester linkage with the adenosyl moiety of tRNA, hence peptide bond formation can occur easily with the assistance of a peptide synthetase enzyme. The GTP along with the aid of the second enzyme, a translocase, is believed to function in the advancement of the messenger RNA by one codon. The GTP in an energy-mediated reaction induces a conformational change in the ribosome so that the transfer from a codon to the adjacent one, a distance of approximately 15 Å, can occur. During the course of this change, GTP is dephosphorylated to guanosine diphosphate, GDP.

In addition to the above requirements, several other factors present in the pH5-soluble fraction are needed for protein synthesis to occur. These have not been completely characterized at this time.

The formation of the polypeptides is accomplished as the messenger "tape" moves along, decoding the information contained at each of the codons. At the terminal position on the polyribosome may be found a polypeptidyl-tRNA; that RNA molecule would correspond to the amino acid of the C-terminal end of the protein.

The efficiency of protein biosynthesis is markedly enhanced by increasing the number of ribosomes attached to the messenger so that several molecules of protein might be simultaneously constructed (Fig. 5.23). Thus at any one time, the polyribosome consists of several ribosomes placed at spaced intervals along the messenger RNA containing polypeptide chains of increasing length (Fig. 5.24), i.e., in different stages of polypeptide synthesis. In this respect, the active polyribosome closely resembles an assembly line for the manufacture of protein.

a. THE WOBBLE HYPOTHESIS. It is clear from the tentative assignments of more than one codon for an amino acid (Table 5.8), that some modification of the rules governing codon-anticodon binding may be required. Is the recognition of anticodon on the messenger RNA by the

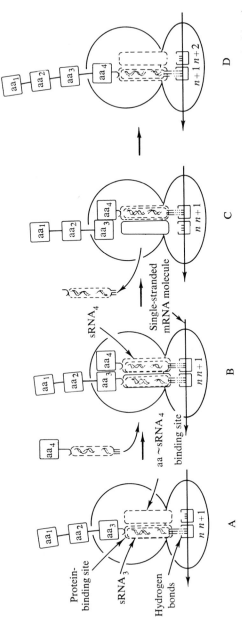

Fig. 5.22. Diagrammatic representation of the stepwise growth of a polypeptide chain. (A) Growing polypeptide chain, attached by the terminal sRNA group to protein-binding site; (B) Attachment of specific aa-sRNA molecule by hydrogen bonding to $(n + 1)$ codon of mRNA chain; (C) Formation of peptide bond between aa_3 and aa_4. Ejection of $sRNA_3$; (D) Movement of growing polypeptide chain from aa-sRNA binding site to protein-binding site. Simultaneous movement of mRNA to place $(n + 2)$ codon at the aa-sRNA binding site. From J. D. Watson. "Molecular Biology of the Gene." Benjamin, New York, 1965.

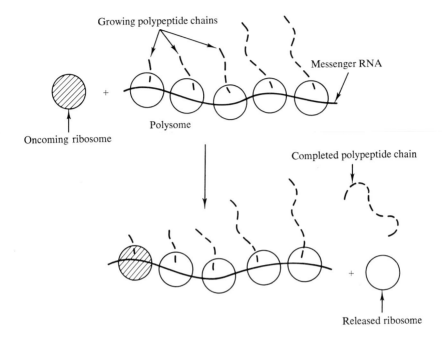

Fig. 5.23. Movement of ribosomes along an RNA chain in protein synthesis.

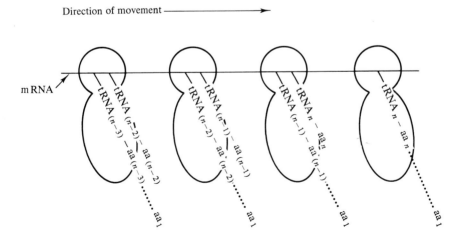

Fig. 5.24. Polyribosome in operation.

codon totally governed by the rules of hydrogen bonding? For example, if the anticodon for phenylalanine is AAA, must the codon be UUU? Since it is known that UUC is also a codon for phenylalanine, must the existence of transfer RNA with an anticodon, AAG, be postulated? An attempt has been made by Crick to circumvent this difficulty. The wobble hypothesis has been proposed in which Crick postulates that the first two bases of the codon must hydrogen bond in the normal fashion to their respective anticodons. In the base pairing of the third base of the codon, more than one type of pairing is possible due to some play, or *wobble*, in the structure. He has computed the possibilities and these are reproduced in Table 5.9. If the third base of the anticodon were A or G, it would recognize U on the codon. If a transfer RNA had guanosine in the third place on the anticodon, then it must recognize uridine or cytidine at that codon site.

b. INITIATION AND TERMINATION OF CHAIN SYNTHESIS. The preceding discussion has been concerned largely with the process of chain extension. An equally weighty problem is the question of chain initiation and chain termination. In chain extension, the substrates for both peptide synthetase and translocase are peptidyl tRNA moieties in which the neutral amide configuration is present. However, in the initiation of a protein chain, both enzymes are presented amino acyl-tRNA with a protonated α-amino group, $-NH_3^+$. In addition, investigations undertaken several years ago revealed an unusual preponderance in bacterial proteins of either methionine or alanine as the N-terminal amino acid. These considerations suggested the possibility for protein synthesis of the initiation process by one or at most two substances. The discovery of N-formylmethionyl-tRNA by Marcker and Sanger (1964) has supplied

TABLE 5.9

Pairing at the Third Position of the Codon[a]

Base on the anticodon	Bases recognized on the codon
U	A
	G
C	G
A	U
G	U
	C
I	U
	C
	A

[a] From F. Crick *J. Mol. Biol.* **19**, 548 (1966).

the clue to this problem of chain initiation. The latter derivative no longer possesses a protonated α-amino group and hence would present a desirable substrate to the enzymes.

The formylation of methionine is carried out by a transformylase enzyme and N^{10}-formyltetrahydrofolic acid *after* the latter has been linked to tRNA. The codon responsible for the binding of the N-formyl-methionyl-tRNA at the initial N-terminal position is AUG or UUG. Once the polypeptide chain has been released from the polysomal surface, the N-formylmethionine group is either enzymically removed in those proteins that *do not* possess a methionine as an N-terminal amino acid or the formyl group may be cleaved. In summary, it is presently believed that methionine, as the N-formyl derivative, is a universal chain initiator.

In certain bacterial strains, several codons do not apparently specify any amino acid. These include UAG and UAA and are referred to as nonsense codons. The nonsense codons are believed to function as chain terminators. It has been proposed that the recognition of the codons for chain termination is conducted by two special transfer RNA molecules, just like the other codons. These transfer RNA's are not esterified by amino acids but carry a specific compound which results in chain termination.

c. CHEMICAL DIRECTION OF PROTEIN SYNTHESIS. The construction of the protein is believed to proceed from the 5′ end of the messenger to its 3′. In this respect, the direction of protein synthesis is identical with direction of the transcription process. Theoretically, it is possible to have protein synthesis occur as the messenger is being elaborated on the genome.

H. Antibiotic Inhibitors of Protein Synthesis

a. STREPTOMYCIN. Streptomycin was originally isolated by S. A. Waksman and colleagues in 1944 from cultures of *Streptomyces griseus*. Its structure has been elucidated and consists of three independent units (Fig. 5.25) joined by glycosidic bonds.

The first indication of the mechanism of action of streptomycin stemmed from the inhibiting effects of the antibiotic upon the inducibility by lactose of β-galactosidase in various bacteria. These studies suggested protein synthesis as the site of action of the antibiotic. The antibiotic also inhibited the incorporation of amino acids into polypeptide material in cell-free systems. The inhibitory effect has since been localized to the ribosomal structure and in particular, to the 30 S subunit, the subunit to which the messenger RNA is attached in the formation of the polyribosome.

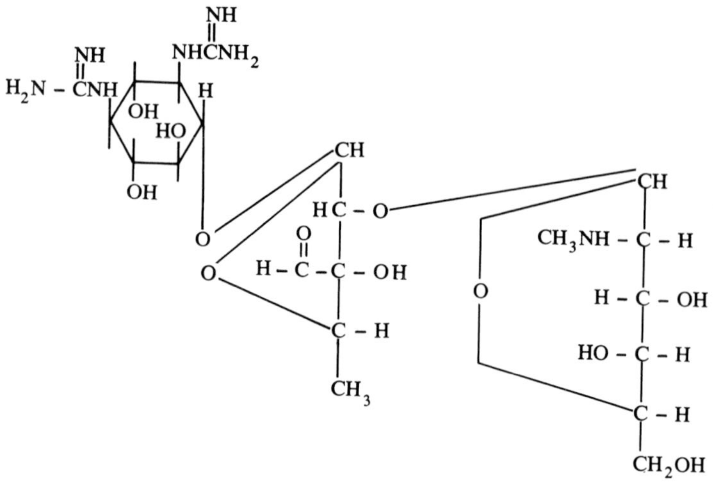

Fig. 5.25. Structure of streptomycin.

Although it was initially felt that streptomycin bound to the ribosome and thus inhibited the formation of the polyribosome, recently the attachment of messenger RNA to streptomycin-bound ribosomes has been demonstrated. The binding of the streptomycin to the 30 S subunit of the ribosome is effected by a chelation of the three cationic groupings of the antibiotic structure, i.e., two guanidine and one methylamine moiety, via magnesium, to some portion of the messenger RNA structure.

Streptomycin has recently been demonstrated to cause the binding of incorrect aminoacyl-transfer RNA to the polyribosome, thus increasing the mistake level in translation, i.e., streptomycin produces "miscoding." The union of the antibiotic to the 30 S ribosomal subunit presumably effects an alteration in the conformation of the ribosomal structure such that the message is incorrectly read. These results have important implications. They indicate that the ribosome, *itself*, is an important factor in regulating protein synthesis, not just a passive bystander.

b. PUROMYCIN. Puromycin was isolated from cultures of *Streptomyces alboniger* in 1952 and has proved a potent inhibitor of protein synthesis in a variety of species. The mechanism of action is dependent upon its structural resemblance to phenylalanyl-tRNA (Fig. 5.26).

Puromycin blocks the incorporation of amino acids from amino acyl-tRNA into microsomal protein and also causes the premature release of incomplete polypeptides into the incubation medium. The antibiotic produces these effects by first, substituting for phenylalanyl-tRNA at the codon present in the ready position on the polyribosome and second, attacking the peptidyl-tRNA at the adjacent codon and cleaving the tRNA

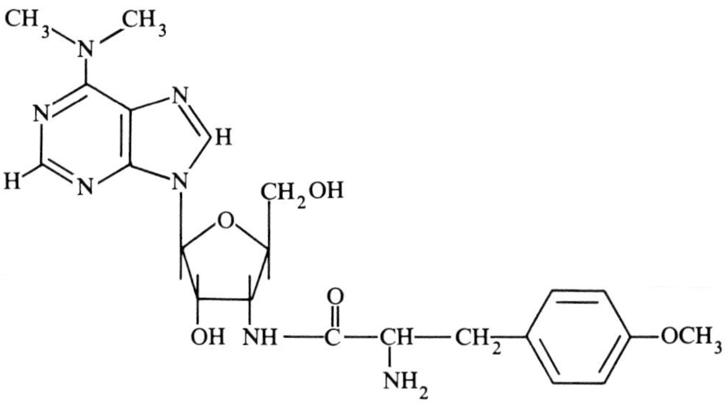

Puromycin

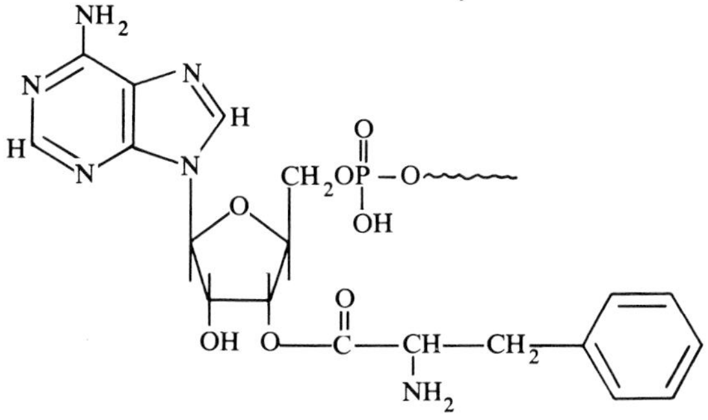

Phenylalanyl-t RNA

Fig. 5.26. Puromycin and phenylalanyl-tRNA.

from the nascent peptide. The resultant puromycyl peptide is released into the medium. Puromycin peptides have been isolated from cell-free systems and are sensitive to peptidases, confirming the existence of a peptide linkage between the antibiotic and the peptide.

c. CYCLOHEXIMIDE (ACTIDIONE). Cycloheximide has recently been introduced into the laboratory as an inhibitor of protein synthesis (Fig. 5.27). The antibiotic was isolated from filtrates of *Streptomyces griseus* and is toxic for fungi, protozoa, and mammalian cells but has no effect upon bacteria. The available evidence suggests that cycloheximide inhibits the readout process, i.e., the movement of the ribosomes along the messenger tape.

d. CHLORAMPHENICOL. Chloramphenicol, isolated from *Streptomyces venezuela* in 1947, is a most interesting agent (Fig. 5.28). The elucidation of its mechanism of action has occupied biochemists for a number of years. The early observations established that protein synthesis in bac-

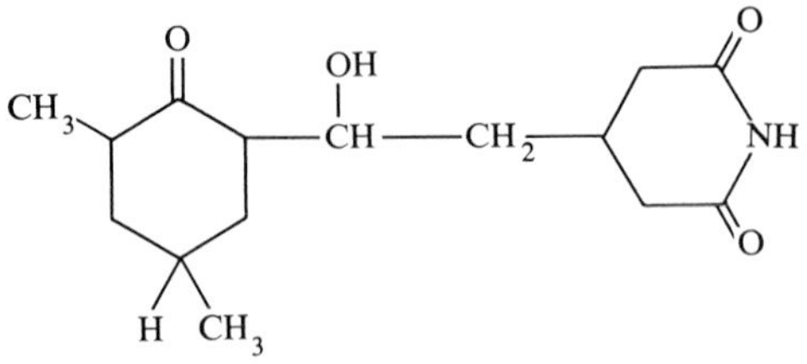

Fig. 5.27. Cycloheximide.

terial systems was particularly sensitive while most mammalian species were resistant to chloramphenicol.

After the development of the polyribosome concept, chloramphenicol was shown to inhibit the formation of this structure but to have no effect upon the subsequent translation process.

In fact, chloramphenicol does have some effect upon certain mammalian tissues, e.g., the hematopoietic organs. The antibiotic is more toxic to bacterial systems because of the ephemeral existence of messenger RNA and consequently of the polyribosome. The formation of bacterial polyribosomes is in a constant state of flux, therefore chloramphenicol can exert its action frequently. In mammalian systems, polyribosomes are longer lived and chloramphenicol is generally without effect. The polyribosomes of the hematopoietic organs, e.g., bone marrow, do not possess as long a life as other mammalian systems and consequently, they are more susceptible to the action of the antibiotic.

Chloramphenicol binds to the 50 S subunit of the ribosome and nuclear magnetic resonance studies have shown that that steric configuration of chloramphenicol bears some resemblance to UMP. Perhaps this resemblance is responsible for the inhibition of polyribosomal formation.

e. TETRACYCLINES AND ERYTHROMYCIN. The tetracyclines and

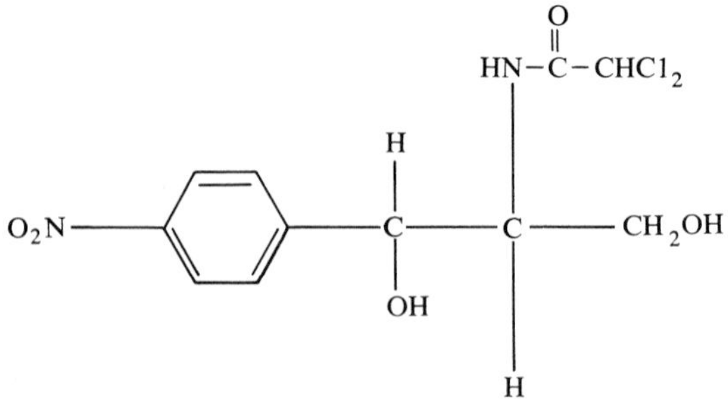

Fig. 5.28. Chloramphenicol.

erythromycin interfere significantly with the binding of aminoacyl-transfer RNA to the polyribosome. These antibiotics are believed to mediate their effect by reaction at the 50 S ribosomal subunit.

I. Control of Protein Synthesis

The concentration of many enzymes within a bacterial or mammalian cell is generally not constant but varies with the composition of the environment. This phenomenon is particularly apparent in microbial physiology where enzymes are classified as either *constitutive* or *induced*. The former is an enzyme which does not respond dramatically to changes in media; the latter represents a category of enzymes whose concentration, and actual synthesis, depend markedly upon the external milieu. In the presence of a suitable *inducer*, the amount of an enzyme may increase manyfold.

Constitutivity and induction are not limited to bacteria but are found in virtually all species. The administration of a variety of barbiturates to rats effects a marked increase in the enzymes functioning in drug metabolism (see p. 276). The surgical removal of 70% of the liver from rodents institutes a process in which the amount of a number of enzymes, i.e., especially those which are concerned with DNA biosynthesis, is increased. Parenteral administration of many hormones or the substitution of a high protein diet for the normal diet may also produce similar changes in the patterns of certain enzymes in a cell.

These mammalian homeostatic mechanisms can also be altered in the other direction. Starvation, irradiation, dietary change, and various agents may decrease the amount of certain enzymes in the tissues. It is clear, then, that the enzymic patterns within a cell must be under some regulation and must be capable of responding within short periods of time to external stimuli. What are the mechanics of these regulatory mechanisms?

The outstanding studies of Jacob and Monod (1961) in bacterial genetics have placed us closer to an answer to this intriguing problem. Most bacterial geneticists have used β-galactosidase in *E. coli* as their model system. In certain strains of *E. coli*, lactose may be utilized only after an initial lag period, i.e., the growth of these bacteria in the presence of lactose is limited for a finite time after which a rapid multiplication ensues. The bacteria are unable to utilize lactose because of the "absence" of the enzyme, β-galactosidase, an enzyme whose function is the catalysis of the following:

$$\text{lactose} \longrightarrow \text{galactose} + \text{glucose}$$

In the presence of the inducer, lactose, however, the wheels of protein machinery begin to turn, and soon a rapid synthesis of β-galactosidase occurs that is followed by logarithmic growth. It is this process, in which an increase in actual enzyme may be observed as a result of the administration of an exogenous agent, that is termed *induction* or *derepression*. During the brief lag period following the administration of the inducer lactose, preparations are being made for the burst in enzyme synthesis to follow.

Similarly, a number of investigators have observed a marked drop in the synthesis of an enzyme in the presence of an end-product metabolite. This process has been called *repression* (Fig. 5.29).

Jacob and Monod (1961) have advanced the following schema to explain the mechanisms of induction and repression. The underlying current of their hypothesis is that induction and repression represent two sides of the same coin (Fig. 5.30).

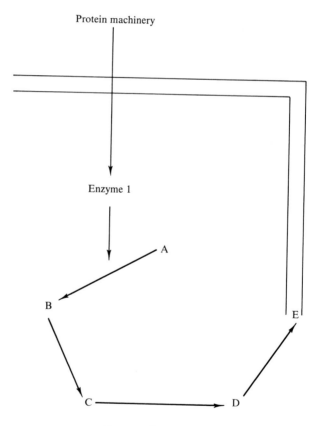

Fig. 5.29. Repression.

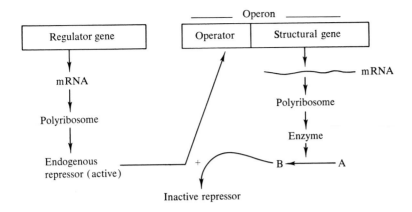

Fig. 5.30A. Induction.

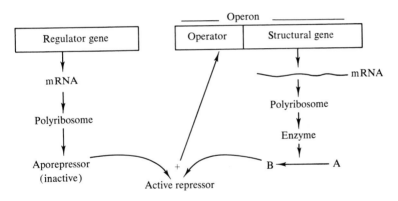

Fig. 5.30B. Repression.

A number of genes are responsible for the elaboration of a single protein molecule; a *regulator gene, operator,* and *structural gene.* The latter two comprise the *cistron.* The operator functions by signaling the structural gene to synthesize an mRNA containing the appropriate codons. If a number of structural genes are under the control of a single operator, the system is referred to as an *operon.* An operon consists of a set of cistrons that are linked sequentially along the gene and are controlled by a common gene, the operator.

After the "go" signal has been relayed to the structural gene, an mRNA is elaborated via RNA polymerase. The mRNA, containing the appropriate codons for a particular protein, is suitably linked into a polyribosomal structure, with the subsequent production of the protein.

The function of the regulator gene in the induction mechanism is the promulgation of an active repressor through the intervention of the polyribosome. The active repressor is inactivated by reaction with the end product, whose synthesis is under the control of the enzyme elaborated by the structural gene. The operator remains in the "go" position and continuously produces enzyme.

On the other hand, if an inactive repressor is elaborated by the regulator (Fig. 5.30B), the reaction of the former with the end product produces an active repressor. The newly formed active repressor may then regulate the activity of the operator by a mechanism that is not understood at present.

In what manner does the active repressor influence the action of the operator? The principal action of the former is upon the synthesis and not upon the function of the mRNA, elaborated by the structural gene. Does the repressor inhibit each small cistronic mRNA of the operon? Martin (see Ames and Martin, 1964) has shown that the operon which is concerned with the synthesis of the enzymes of the histidine biosynthetic pathway consists of ten cistrons and further, these cistrons give rise not to ten different mRNA molecules, but to a polycistronic mRNA containing approximately 10,000 nucleotides (molecular weight $= 4 \times 10^6$). Thus, contained within the polycistronic messenger is the information for the synthesis of *all* the enzymes of the histidine pathway.

The polycistronic message is translated beginning at its operator end. Transcription of DNA also starts at the operator end of the operon. Since protein synthesis is initiated by the assembly of amino acids from the N-terminal to the C-terminal end (see page 259), it can be concluded that the genes are arranged so that the cistron nearest the operator codes for the N-terminal end of the polypeptide.

A control mechanism also occurs at the level of the substrate rather than at the level of the enzyme. This mechanism, end-product inhibition, is demonstrated in Fig. 5.31.

In this regulatory mechanism, enzymic activity is automatically inhibited when the concentration of the end product approaches a certain value. Enzymic activity may be inhibited by the end product by either of two mechanisms: If the inhibitor structurally resembles one of the substrates, the former may bind to the same site on the enzyme surface as the latter and in this manner effect an inhibition. Since the inhibitor in this case bears a structural resemblance to the substrate, the inhibition is of the *isosteric type*. Generally, however, the end product metabolite bears little resemblance to the substrate of an earlier reaction. In this case, the binding of the inhibitor occurs at a site other than the substrate site. This mechanism has been termed an *allosteric inhibition*. The

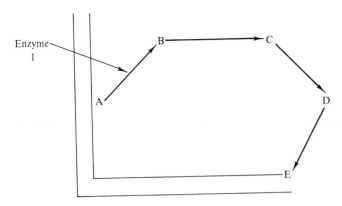

Fig. 5.31. End-product inhibition.

inhibitor site may be adjacent to or indeed quite remote from the substrate site. The formal union of the inhibitor to the enzyme then leads to a distortion of the conformation at the substrate site, inhibiting the formation of the enzyme-substrate complex.

One can imagine an example of positive feedback inhibition in which the binding of an effector to the allosteric enzyme will lead to a conformational change more favorable for catalytic activity. The enzymic activity will be enhanced. It is possible that certain hormones may act in this fashion.

Translational Control. In higher organisms there is evidence of the occurrence of another type of regulatory mechanism, operative at the translational level. It is known from studies on amphibian embryogenesis that the messenger RNA may be long-lived in the unfertilized egg and occur in a "masked" form, i.e., a form protected from catabolism but incapable of transcription until activated. The "unmasking" is believed to take place upon fertilization and is thought to involve a proteolytic digestion of a proteinaceous protector substance on the messenger RNA. The processes of "masking" and "unmasking" fall into the category of translational control. The presence or absence of the protector substance could be a function of the stage and development and differentiation of the organism.

Regulation of Catabolism. In Mammalian cells, the effective concentration of an enzyme is the net result of its rate of synthesis and degradation. Thus, it is not necessary to *increase* enzyme synthesis in order to increase the effective concentration. The rate of degradation may be diminished which will lead to the same result—an increase in the amount of intracellular enzyme. One must therefore consider *both* the rate of synthesis and of degradation of an enzyme.

IV. ADDITIONAL ENZYMIC ACTIVITIES OF THE ENDOPLASMIC RETICULUM

Microsomal preparations contain enzymes which catalyze reactions in all the major areas of metabolism. Many of these reactions take place in the "smooth" or agranular endoplasmic reticulum. In this section, we will consider several of these processes to indicate the breadth of metabolic activity manifest within this organelle. The chapter will conclude with a brief discussion of several drugs that specifically affect the endoplasmic reticulum and its enzymic complement.

1. CARBOHYDRATE METABOLISM

a. GLUCOSE-6-PHOSPHATASE. Glucose-6-phosphatase is located exclusively within the microsomes of specific tissues, i.e., liver, brain, kidney, heart; appears to be tightly bound to the membranous structures; and is absent from the ribosomes. The enzyme catalyzes the terminal reaction in glycogenolysis, the dephosphorylation of glucose 6-phosphate,

$$\text{glucose 6-phosphate} \longrightarrow \text{glucose + phosphate.}$$

Its importance lies in the release of a potential energy source from the liver to the peripheral tissues.

Glucose-6-phosphatase activity is markedly affected during the ontogenesis from fetal to adult liver, i.e., developmental transitions. In fetal liver, glucose-6-phosphatase activity is undetectable, hence, glycogen accumulates. At term, and shortly thereafter, enzymic activity increases rapidly, facilitating the rapid degradation of the stored glycogen to glucose which can be utilized for the production of energy at a time when the intake of nutrient material is limited. Glucose-6-phosphatase has additional significance for the biochemist; the enzyme serves as a "marker" for membranous structures.

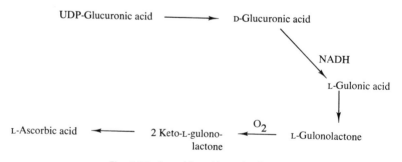

Fig. 5.32. Ascorbic acid synthesis.

b. TRANSFERASES AND ASCORBIC ACID BIOSYNTHESIS. Localized specifically within the liver microsomes are the glucuronyl transferases that function in the disposition of many substances via conjugation and the subsequent elimination by the liver. The initial enzymes that catalyze the reactions leading to the synthesis of the biological conjugating agents, uridine diphosphate glucose (UDP-glucose) and UDP-glucuronic acid (see reactions 1 and 2), are probably present in the soluble portion of the cytoplasm.

(1) UTP + glucose 1-phosphate $\xrightarrow{\text{UDP-glucose} \atop \text{pyrophosphorylase}}$ UDP-glucose + pyrophosphate.

(2) 2 UDP-glucose + NAD $\xrightarrow{\text{UDP-glucose} \atop \text{dehydrogenase}}$ 2 UDP-glucuronic acid + 4 NADH.

(3) UDP-glucuronic acid + acceptor $\xrightarrow{\text{glucuronyl} \atop \text{transferase}}$ UDP + acceptor-glucuronide.

The glucuronide-forming enzyme system is of vital importance in the transformation of bilirubin to an excretable form, bilirubin diglucuronide. The glucuronide-conjugating activity also undergoes marked changes during the development of mammalian liver. At midgestation in the rat, glucuronyl transferase is virtually absent in the liver. The newborn liver possesses about one-fifth the adult activity while at 15 − 20 days of life, enzymic activity equals the level found in the adult liver.

Glucuronic acid, produced from UDP-glucuronic acid by a specific phosphatase, may serve as a precursor in the formation, within the endoplasmic reticulum, of the vitamin, L-ascorbic acid (see Fig. 5.32).

2. ELECTRON TRANSPORT

The microsomal fraction of rat liver also contains a specific cytochrome pigment not found in mitochondria. Subsequent purification and examination of the properties of the microsomal cytochrome system, cytochrome b_5, have established the following properties:

1. In the oxidized state, the cytochrome shows a spectrum with maxima at 530 and 565 mμ and an intense Soret band at 413 mμ. When reduced by dithionite, the Soret band shifts to 423 mμ.

2. The molecular weight is about 17,000 with one iron protoporphyrin prosthetic unit per molecule of apoprotein. No flavin or nonheme iron is present.

3. The purified cytochrome b_5 functions as a single univalent electron acceptor, being rapidly reduced by NADH in the presence of a specific microsomal reductase.

The enzyme which brings about the reduction in the presence of

NADH, cytochrome b_5 reductase has been isolated and possesses a flavin adenine dinucleotide (FAD) prosthetic group tightly bound to the aporeductase. Also associated with the apoenzyme in a loose bond is a reduced pyridine nucleotide, NADH (NADH reacts very slowly with this enzyme). The flavin-dependent, NADH-cytochrome b_5 reductase is completely insensitive to antimycin A, one of several distinguishing features between the microsomal and mitochondrial cytochrome enzyme systems (see Chapter III).

In isolated systems, reduced cytochrome b_5 can reduce cytochrome c or c_1 in the presence of the appropriate NADPH-specific, microsomal cytochrome c or c_1 reductase. The significance of the latter reaction, however, is still obscure since neither cytochrome c or c_1 has been identified in microsomes or in the cytoplasm of the cell.

Since both cytochrome b_5 and its reductase are quite firmly bound to the isolated microsomal fraction presumably in the "smooth" endoplasmic reticulum, and are distributed in a variety of mammalian tissues, it would appear that the highly active endoplasmic reticulum of the intact cell possesses at least part of a system that can actively transport electrons. What possible physiological significance might this suggest? A number of ideas have been put forth but the absence of any convincing evidence renders most of them of dubious importance. These include, e.g., involvement in terminal oxidative processes of protein synthetic reactions; the storage of hemin or cytochrome or precursor components.

Recently, more plausible explanations for the functional significance of the microsomal cytochrome system have gained prominence. The system may participate in a number of "reductive" synthetic reactions, shunting electrons from NADH to various acceptors. The synthesis of steroids, for example, requires a hydroxylation step mediated by NADPH and molecular oxygen. A similar microsomal pyridine nucleotide-oxygen enzyme system is also required for a wide variety of metabolizing systems involved in the biotransformation of many pharmacological and physiological compounds, e.g., barbiturates, norepinephrine. The location of the microsomal cytochrome system together with the low redox potential of the cytochrome b_5 make the system a likely candidate for the transfer of electrons in such reactions. Much more evidence is necessary, however, before such a functional role may finally be accepted.

Another interesting possibility regarding the significance of the microsomal cytochrome content revolves around active transport processes. Among the many theories regarding the mechanism of carrier-linked active transport of substances against concentration gradients, one which was proposed many years ago requires the transfer of electrons as an essential step. Recently the discovery of an enzyme system which hydrolyzes ATP maximally in the presence of $Na^+ + K^+ + Mg^{2+}$ in the microsomal fraction of most mammalian cells recalls to mind the electron-requiring transport theory. This "membrane" ATPase system fulfills many of the requirements for an enzymic active transport system (see Chapter VII), and since the ATPase appears to be localized in the same region of the cell as is the microsomal cytochrome system, an association between electron movements and active transport becomes a possibility. However, in this area as well, definitive proof is lacking.

3. LIPID METABOLISM

It has long been known that the microsomes contain large amounts of lipid material, i.e., 30–50% of its total weight, over 50% of which is phospholipid. In view of the membranous constitution of the endoplasmic reticulum, it is not surprising to find this relatively large amount of phospholipid. The amounts of phospholipid in smooth or rough endoplasmic reticulum of at least liver tissue do not significantly differ (Table 5.10).

Lecithin and cephalin comprise from 50 to 90% of the total phospholipids of liver, heart, or brain microsomes; phosphatidylserine is in low amount. The lecithin/cephalin ratio is much higher in microsomes than in other organelles, suggesting a function of the endoplasmic reticulum in either the synthesis or the concentration of this phospholipid.

a. FATTY ACID METABOLISM. Although the mitochondria appear to contain all the enzymes required for fatty acid oxidation and nearly all that are necessary for the incorporation of acetate into the fatty acid structure, the microsomal fraction does facilitate the reductive synthesis

TABLE 5.10

Composition of Smooth and Rough Endoplasmic Reticulum[a]

	Amount/gm liver	
	Rough	Smooth
Phosphorus (μg)	202	275
Nitrogen (mg)	1.2	1.6
Protein (mg)	7.0	8.9
Lipid-phosphorus (μg)	90	106
RNA (mg)	1.3	0.4
Cytochrome b_5 (mμmoles)	4.3	8.2

[a]The data presented in the table have been abstracted from H. Remmer and J. H. Merker, *Ann. N. Y. Acad. Sci.* **123**, 79 (1965).

of short-chain fatty acids (see below), the formation of long chain fatty acids, and of sphingosine (by brain microsomes).

$$\text{crotonyl - CoA} + \text{NADPH} \longrightarrow \text{butyryl-CoA} + \text{NADP}$$
$$\text{ATP} + \text{CoA} + \text{fatty acid} \longrightarrow \text{fatty acyl-CoA}$$
$$\text{palmityl CoA} + \text{serine} + \text{NADPH} \longrightarrow \text{sphingosine} + \text{CoA} + \text{NADP}$$

b. PHOSPHOLIPID METABOLISM. The involvement of the liver in the synthesis of phospholipids has long been known. The functions of the phospholipids fall into two categories, structural and dynamic. Phospholipids as constituents of the lipoprotein foundation of various organelles, e.g., membranes, fulfill a vital role in the development of the cell. This category of phospholipid is characterized by a long, and indeed for the particular organelle, an eternal life. The second class of phospholipids, on the other hand, is formed more rapidly, and is active in the transport of fatty acids and in the formation of triglycerides.

Synthesis (see Fig. 5.33). A great number of substances form part of the phospholipids: glycerol in the *glycerophosphatides*; the *sphingomyelins*, composed of the alcohol, sphingosine, and phosphoric acid, fatty acid and chlorine; the *phosphoinositides* in which inositol is an important constituent.

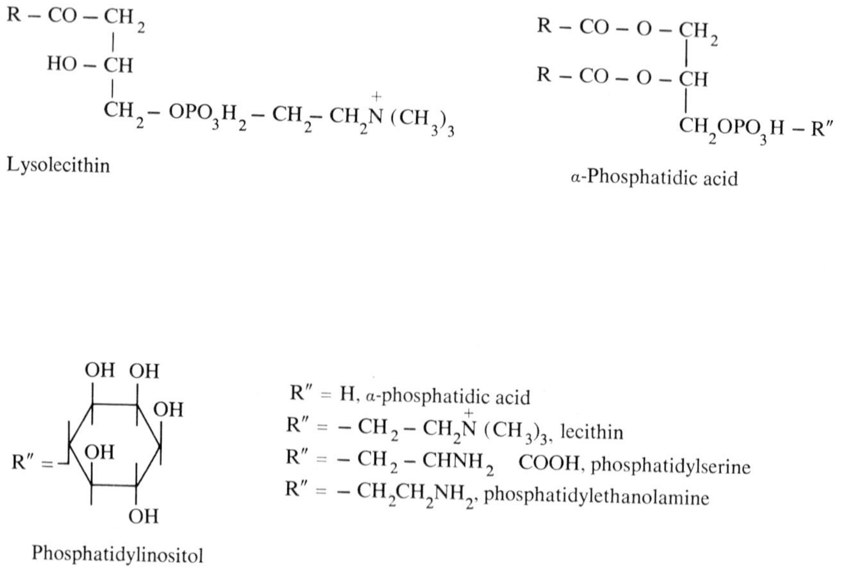

R – CO – CH$_2$
|
HO – CH
|
CH$_2$ – OPO$_3$H$_2$ – CH$_2$– CH$_2$N$^+$(CH$_3$)$_3$

Lysolecithin

R – CO – O – CH$_2$
|
R – CO – O – CH
|
CH$_2$OPO$_3$H – R″

α-Phosphatidic acid

R″ =

OH OH

OH

OH

OH

Phosphatidylinositol

R″ = H, α-phosphatidic acid

R″ = – CH$_2$ – CH$_2$N$^+$(CH$_3$)$_3$, lecithin

R″ = – CH$_2$ – CHNH$_2$ COOH, phosphatidylserine

R″ = – CH$_2$CH$_2$NH$_2$, phosphatidylethanolamine

Fig. 5.33. Structural formulas of phospholipid derivatives.

Although phospholipid synthesis can take place in all tissues, only the liver appears to be the source of the plasma phospholipids and further, this organ is the major site of degradation or metabolism of the phospholipids.

One of the most important contributions in this area was the elucidation of the mechanisms of phospholipid synthesis by E. P. Kennedy and co-workers (1955, 1956). Kennedy observed that the nucleotide, cytidine triphosphate (CTP) was an important cofactor in the biosynthetic pathway and the basis of the mechanism was a phosphoryl exchange. The over-all schema has been summarized in Fig. 5.34. The formations of CDP-ethanolamine, CDP-choline, CDP-diglyceride occur by a common mechanism (Fig. 5.35).

An additional route of synthesis of phospholipid occurs via phosphatidylserine and was initially proposed by Bremer et al. in 1960 (Fig. 5.36).

The conversion of phosphatidylethanolamine to lecithin is accomplished by means of a series of methylation reactions, in which S-adenosylmethionine contributes the methyl groups. Indeed, the major route for lecithin synthesis is believed to be the latter pathway rather than the formation from CDP-choline.

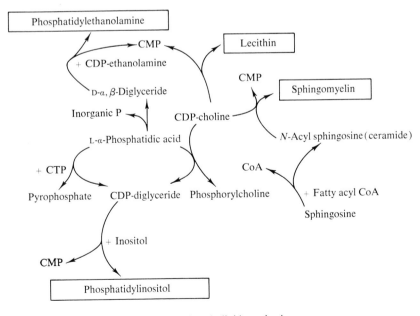

Fig. 5.34. Phospholipid synthesis.

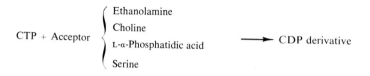

$$\text{CTP + Acceptor} \begin{cases} \text{Ethanolamine} \\ \text{Choline} \\ \text{L-}\alpha\text{-Phosphatidic acid} \\ \text{Serine} \end{cases} \longrightarrow \text{CDP derivative}$$

Fig. 5.35. Cytidine diphosphate derivatives.

The endoplasmic reticulum plays an important role in the synthesis of phospholipids since the location of the enzymes responsible for the formation of lecithin, the methylation of phosphatidylethanolamine, the acylation of glycerophosphate to form phosphatidic acid, the formation of phosphatidic acid from ATP and monoglycerides (in brain), as well as the decarboxylation of phosphatidylserine to form phosphatidylethanolamine all occur in this organelle. The metabolic routes from CDP-ethanolamine and CDP-diglyceride are found both in the microsomes and in the mitochondria.

The importance of the microsomes lies not only in the synthesis of the phospholipids but in the apportionment of the proper distribution among the phosphatidylcholine, phosphatidylserine, and phosphatidylethanolamine. Siekevitz (1963) in his review of the endoplasmic reticulum and microsomes has suggested ". . . perhaps the properties of the membranes of the endoplasmic reticulum depend on the relative amounts of these three different phosphatides, and the microsomal enzymes have as one of their purposes a redistribution, under certain conditions, of the relative amounts of the phosphatides in the membranes, thus conferring different properties on the membrane."

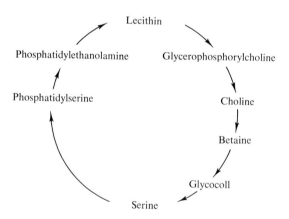

Fig. 5.36. Phospholipid cycle.

c. STEROID METABOLISM. Microsomes perform a number of important functions in steroid biochemistry, among which may be included the synthesis of cholesterol, i.e., the conversion of mevalonic acid to squalene or sterol, and various oxidations and reductions, e.g., C-20 keto reduction, C-11 oxidation of cortisol, reduction of 4,5-double bond of ring B. Steroid metabolism consists of a number of hydroxylations, the bulk of which occur in the microsomes of the liver. The mechanism underlying these hydroxylations will be discussed in the following section.

Virtually all the bile acids present in the bile are conjugated with either taurine or glycine. The conjugation takes place in two steps, (1) activation of the bile acid, cholic acid,

$$\text{cholic acid} + \text{CoA} + \text{ATP} \xrightarrow{\text{Mg or Mn}} \text{cholyl-CoA} + \text{AMP} + \text{pyrophosphate}$$

(2) conjugation with amino acid or derivative,

$$\text{cholyl-CoA} + \begin{matrix} \text{glycine} \\ \text{or} \\ \text{taurine} \end{matrix} \longrightarrow \text{CoA} + \begin{matrix} \text{glycocholic acid} \\ \text{or} \\ \text{taurocholic acid} \end{matrix}$$

The first reaction is catalyzed by a microsomal enzyme while the intracellular site of the conjugating enzyme has not definitely been established although evidence has been gathered favoring the microsome.

4. DRUG METABOLISM

Liver microsomes as well as microsomes from other tissues, e.g., kidney, lung, contain a number of enzymes that function in the modification of pharmacological as well as physiological agents, a process referred to as *biotransformation*. The intracellular site of these enzymes has recently been defined as the "smooth" endoplasmic reticulum. The structural transformations are accomplished by the microsomes with the purpose of (1) catabolism of drugs, i.e., detoxification, (2) solubilization of drugs prior to excretion from the host, (3) solubilization of drugs to aid in transport from liver to the peripheral tissues, and (4) anabolism to substances to be used by the peripheral tissues.

The anatomical cartography of the intestine and kidney tubule favors the absorption and hence utilization of lipid-soluble substances. The physical properties of the drugs strongly influence the degree of effectiveness in accomplishing their prescribed functions. Although lipid solubility is an important property in the utilization of drugs and indeed of hormones, this property proves an obstacle in the elimination of the substance. The latter may be best exemplified by a statement made by Brodie (1964):

For simplicity, let us assume that a drug is evenly distributed throughout the body water. If the compound has a low lipid solubility, about five hours would elapse before half the substance is lost from the body; if the drug is also excreted by the tubules (kidney), this time will be shortened to as little as one hour. However, if the drug is lipid-soluble, the excretion rate will be drastically reduced by its back diffusion in the plasma from the tubular segment where the urine is concentrated. About 30 days would elapse before half the drug leaves the body—a possible therapeutic advantage with an antibacterial agent but of doubtful value with an anesthetic agent. If, in addition, the drug—atabrine or thiopental for example—is highly localized in tissues, the half life in the body would be about 100 years—considerably longer than that of the physician and the patient combined!

The types of enzymic reactions occurring in the smooth endoplasmic reticulum include the oxidative reactions, glucuronide-forming systems, and the reduction of nitro and azo bonds. The oxidative reactions, summarized in Tables 5.11 and 5.12, possess two common features, the requirements for NADPH and oxygen.

Recently, the presence of a new cytochrome has been reported in microsomal fractions. The cytochrome is unusual in that the absorbancy spectra of the oxidized and reduced materials are identical. It also differs from the mitochondrial cytochromes in its relative inertness to cyanide. However, the reduced form rapidly reacts with carbon monoxide to yield a complex with a new absorbancy maximum at 450 mμ. This cytochrome was initially called the "carbon-monoxide-binding" pigment but now has been dubbed cytochrome P_{450}. Considerable evidence has accumulated implicating cytochrome P_{450} in the processes of biotransformation. The reactions involved may proceed as follows:

$$NADPH + \text{flavoprotein} \longrightarrow \text{reduced flavoprotein} + NADP$$
$$\text{reduced flavoprotein} + \text{cyt. } b_5 \longrightarrow \text{reduced cyt. } b_5 + \text{flavoprotein}$$
$$\text{reduced cyt. } b_5 + \text{cyt. } P_{450} \longrightarrow \text{reduced cyt. } P_{450} + \text{cyt. } b_5$$
$$\text{reduced cyt. } P_{450} + O_2 \longrightarrow \text{cyt. } P_{450} \cdot O_2 + H_2O$$
$$\text{or}$$
$$(\text{"active" oxygen})$$
$$\text{"active" oxygen} + \text{drug} \longrightarrow \text{oxidized drug} + \text{cyt. } P_{450}$$

Factors Affecting Drug Metabolism. Recently, it has been shown that the metabolism of various lipid-soluble drugs within the liver microsomes can be enhanced by the pretreatment of the animals with numerous unrelated compounds (see Conney and Burns, 1962). Among the agents eliciting this effect may be included barbiturates such as phenobarbital and barbital; oral antidiabetic agents such as tolbutamide; antihistaminic agents such as Benedryl; anti-inflammatory drugs such as phenylbutazone; and polycyclic hydrocarbons, e.g., 3-methylcholanthrene.

The increase in drug metabolism correlates well with an increase in the synthesis of drug-metabolizing enzymes. It appears probable that the above-mentioned agents stimulate the production of new enzyme

TABLE 5.11

Oxidative Reactions Catalyzed by the Drug-Metabolizing Enzymes

Type	Example
Hydroxylation of aromatic rings	Zoxazolamine $\longrightarrow$ hydroxyzoxazolamine
	3,4-Benzpyrene $\longrightarrow$ 8- or 10-hydroxy-3,4-benzpyrene
N-Hydroxylation	2-Acetylaminofluorene $\longrightarrow$ *N*-hydroxy-2-acetylaminofluorene
Side-chain oxidation	Phenylbutazone $\longrightarrow$ "hydroxy"-phenylbutazone
N-Dealkylation	Imipramine $\longrightarrow$ demethylimipramine
O-Dealkylation	
S-Dealkylation	6-Methylthiopurine $\longrightarrow$ 6-thiopurine
Sulfoxidation	Chlorpromazine $\longrightarrow$ chlorpromazine sulfoxide
Deamination	Zoxazolamine $\longrightarrow$ chlorzoxazone

TABLE 5.12

Other Microsomal Enzyme Reactions

Type	Example
Hydrolysis of esters	Procaine $\longrightarrow$ *p*-aminobenzoic acid
Reduction of nitro groups	Chloramphenicol $\longrightarrow$ amine
Reduction of azo groups	Prontosil $\longrightarrow$ sulfanilamide
Glucuronide formation	Phenol $\longrightarrow$ phenol glucuronide

protein by a mechanism involving enhanced synthesis of messenger RNA.

The livers of rats pretreated with phenobarbital have been studied qualitatively with the electron microscope and with more quantitative techniques by Remmer and Merker (1965). A striking proliferation of the smooth endoplasmic reticulum of the liver cell was observed in the treated animals with no apparent change in the rough type. These cytological changes corresponded with increases in the total nitrogen, phosphorus, RNA, and protein in the isolated smooth endoplasmic reticulum of the phenobarbital-treated rats.

These combined morphological and biochemical observations clearly indicate the necessity for an accurate clinical history of patients and conservatism in the prescription of medication. Today, in the age of habitual usage of sleeping potions, of insecticides, of food coloring agents, and, indeed, of aspirin, the question of drug metabolism is a critical one.

Inhibition of the metabolism of drugs has also been observed with certain agents. Carbon tetrachloride, which produces a very profound effect upon the morphology of the smooth endoplasmic reticulum, markedly diminishes the activity of the drug-metabolizing enzymes. The drug-metabolizing enzymes undergo marked developmental changes as well. The level of enzymic activity in rat fetal liver is barely detectable, at *partum*, the activity begins to rise and by a week or so after birth, adult values are reached. A corollary of these changes is the marked sensitivity of the fetus and indeed of the newborn to the pharmacological effects of drugs, i.e., these agents cannot be metabolized. The ontogenetic biochemical changes just described correspond well with the observed morphological transitions. The smooth endoplasmic reticulum is not very pronounced in fetal liver but proliferates after birth.

V. CONCLUSIONS

The endoplasmic reticulum is intimately involved in protein synthesis, electron and ion transport, steroid and fat metabolism, various oxidative reactions and in drug metabolism. The canalicular appearance of this organelle certainly suggests a feeding of the products of these reactions to the other parts of the cell and indeed the elimination from the cell.

GENERAL REFERENCES

Bourne, G. H. (1962). "Division of Labor in Cells." Academic Press, New York.
Cameron, G. B. (1964). *In* "The Liver" (C. H. Rouiller, ed.), Vol. 2, pp. 92-134. Academic Press, New York.
Combs, B. (1964). *In* "The Liver" (C. H. Rouiller, ed.), Vol. 2, pp. 1-36. Academic Press, New York.
Dallner, G. (1963). *Acta Pathol. Microbiol. Scand.* Suppl. 166, 1-94.
DeRobertis, E. D. P., Nowinski, W. W., and Saez, F. A. (1965). "Cell Biology." Saunders, Philadelphia, Pennsylvania.
Ernster, L., Siekevitz, P., and Palade, G. E. (1962). *J. Cell Biol.* 15, 541-562.
Haguenau, F. (1958). *Intern. Rev. Cytol.* 7, 425-483.
Porter, K. R. (1961). *In* "The Cell" (J. Brachet and A. E. Mirsky, eds.), Vol. 2, pp. 621-675. Academic Press, New York.
Siekevitz, P. (1963). *Ann Rev. Physiol.* 25, 15-40.
Sjöstrand, F. (1964). *In* "Cytology and Cell Physiology" (G. H. Bourne, ed.), pp. 311-376. Academic Press, New York.
Taylor, J. H., ed. (1964). "Molecular Genetics," Part 1. Academic Press, New York.
Watson, J. D. (1965). "Molecular Biology of the Gene." Benjamin, New York.
Zamecnik, P. C. (1960). *Harvey Lectures* 54, 256-281.

SPECIFIC REFERENCES

PROTEIN SYNTHESIS

Ames, B. N., and Martin, R. G. (1964). *Ann. Rev. Biochem.* **33**, 235-258.

Byrne, R., Levin, J. G., Bladen, M. W., and Nirenberg, M. W. (1964). *Proc. Natl. Acad. Sci. U.S.* **52**, 140-148.

Chapeville, F., Lipmann, F., Ehrenstein, G., Weisblum, B., Ray, W. J., Jr., and Benzer, S. (1962). *Proc. Natl. Acad. Sci. U.S.* **48**, 1086-1095.

Cohen, S. S. (1948). *J. Biol. Chem.* **174**, 281-293.

Crick, F. (1963). *Progr. Nucleic Acid Res.* **1**, 163-217.

Crick, F. H. C. (1966). *J. Mol. Biol.* **19**, 548-555.

Dunn, D. B. (1959). *Biochim. Biophys. Acta* **34**, 287-288.

Eikenberg, E. F., and Rich, A. (1964). *Proc. Natl. Acad. Sci. U.S.* **51**, 810-818.

Hall, B. D., and Spiegelman, S. (1961). *Proc. Natl. Acad. Sci. U.S.* **47**, 137-163.

Henshaw, E. C., Revel, M., and Hiatt, H. H. (1965). *J. Mol. Biol.* **14**, 241-256.

Hershey, A. D. (1953). *J. Gen. Physiol.* **37**, 1-23.

Jacob, F., and Monod, J. (1961). *J. Mol. Biol.* **3**, 318-356.

Lengyel, P. (1966). *J. Gen. Physiol.* **49**, 305-330.

Lipmann, F. (1963). *Progr. Nucleic Acid Res.* **1**, 135-162.

Littlefield, J. W., Keller, E. B., Gross, J., and Zamecnik, P. C. (1955). *J. Biol. Chem.* **217**, 111-123.

McConkey, E. H., and Dubin, D. T. (1965). *J. Mol. Biol.* **15**, 102-110.

McQuillan, K. (1962). *Progr. Biophys. Biophys. Chem.* **12**, 67-106.

Marcker, K., and Sanger, F. (1964). *J. Mol. Biol.* **8**, 835-840.

Marks, P. A., Rifkind, R. A., and Danon, D. (1963). *Proc. Natl. Acad. Sci. U.S.* **50**, 336-342.

Nirenberg, M. W., Matthaei, J. H., Jones, O. W., Martin, R. G., and Barondes, S. H. (1963). *Federation Proc.* **22**, 55-61.

Noll, H. (1965). *In* "Developmental and Metabolic Control Mechanisms," pp. 67-113. Williams & Wilkins, Baltimore, Maryland.

Noll, H. (1966). *Science* **151**, 1241-1245.

Ochoa, S. (1963). *Federation Proc.* **22**, 62-74.

Perry, R. P. (1963). *Proc. Natl. Acad. Sci. U.S.* **48**, 2179-2186.

Revel, M., and Hiatt, H. H. (1964). *Proc. Natl. Acad. Sci. U.S.* **51**, 810-818.

Revel, M., and Littauer, Z. (1965). *J. Mol. Biol.* **15**, 389-394.

Spencer, M., Fuller, W., Wilkins, M. H. F., and Brown, G. L. (1962). *Nature* **194**, 1014-1020.

Stent, G. S. (1964). *Science* **144**, 816-820.

Volkin, E. (1962). *Federation Proc.* **21**, 112-119.

Warner, J. R., Knopf, P. M., and Rich, A. (1963). *Proc. Natl. Acad. Sci. U.S.* **49**, 122-129.

Wettstein, F. O., Staehelin, T., and Noll, H. (1963). *Nature* **197**, 430-435.

Yankovsky, S. A., and Spiegelman, S. (1963). *Proc. Natl. Acad. Sci. U.S.* **49**, 538-544.

ANTIBIOTICS

Goldberg, I. H. (1965). *Am. J. Med.* **39**, 722-752.

Nathans, D. (1964). *Federation Proc.* **23**, 984-989.

Weisburger, A. S., and Wolfe, S. (1964). *Federation Proc.* **23**, 976-983.

TRANSPORT AND ION MOVEMENTS

Garfinkel, D. (1963). *Comp. Biochem. Physiol.* **8**, 367-379.

Skou, J. (1961). *1st Intern. Congr. Pharmacol., Stockholm,* 1962.
Strittmatter, C. F. (1961). *J. Biol. Chem.* **236**, 2326-2341.

LIPID METABOLISM

Bremer, J., Figard, P. H., and Greenberg, D. M. (1960). *Biochim. Biophys. Acta.* **43**, 477-488.
Kennedy, E. P., and Weiss, S. B. (1955). *J. Am. Chem. Soc.* **77**, 250-251.
Kennedy, E. P., and Weiss, S. B. (1956). *J. Biol. Chem.* **222**, 193-214.

DRUG METABOLISM

Boyland, E., and Booth, J. (1962). *Ann. Rev. Pharmacol.* **2**, 129-142.
Brodie, B. B. (1964). *Pharmacologist* **6**, 12-26.
Conney, A. H., and Burns, J. J. (1962). *Advan. Pharmacol.* **1**, 31-58.
Ernster, L., and Orrenius, S. (1965). *Federation Proc.* **24**, 1190-1199.
Fouts, J. R. (1963). *Ann. N. Y. Acad. Sci.* **104**, 875-880.
Remmer, H., and Merker, J. H. (1965). *Ann. N. Y. Acad. Sci.* **123**, 79-97.

RIBOSOMES

Attardi, G., Huang, P. C., and Kabat, S. (1965). *Proc. Natl. Acad. Sci. U.S.* **54**, 185-192.
Goodman, H. M., and Rich, A. (1962). *Proc. Natl. Acad. Sci. U.S.* **48**, 2101.
Hirsch, C. A., and Hiatt, H. H. (1967). *J. Biol. Chem.* (in press).
Loeb, J. N., Howell, R. R., and Tomkins, G. M. (1965). *Science* **149**, 1093-1094.
Perry, R. P. (1965). *Natl. Cancer Inst. Monograph* **18**, 325-340.
Petermann, M. L., and Pavlovec, A. (1966). *Biochim. Biophys. Acta.* **114**, 264-276.
Ritossa, F. M., and Spiegelman, S. (1965). *Proc. Natl. Acad. Sci. U.S.* **53**, 737-745.
Steele, W. J., and Busch, H. (1967). *In* "Methods in Cancer Research" (H. Busch, ed.), Vol. 3. Academic Press, New York.
Vaughan, M. H., Warner, J. R., and Darnell, J. E. (1967) *J. Mol. Biol.* **25**, 235-251.
Warner, J. R. (1966). *J. Mol. Biol.* **19**, 383-398.

TRANSFER RNA

Carbon, J. A., Hung, L., and Jones, D. (1965). *Proc. Natl. Acad. Sci. U.S.* **53**, 979-986.
Holley, R. W., Apgar, J., Everett, G. A., Madison, J. T., Marquisee, M., Merrill, S. H., Penswick, J. R., and Zamir, A. (1965). *Science* **147**, 1462-1465.
Lipsett, M. N. (1965) *Biochem. Biophys. Res. Commun.* **20**, 224-229.
Ritossa, F. M., Atwood, K. C., and Spiegelman, S. (1966). *Genetics* **54**, 663-676.
Zachau, H. G., Duttling, D., and Feldman, H. (1966). *Angew Chem. Intern. Ed.* **5**, 422.

Chapter VI

OTHER CYTOPLASMIC STRUCTURES: LYSOSOMES AND THE GOLGI SYSTEM

I. LYSOSOME

The *lysosomes,* the latest addition to the list of subcellular organelles, were identified in rat liver cells by de Duve in 1955 and are unique in regard to their pleomorphism, i.e., many shapes and multi-functions. Subsequent studies have established the widespread occurrence of lysosomes in the cells of vertebrates and in unicellular organisms. The lysosomes are present in large numbers in macrophages and leukocytes, in which they appear in electron micrographs as large dark objects.

A chance observation by de Duve in 1949 was ultimately responsible for the recognition of this cellular constituent. De Duve was studying acid phosphatase activity in fractions of rat liver preparations that after "mild" homogenization had been isolated by differential centrifugation techniques. The enzymic activity of these fractions was only 10% of the activity recoverable after more drastic treatment of tissues, e.g., in a Waring blender. Furthermore, after "aging" of the liver preparations at 0°C, the acid phosphatase activity rose markedly. The "latent" enzymic activity and its subsequent augmentation were especially apparent with the crude mitochondrial fraction.

De Duve postulated that acid phosphatase was compartmentalized within a specific cellular entity bounded by a surface membrane; the subcellular component sedimented with the mitochondrial fraction. The function of the membrane was not only to prevent the escape of the enzyme but also to restrict the passage of phosphate esters from the cytoplasm and thus avoid the resultant degradation of the ester. During the

281

preparation of the rat liver homogenates by the newer and milder methods, the integrity of the subcellular particle was maintained, and the subsequent release of contained acid phosphatase was prevented.

A. Pleomorphism

It is now appreciated that the lysosomes do not represent a single entity but a class of cellular organelles whose condition within a tissue may vary under different stimuli. A list of the pleomorphic forms of lysosomes and their close relatives are presented in Table 6.1.

1. LYSOSOMES

Primary lysosomes or "storage-granule" form, generally present in leukocytes and macrophages, may be defined as a cellular structure bounded by a single unit membrane (see Chapter II) and containing certain lytic, i.e., degradative, enzymes (e.g., acid phosphatase). The lysosomal membrane may merge with the membranes of endocytic vacuoles or phagosomes which had been formed as invaginations of the cell membrane in the ingestion of foreign proteins or other macromolecular structures (see p. 291). The resultant structure is the digestive vacuole or phagolysome. Unlike the latter, the endocytic vacuole is not a true lysosomal structure and does not possess the complement of lysosomal enzymes.

2. AUTOPHAGIC VACUOLE

Another structure in this category, the "autophagic vacuole" was first described in kidney after the ligation of the ureter and later, in liver, after perfusion of the organ with glucagon. The structure was originally called a *cytolysome*. The autophagic vacuole appears whenever a cell must utilize a part of its own cytoplasm as would occur in fasting and anoxia or during intense endocytic activity (see p. 292). The autophagic

TABLE 6.1
More Commonly Found Pleomorphic Forms of the Lysosomes and Close Relatives

Autophagic vacuole or cytolysome, or autophagosome,
　cytosegresome, autolytic vacuole
Cytosome
Residual bodies
Multivesicular bodies
Microbodies
Storage granules or digestive vacuole or phagolysome

vacuole represents a vacuole lined with a single membrane and containing morphologically recognizable cytoplasmic constituents.

3. CYTOSOME

The cytosome is an ill-defined structure that may represent any cellular entity possessing a single unit membrane. In most cases, the cytosome proved to be a lysosomal form.

4. MICROBODIES

Microbodies are structures described in the cells of the liver and kidney, bounded by a single unit membrane, containing a finely granulated interior and occasionally, a dense center of regular structure. The *microbodies*, although not a true form, are a close relative of the lysosome.

5. RESIDUAL BODIES

The residual bodies are lysosomes that have conducted various digestive functions and appear stuffed with an undigested lipoid debris. The residual bodies can assemble into membranous arrangements such as the myelin figures observed in ultrastructural studies of nervous tissues. In the latter, the enclosed residue consists of a phospholipid-cholesterol complex. In the liver, residual bodies are often found enclosing ferritinlike particles. The undigested fragments are extruded from the cell via the residual bodies by the process of *exoplasmosis* (see p. 291).

6. MULTIVESICULAR BODIES

These are single-membraned structures that arise from the fusion of endocytic vesicles with some components of the Golgi apparatus (p. 295). The function of these lysosomal forms is not clear.

B. Visual Identification of Lysosomes

Although a number of enzymes are present within the lysosome, acid phosphatase has been the most frequently and successfully employed aid in the visual identification of this cytoplasmic organelle. The cytochemical technique for the demonstration of acid phosphatase activity is the Gomori method (see Chapter I), introduced in 1952. The tissue section is incubated in a medium in which are included the substrate, β-glycerophosphate, acetate buffer, pH 5.0, and a lead salt as a trap for the enzymically released phosphate. The insoluble lead phosphate is converted to the black lead sulfide which precipitates at the site of enzymic activity. Since the lead sulfide has a high electron density, the method may be utilized in both light and electron microscopy. Examples of the utilization of the Gomori method in light and electron microscopy are presented in Figs. 6.1 and 6.2. In the former, acid phosphatase activity is present in a preparation from normal human prostate. In Fig. 6.2 is offered an electron micrograph of a secretory cell from a large axillary sweat gland from a human. Several lysosomes are present in this field and the acid phosphatase activity is principally at the periphery of these structures. Recently, the azo-dye technique described in Chapter I has been introduced as another method for the demonstration of lysosomal activity.

In suitably prepared liver preparations, the lysosome appears as a spherical organelle (see Fig. 6.2) with a mean diameter of 0.4 μ and an average density of 1.20 gm/ml.

C. Concentration of Lysosomes

In the early studies, acid phosphatase activity was localized within the crude mitochondrial or "heavy" microsomal fractions. Investigations by de Duve led to the recognition and acceptance of a distinct organelle

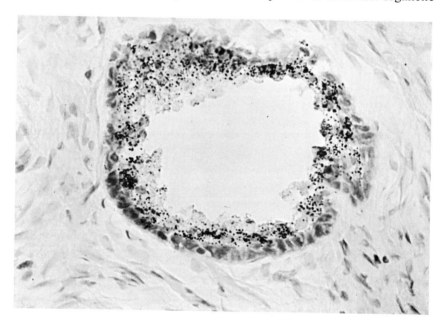

Fig. 6.1. Acid phosphatase by the Gomori method. A section of normal human prostate has been treated by the Gomori method for the localization of acid phosphatase activity. × 1000. Note the intense black stippling. Courtesy of F. Gjörky.

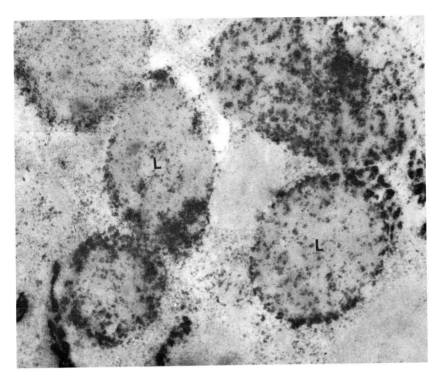

Fig. **6.2.** Electron micrograph of acid phosphatase activity in lysosomes (L). The secretory cells from human axillary sweat glands were fixed in glutaraldehyde, processed by the Gomori method, and stained with OsO_4. Intense enzymic activity was demonstrable mainly at the periphery of the cells. × 51,000. Courtesy of L. Biempica and L. F. Montes.

in which was present all the acid phosphatase activity and which could be separated from the mitochondria by the more elegant density gradient techniques. The mitochondrial fraction obtained by the Schneider procedure is layered in a test tube on a discontinuous gradient of sucrose solutions with densities from 1.17 to 1.24 gm/ml and is then centrifuged at 100,000 g for 3 hours (Fig. 6.3). Two distinct fractions may be recovered, one settling at a density of sucrose ranging from 1.18 to 1.20, and another at 1.21 to 1.23 gm/ml. The former was identified as mitochondria and was devoid of acid phosphatase activity. The latter, the lighter fraction, contained not only acid phosphatase but an entire group of enzymes capable of splitting important biological materials. The enzymes contained within this lighter particle functioned in a digestive or *lytic* capacity, hence, de Duve coined the name *lysosomes* to describe this particle. In the liver cell, one lysosome may be found for every 100 mitochondria, while in the cells of the proximal convolutions of the kidney tubule, the ratio is closer to 1/10.

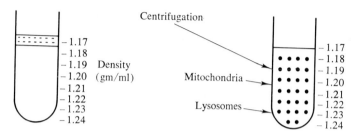

Fig. 6.3. Preparation of lysosomal fraction.

D. Enzyme Activity

The lysosomal membrane may be ruptured in a Waring blendor, with sonic vibration, by repeated freezing and thawing in hypotonic media, by bile salts, digitonin, or other surface-active agents (Triton), by proteolytic or lipolytic enzymes, or by autolysis at 37°C at pH 5.0. Disruption of the lysosome is accompanied by the emergence of a wide range of enzymic activity into the soluble fraction (see Fig. 6.4).

The common properties of the released enzymes included the following: (1) all enzymes were hydrolases, i.e., the lytic property required water, and (2) all enzymes exhibited an acid pH optimum, pH 5.0. As long as the integrity of the lysosome was maintained, the enzymic digestion of the substrates was restricted within the lysosome. As soon as the lysosome was ruptured, the enzymes spilled into the cytoplasm and digestion proceeded. These properties were responsible for de Duve's reference to the lysosomes as "suicide bags."

The substrates for the lysosomal hydrolytic enzymes comprise the nucleic acids, proteins, polysaccharides, e.g., glycogen, mucopolysaccharides, and phospholipids. In addition to enzymes cited in Fig. 6.4, lysozyme, phospholipase, phosphatidic acid phosphatase, nonspecific esterases, and hyaluronidase are located within the lysosome.

Embodied within the lysosomal structure are a number of other proteinaceous substances, the functions for which have not entirely been deduced. These materials are tabulated in Table 6.2.

E. Composition of Lysosomal Material

The lysosome possesses a high phospholipid content as well as periodic acid-Schiff (PAS)-positive material, i.e., glycoprotein. Koenig (1962), employing histochemical techniques, has made some interesting observations on the nature of the lysosomes isolated from the brain. The particles stained intensely for glycolipoproteins and Koenig has proposed that the

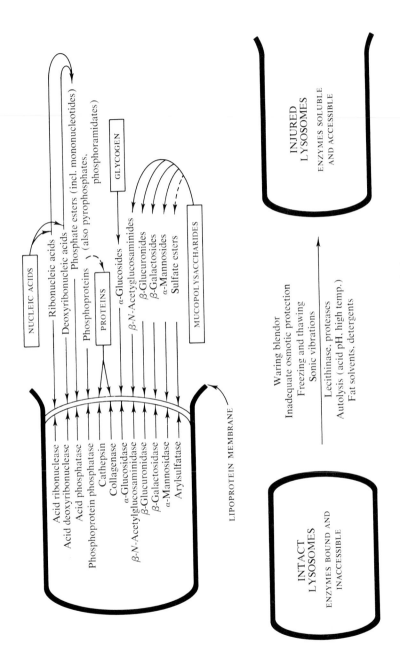

Fig. 6.4. Lysosomes as a biochemical concept.

TABLE 6.2

Other Lysosomal Constituents

Cationic protein that induces inflammation
Phagocytin, antibacterial agent
Tissue plasminogen activator
Hemolysin
Mucopolysaccharide or glycoprotein matrix
Proteaselike permeability factor

---

glycolipid portion is a ganglioside, an acidic, hydrophilic glycosphingo-lipid containing *N*-acetylsialic acid. A brief discussion of the ganglio-sides is presented below. Within the ganglioside meshwork are entrapped the conglomeration of lysosomal enzyme activities. Koenig's hypothesis has not been fully established but the concept has some merit in account-ing for the ease of rupture of the lysosomes under various physiological as well as pathological conditions.

1. BIOCHEMISTRY OF GANGLIOSIDES

The gangliosides are glycolipids containing 60% carbohydrate, of which 24% is hexose, 15% hexosamine, and 21% a sialic acid derivative. A typical ganglioside structure which has been recently postulated by Faillace and Bogoch (1962) is presented in Fig. 6.5.

2. SIALIC ACIDS

The sialic acids are a class of sugars derived from neuraminic acid. The currently accepted structures for neuraminic and *N*-acetylneuraminic acids are given in Fig. 6.6.

Neuraminic acid and its *N*-acetyl derivative are conjugates of pyruvic acid and sugar, D-mannosamine. It is noteworthy that this example is

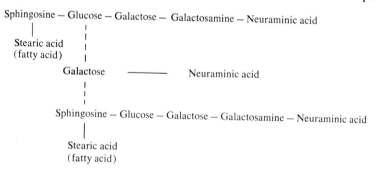

Fig. 6.5. Ganglioside ("double unit" structure).

COOH
|
C = O
|
CH₂
|
CHOH
|
NH₂ – C – H
|
HO – C – H
|
CHOH
|
CHOH
|
CH₂OH

Neuraminic acid

COOH
|
C = O
|
CH₂
|
CHOH
|
CH₃ – C – NH – C – H
‖O
HO – C – H
|
CHOH
|
CHOH
|
CH₂OH

N-Acetylneuraminic acid

Fig. 6.6. Formulas for neuraminic acid derivatives.

the first indication of the natural occurrence of D-mannosamine. Much of our knowledge of the intracellular synthesis of the sialic acids has evolved from the studies of Roseman and his colleagues (1962) who have proposed the schema presented in Fig. 6.7.

3. GANGLIOSIDES

The gangliosides are formed from N-acetylneuraminic acid via the intervention of a cytidine monophosphate derivative. N-Acetylneuraminic

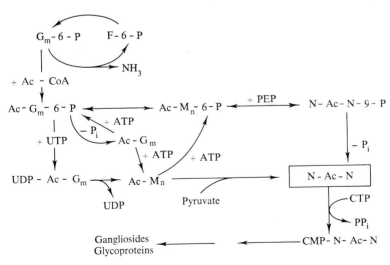

Fig. 6.7. Synthesis of gangliosides, Gₘ-6-P, glucosamine 6-phosphate; F-6-P, fructose 6-phosphate; Ac-CoA, acetyl-CoA; Mₙ-6-P, mannosamine 6-phosphate; N-Ac-N-9-P, N-acetylneuraminic acid 9-phosphate; Pᵢ, inorganic phosphate; PPᵢ, pyrophosphate; PEP, phosphoenolpyruvate.

acid may be produced by two pathways, (1) the racemization of uridine diphosphate-N-acetylglucosamine to the mannosamine derivative and the subsequent addition of pyruvate, and (2) the racemization of acetyl-glucosamine 6-phosphate to the mannosamine derivative, the addition of phosphoenolpyruvate, and the dephosphorylation of the resultant N-acetylneuraminic acid 9-phosphate.

F. Intracellular Construction of the Lysosome

The lysosome is believed to originate through the combined activities of the endoplasmic reticulum and the Golgi apparatus (see Section II). The acid hydrolases, like all other proteins, are elaborated as the result of polyribosomal activity, and first appear in the cisternae of the endo-plasmic reticulum and are then transferred to the Golgi system. The hydrolases are finally released as lysosomes from the Golgi within vesicles derived from the latter. The processes closely resemble the re-lease of zymogen granules in pancreatic secretions. The evidence for the derivation of the lysosome from the Golgi is based upon (1) the positive acid phosphatase reaction of many Golgi systems and (2) the appearance of phagocytic vacuoles first in the neighborhood of the Golgi.

The acceptance of this hypothesis is by no means universal. Indeed, several laboratories have postulated the existence of two mechanisms which depend upon the particular cell. In cells possessing a very en-riched smooth endoplasmic reticulum, the Golgi apparatus participates in the elaboration of the lysosome; in cells rich in rough endoplasmic reticulum, e.g., rat prostate and seminal vesicles, only the latter structure appears responsible.

G. Functions of the Lysosome

From the earliest studies of de Duve and his colleagues, it was ap-parent that the rupture of the lysosomal membrane and its consequent release of the hydrolytic enzymes were associated with a number of autolytic processes. The escape of the hydrolases into the cell could effect the dissolution of nucleic acids, proteins, or polysaccharides. From these facts emerged a picture in which the lysosome is viewed as the principal performer in a number of physiological as well as pathological states. Let us consider examples of lysosomal activity in these conditions.

1. ACTIVITY UNDER PHYSIOLOGICAL CONDITIONS

a. ENDOCYTOSIS. Lysosomes play an active role in the storage, pro-cessing, and digestion of extracellular materials incorporated into the cells. The nature of these materials may vary from whole cells to macromolecular constituents. The extracellular material is taken into the cell by some form of vesicle formation at the cell membrane, endocytosis. When incorporated into the cell, the ingested material is often found within a membrane-bound vacuole, the *phagosome* (see Fig. 6.8), a lysosomelike structure but without enzymic activity.

The phagosome collides with a lysosome within the cell, a fusion of their respective membranes takes place, with the production of a *digestive vacuole* or *heterolysosome*. The coalescence of the membranes is believed to occur by a process of *exoplasmosis*, similar to a reverse pinocytosis. The ingested material is thus exposed to the hydrolytic enzymes and may be degraded to nutrients for the cell.

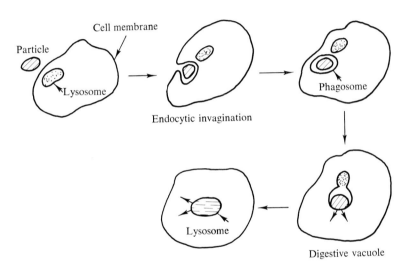

Fig. 6.8. Endocytosis.

b. AUTOLYTIC PROCESSES (AUTOPHAGY). The digestion of large amounts of a cell's own cytoplasm is very well documented. The mechanism has been referred to as *focal autolysis* or *cellular autophagy*. Examples include the changes that occur during the differentiation of the newborn kidney, in metamorphosing insect salivary glands, during the regression of the Müllerian ducts, etc. In the latter case, the Müllerian ducts occur as paired, symmetrical embryonic organs in both sexes of the chick. In genetic males, these structures possess only a temporary function, while in females, the ducts form the foundation for future oviducts of the hen. At the ninth day of incubation shortly after the morphological differentiation of the gonads, a regression of the Müllerian ducts takes place in the male embryo which is complete within 48 hours. The regression is the result of intense lysosomal activity.

In the mammal, cellular autophagy is observed in the atretic ovarian follicle, in the rat prostate after castration, in the liver of starved animals, in the adipose cells after stressing by exposure to cold and fasting, in the heart during aging, and in the hypoxic brain.

The enclosed cytoplasm which may contain mitochondria, endoplasmic reticulum, and other cellular structures, is surrounded by a single unit membrane and will exhibit a positive acid phosphatase reaction. The resultant structure is called the *cytolysome* or *autophagic vacuole*. The origin of the membrane surrounding the autophagic vacuole is the subject of much conjecture. Although several investigators have considered the synthesis to occur *de novo*, most workers hold to the derivation of the membrane from preexisting intracellular membranes, perhaps arising from the endoplasmic reticulum.

c. EXOCYTOSIS. In many protozoa, the normal fate of the residue remaining within a lysosome is the elimination from the cell by *exocytosis*, i.e., the contents of the vacuoles are released by a coalescence of its membrane, with the plasma membrane mimicking a reverse phagocytosis. A similar mechanism has been observed in higher animals.

d. LYSOSOMES AND MITOSIS. Evidence has accumulated suggesting a function for the lysosomes during the processes of cell division or mitosis (see Chapter VIII). During the prophase, a mobilization of the lysosomes from their usual positions to the perinuclear region has been observed, prior to the dissolution of the nuclear membrane.

e. HETEROPHAGIC FUNCTIONS. Tissue destruction is often mediated by an invasion of phagocytic cells; this event represents a heterophagic process. The phagocytic cells are the principal destroyers of all types of invading organisms, e.g., microorganisms and viruses, foreign particles, blood clots, and thrombi.

The physiological functions of the lysosome in the proper maintenance of the cell and of the organism are summarized in Tables 6.3 and 6.4 which have been taken from the review by de Duve and Wattiaux (1966).

2. ACTIVITY UNDER PATHOLOGICAL CONDITIONS

a. METABOLIC DISEASES. The lysosomes have been shown to be the site of primary or secondary injury in only a few human or experimental diseases. The best example of a true inborn lysosomal disease was discovered by Hers and colleagues who identified in children a glycogen storage disease type II in which an absence of the lysosomal enzyme,

TABLE 6.3
Functions of Lysosomes in the Economy of the Cell[a]

I. Functions dependent on heterophagy or heterolysis
    A. Heterotrophic nutrition by intracellular digestion and, under special ecological conditions, by extracellular digestion
    B. Defense against bacteria and other microorganisms, viruses, and informational or toxic macromolecules
    C. Invasion by lysis of obstructing structures

II. Functions dependent on autophagy or autolysis
    A. Nutrition under unfavorable conditions of food supply through piecemeal self-digestion
    B. Cellular differentiation and metamorphosis
    C. Intracellular scavenging as part of the self-rejuvenation of long-lived cells?
    D. Other functions: mitosis?
    E. Self-clearance of dead cells

[a] From C. de Duve and R. Wattiaux, *Am. Rev. Physiol.* **15**, 435 (1966).

α-glucosidase, was noted. In the absence of this enzyme, glycogen is deposited in the liver granules.

Another example is metachromatic leukodystrophy which is associated with a defect in lysosomal arylsulfatase A, capable of degrading cerebroside sulfates. The defect is associated with an accumulation of sulfated mucopolysaccharides in nervous tissue.

A condition in which the induction of new lysosomes occurs is potassium depletion: the appearance of lysosomes in renal papillas has been noted in rats rendered deficient in potassium. The condition has also been observed in human kidney biopsy specimens from patients with clinical evidence of potassium deficiency. However, it is not known whether the increase in lysosomal activity is responsible for the subsequent vulnerability of the kidneys to pyelonephritis.

b. CONNECTIVE TISSUE. Circumstantial evidence has been presented attributing certain diseases of the connective tissue to abnormal fragilities of lysosomes in synovium, vascular pericytes, or mesenchymal cells. The activity of the lysosomes in rheumatoid arthritis has been debated for several years.

3. STABILIZATION AND LABILIZATION OF LYSOSOMES

The lysosome is surrounded by a single lipoprotein unit membrane, and accordingly, should be affected by a number of membrane-active agents. A list of agents that protect or labilize the lysosomal structure

TABLE 6.4
Functions of Lysosomes in the Economy of the Organism[a]

I. Functions associated with autonomous cell life

II. Functions associated with programmed cellular breakdown
   A. Fertilization, early embryological development, and nidation?
   B. Developmental processes, metamorphosis, regression, involution, atresia, keratinization, sexual cycles, etc. . . .
   C. Apocrine and holocrine secretion

III. Functions associated with food processing
   Participation in digestion in gastrointestinal tract

IV. Functions associated with breakdown of secreted macromolecules
   A. Reabsorption in kidney and bladder, possibly in other excretory outlets
   B. Absorption and digestion in epithelial, mesothelial, and endothelial cells
   C. Processing of secretory products in gland cells (thyroid, possibly others)
   D. Resorption of bone and of connective tissue fibers

V. Functions associated with immunity, scavenging, and detoxication
   A. Destruction of aged erythrocytes, dead cells, cell debris by macrophages, and reticuloendothelial cells
   B. Destruction of microorganisms, viruses, foreign particles, and macromolecules by leukocytes, macrophages, reticuloendothelial cells, possibly hepatic parenchymal cells.
   C. Processing of antigens and possibly role in lymphocyte transformation

VI. Functions in thrombolysis
   A. Platelet aggregation and lysis?
   B. Activation of plasminogen

---

[a]From C. de Duve and R. Wattiaux, *Am. Rev. Physiol.* **15**, 435 (1966).

either *in vitro* or *in vivo* is presented in Table 6.5. This table has been adapted from the review by Weissmann (1965).

Only a few of the compounds affecting the stability of the lysosomes *in vitro* exert unequivocal actions upon these structures *in vivo*. In this regard, vitamin A, the streptolysins, ultraviolet, and X-irradiations produce histochemical as well as biochemical lesions in lysosomes in the living cell. Several other agents only exert their effects in the living cell and have little or no action *in vitro*. Thus, 2,4-dinitrophenol and excess oxygen must labilize the lysosome through a mechanism operative only in the intact cell.

The stabilizers of lysosomal structure are fewer in number. Among these may be included the anti-inflammatory drugs, cortisone, cortisol, prednisone, and β-methasone. The glucocorticoids may owe many of their effects against cellular injury to this ability to stabilize the lysosomal structure against disruption.

II. GOLGI SYSTEM

The Golgi system was originally described in 1898 by C. Golgi as a concentric network about the nucleus of the Purkinje cells in the cerebellar cortex of the barn owl. Concern and speculation as to the function of this "internal reticular apparatus" have persisted for many years. The trend at the turn of the twentieth century favored the acceptance of the Golgi system only as an artifact of the method of preparation of the tissue.

TABLE 6.5
Labilization and Stabilization of Lysosomes[a]

	In Vitro	In Vivo
Labilization	Hypotonic conditions	Vitamin A
	Freezing and thawing	Endotoxin
	Low pH	Streptolysins O and S
	Nonionic detergents	UV and X-irradiation
	Blendorization	2,4-Dinitrophenol
	Deoxycholate	Antigen-antibody reactions
	UV and X-irradiation	High oxygen excess
	Phospholipase and proteases	Ischemia
	Phospholipase and proteases	Anoxia
	Streptolysins O and S	Shock
	Progesterone, testosterone	Prophase of mitosis
	Progesterone, testosterone	Endocytosis
	Diethylstilbesterol	Metamorphosis
	5-β-H bile acids	Virus infection
	Vitamin A	Starvation
	Cysteine	Thyrotropin in thyroid
	Lysolecithin	
Stabilization	Cortisol and cortisone	Cortisone and analogs
	Cholesterol	Chloroquine
	Chloroquine	Serum factors in autoimmunity
	Antihistamines	Serum factors in autoimmunity
	Serum factors in autoimmunity	"Tolerance" to endotoxin
	Prednisone	
	β-Methasone	

[a] Adapted from G. Weissmann, *New Engl. J. Med.* **272**, 1084 (1965).

The skepticism as to the existence of this organelle was founded in the method chosen for its microscopic identification, i.e., the classic silver impregnation technique. In this method, silver nitrate is reduced to free silver by the Golgi system and the metal is deposited as a dark opaque area. Only those vacuoles that would stain supravitally with neutral red, however, were recognized as structures "naturally" occurring within the cell; these structures were termed "vacuomes" by Parat and Poinslevé in 1924. Paradoxically, the existence of the "dictyosome" in invertebrate tissues was readily accepted by most investigators; this spherical or concave discoid-shaped subcellular constituent is identical with the vertebrate Golgi system. With the advent of electron microscopic techniques, this controversy was resolved by Dalton in 1951 and the use of classic silver impregnation techniques as an aid in the identification of the Golgi was vindicated.

A. Identification of the Golgi Apparatus

A cytoplasmic region that can readily reduce silver nitrate may be demonstrated in a majority of cell types. The Golgi system is variable in form, size, and in location within the cell. Under the electron microscope, after "staining" with osmium tetroxide, the Golgi system of highly differentiated, functional cells may appear as three distinct components (Fig. 6.9): (1) Large vacuoles. (2) Paired membranes, i.e., lamellae, $60-70$ Å each in thickness with an interspace about 70 Å. No particles are attached to the lamellae. (3) Small granules or microvesicles which range from 40 Å in diameter to the size of zymogen granules.

The most consistent morphological form in certain inactive secretory and in embryonic secretory cells is the lamellar structure. In many cell types, e.g., chick pancreas, tumor cells, a close spatial relationship is apparent between the nuclear membrane and the Golgi regions, with the open nuclear pores closest to these regions.

B. Functions

Evidence from morphological sources suggested that the Golgi system may function in various secretory phenomena as a segregation and concentration area. Indeed, the location of the Golgi apparatus immediately surrounding the contractile vacuole of certain protozoa implied an involvement in the control of the water balance in the protozoa. The amount and distribution of the Golgi system was dependent upon the changes in secretory activity of various glandular cells. This concept of the Golgi function was first nurtured in 1929 by Bowen who stated:

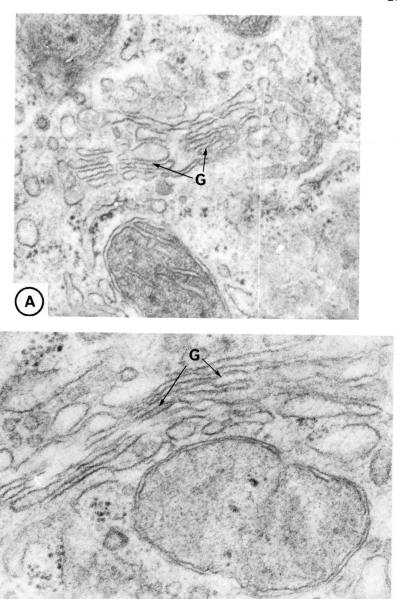

Fig. 6.9. Golgi apparatus in rat liver and Walker tumor. Sections were stained with OsO_4. A. Golgi (G) in rat liver. × 60,000. B. Golgi in Walker tumor. × 76,000. Courtesy of H. Busch and K. Shankar.

Secretion is in essence a phenomenon of "granule" or droplet formation. Starting with a single such secretory droplet about to be expelled from the cell, we find it possible to trace its origin step by step to a minute vacuole, which has thus from the beginning served as a segregation center for a specific secretion material. The primordial vacuole is found to arise in that zone of the cell characterized by the presence of the Golgi apparatus, and the evidence indicates, if it does not demonstrate, that the primary vacuole arises through the activity of the Golgi substance and undergoes a part at least of its development in contact with or embedded in, the Golgi apparatus.

The Golgi system may comprise a membrane system with the endoplasmic reticulum extending throughout the cytoplasm. The formation of the secretory product would occur in the endoplasmic reticulum and the new product would appear in the Golgi. The agranular (smooth) endoplasmic reticulum is flecked with certain "blebs" at regions bordering upon the Golgi apparatus and it may be these "blebs" which facilitate the transport of substances into the Golgi zone.

The Golgi apparatus has been implicated in the intracellular transport and storage of enzymes elaborated by the pancreatic exocrine cell. Palade and his associates (1961) have compiled evidence that identifies four stages in the intracellular pathway leading from the *de novo* synthesis of the pancreatic enzymes at a specific intracellular site to their exit from the cell. These stages are depicted schematically in Fig. 6.10.

The first stage follows the synthesis of the enzyme at the ribosome on the endoplasmic reticulum. Within $1-3$ minutes after the *in vivo* administration of a radioactive amino acid precursor, labeled α-chymotrypsinogen may be recovered from the attached ribosomes in pancreas.

The second stage takes place in the cavities of the cisternae of the rough endoplasmic reticulum. The location was originally suggested by the presence of granules, and later by the isolation of a concentrate of these structures that possessed high enzymic activity from pancreatic microsomes.

The third stage is localized by electron microscopic autoradiography to the large vacuoles of the Golgi apparatus wherein a rapid appearance of radioactive material may be noted shortly after the administration of tritiated leucine. The autoradiographic grains are situated over the Golgi zone, with most of them associated with the large vacuoles. The latter structures appear bounded by a smooth-surfaced membrane and filled with enzyme granules.

The granules themselves represent the fourth stage. Gradually the vacuole (third stage) is filled with its complement of enzymes, and finally will progress toward the apical pole of the cell. The digestive enzymes will appear in this stage in less than 45 minutes and they may be stored in this granule for several hours. The contents of the filled granule may be discharged after a fusion of its membrane with the plasma membrane has occurred. Thus, the processes leading from production of newly synthesized enzymes to the extracellular secretion have been reconstructed.

Formation of Mucus. A major function of the Golgi apparatus is related to the synthesis of mucus in the mucus cells. Long before the existence of the Golgi apparatus was officially recognized, Ramon y Cajal suggested the formation of the mucigen granules of the goblet cells in that structure. Electron microscopy has sufficiently confirmed this hypothesis. A schematic representation of the surface goblet cell of the colon

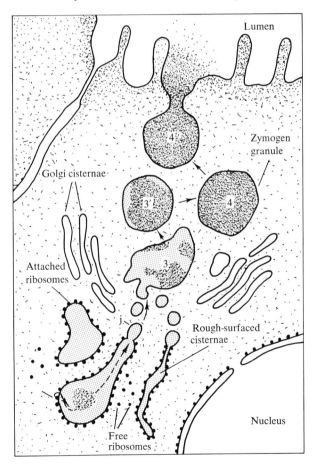

Fig. 6.10. The endocrine pancreas. From G. E. Palade, P. Siekevitz, and L. G. Caro, *in* "The Exocrine Pancreas." (A. V. S. de Reuck and M. P. Cameron, eds.). Little, Brown, Boston, Massachusetts, 1961.

from young rats is presented in Fig. 6.11. The surface goblet cells are abundant and are easily located at the apex and are surrounded by dense cytoplasm which contains the rough endoplasmic reticulum and some mitochondria. An extensive Golgi apparatus is found in the supranuclear region. The Golgi in this region are composed of membrane-bounded cisternae saccules, arranged in several strata. The space above the Golgi is filled with mucigen granules.

Studies with labeled glucose and sulfate-^{35}S have established the synthesis of the mucopolysaccharide and glycoproteins (or part of the glycoproteins) in the Golgi cisternae of the goblet cells. The sequence of events involves the conversion of glucose into the uridine- and guanosine diphosphate sugar derivatives presumably in the endoplasmic reticulum, the synthesis of the protein moiety in the rough endoplasmic reticulum, and the coupling into glycoprotein in the Golgi. It seems unlikely that the mucopolysaccharides and glycoproteins would diffuse into the mucigen granules but that the latter structures are normal outgrowths of the Golgi.

After release of the mucigen granules from the Golgi, they become displaced upward by the continuous synthesis of new structures. They finally are released through gaps in the upper cytoplasmic layer and when outside, the granules burst, liberating mucus.

The participation of the Golgi apparatus in the production of lysosomes has already been presented (p. 290) and will not be discussed here.

C. Isolation of the Golgi System

The isolation has been achieved from the rat epididymis, a tissue which possesses a very extensive Golgi system in the highly convoluted ducts. In collaboration with Schneider, Dalton isolated Golgi membranes by means of a discontinuous sucrose gradient with densities 1.15 (bottom of tube) to 1.04 in 0.34 M saline solution. The homogenate of the epididymis was layered on the top of the gradient and was centrifuged at 145,000 g for 1 hour. The characteristic structures were observed by phase microscopy at the interface between densities 1.09 and 1.13 gm/ml. In the early studies, the Golgi membranes isolated by this procedure contained 6% of the total nitrogen of the homogenate, a considerable amount of phospholipid, alkaline and acid phosphatase, ATPase and AMPase activities and a high concentration of RNA (approximately 20% of the entire amount in the homogenate). The presence of the RNA and the acid phosphatase implied considerable contamination in the isolated material.

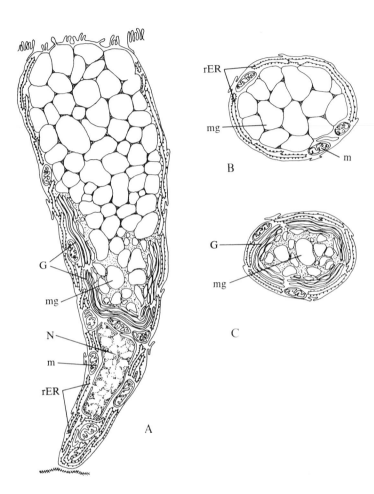

Fig. 6.11. Semischematic drawings based on electron micrographs of surface goblet cells of colon of 10-gm rats. (A) In longitudinal section, the Golgi complex (G) forms a U above the nucleus (N). It is composed of several stacks of saccules. Each stack includes 7 to 12 saccules (only 4 of which are depicted here). There seem to be transitions between the most central saccules and the loosely packed mucigen granules (mg) which occupy the central portion of the supranuclear area. Above, closely packed mucigen granules occupy the cell apex (m, mitochondria; rER, rough-surfaced endoplasmic reticulum or ergastoplasm). (B) Tranverse section through the supranuclear region. The Golgi complex (G) forms a ring around mucigen granules (mg). The Golgi complex in turn is rimmed by a narrow margin of cytoplasm. (C) Transverse section above the Golgi complex. The group of apical mucigen granules (mg) is rimmed by cytoplasm. Rough endoplasmic reticulum (rER) and mitochondria (m) may be distinguished. From M. Neutra and C. P. LeBlond. *J. Cell Biol.* **30**, 119 (1966).

The isolation procedure has been subsequently modified by placing the extract of final density 1.15 at the *bottom* of the centrifuge tube containing the discontinuous gradient from densities 1.15 (bottom) to 1.04. After centrifugation a band could be obtained at the 1.09/1.13 interface which consisted entirely of Golgi membranes whose composition included only 1% of the total nitrogen of the homogenate, trace amounts of RNA, phospholipid, NADH-cytochrome c reductase activity, acid and alkaline phosphatases. Perhaps future investigations will prove successful in the isolation of the Golgi system from other tissues and in this manner, the properties of the Golgi regions from the various cells may be compared.

D. Enzymic Activity of the Golgi System

The existence in the Golgi system of one or two enzymes which catalyze the pyrophosphorolysis of various nucleoside 5'-diphosphates and the phosphorolysis of cocarboxylase, i.e., thiamine pyrophosphate, has recently been demonstrated (Allen, 1963; Novikoff and Goldfischer, 1961).

$$\text{nucleoside - P - P} \longrightarrow \text{nucleoside - P} + P_i$$
$$\text{thiamine - P - P} \longrightarrow \text{thiamine - P} + P_i$$

The number and identity of other enzymes comprising the Golgi system are uncertain, but the existence of the nucleoside diphosphatase and thiamine pyrophosphatase within this organelle is beyond reasonable doubt. Thiamine pyrophosphatase has a limited distribution in the Golgi apparatus from various cell types whereas nucleoside diphosphatase is more ubiquitous. In plant cells, only the latter is found.

The utilization of thiamine pyrophosphatase in the identification of the Golgi apparatus is illustrated in Fig. 6.12. In the latter, the Golgi stained well when thiamine pyrophosphate was employed as substrate, showing clear supranuclear threads.

Allen has separated the thiamine pyrophosphatase and the nucleoside diphosphatase by means of electrophoretic techniques and has determined the substrate specificities. Nucleoside diphosphatase was maximally reactive with uridine 5'-diphosphate (UDP) and inosine 5'-diphosphate (IDP), moderately active with guanosine 5'-diphosphate (GDP) and slightly active with cytidine 5'-diphosphate (CDP) and thiamine pyrophosphate; very low activity was manifested with adenosine 5'-diphosphate (ADP). Thiamine pyrophosphatase was most active with thiamine pyrophosphate as substrate while only limited reactivity was exhibited with CDP, GDP, and IDP.

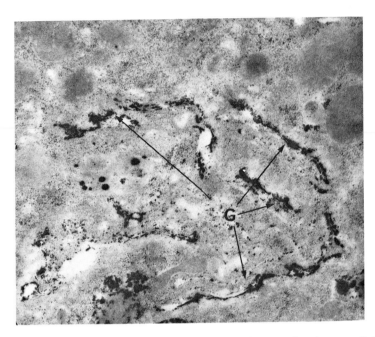

Fig. 6.12. Golgi apparatus (G) demonstrable by thiamine pyrophosphatase activity. A section of human large axillary sweat glands was incubated in a medium containing thiamine pyrophosphate and the enzymic activity was localized by the release of inorganic phosphate. × 25,500. Courtesy of L. Biempica and L. F. Montes.

Both enzymes have been localized and restricted by electron microscopy to the Golgi lamellae. Nucleoside diphosphatase activity is also present in the endoplasmic reticulum although no information is available on the identity of the enzymes in these membranous structures. The functions of these enzymes in the overall plan and activities of the cell are not known at this time. What is also not known is the relationship of these enzymes to the processes of segregation, concentration, and secretion which occur with the Golgi apparatus.

GENERAL REFERENCES

LYSOSOMES

de Duve, C. (1959). *In* "Subcellular Particles (T. Hayashi, ed.), pp. 128-159. Ronald Press, New York.

de Duve, C. (1961). *In* "Biological Approaches to Cancer Chemotherapy" (R. J. C. Harris, ed.), pp. 101-112. Academic Press, New York.

de Duve, C., and Wattiaux, R. (1966). *Am. Rev. Physiol.* 15, 435-492.

de Reuck, A. V. S., and Cameron, M. P. (1963). *Ciba Found. Symp., Lysosomes*

Novikoff, A. B. (1961). *In* "The Cell" (A. E. Mirsky and J. Brachet, eds.), Vol. 2, pp. 423-488. Academic Press, New York.
Weissmann, G. (1965). *New Engl. J. Med.* 272, 1084-1090 and 1143-1149.

GOLGI SYSTEM

Brachet, J. (1957). "Biochemical Cytology." Academic Press, New York.
Dalton, A. J. (1960). *In* "Cell Physiology of Neoplasia," pp. 161-184. Univ. of Texas Press, Austin, Texas.
Dalton, A. J. (1961). *In* "The Cell" (A. E. Mirsky and J. Brachet, eds.), pp. 603-620. Academic Press, New York.
Kurosumi, K. (1961). *Ann. Rev. Cytol.* 11, 1-124.
Neutra, M., and LeBlond, C. P. (1966). *J. Cell Biol.* 30, 119-136 and 137-150.
Novikoff, A. B. (1964). *Federation Proc.* 23, 1010-1022.
Sjöstrand, F. S., and Hanzon, V. (1954). *Exptl. Cell Res.* 7, 415-429.

SPECIFIC REFERENCES

LYSOSOMES

Faillace, L. A., and Bogoch, S. (1962). *Biochem. J.* 82, 527-530.
Gomori, G. (1952). *In* "Microscopic Histochemistry, Principles and Practice." Univ. of Chicago Press, Chicago, Illinois.
Koenig, H. (1962). *Nature* 195, 782-784.
Roseman, S. (1962). *Federation Proc.* 21, 1075-1083.
Straus, W. (1956). *J. Biophys. Biochem. Cytol.* 2, 513-554.

GOLGI SYSTEM

Allen, J. M. (1963). *J. Histochem. Cytochem.* 11, 529-541 and 542-552.
Allen, J. M., and Slater, J. (1958). *Anat. Record* 130, 731-745.
Bowen, R. H. (1929). *Quart. Rev. Biol.* 4, 299-324.
Caro, L. G. (1961). *J. Biophys. Biochem. Cytol.* 10, 37-44.
Dalton, A. J. (1951). *Nature* 168, 244-245.
Kuff, E. L., and Dalton, A. J. (1959). *In* "Subcellular Particles" (T. Hayashi, ed.), pp. 114–127. Ronald Press, New York.
Novikoff, A. B., and Essner, E. (1962). *Federation Proc.* 21, 1130-1142.
Novikoff, A. B., and Goldfischer, S. (1961). *Proc. Natl. Acad. Sci. U.S.* 47, 802-810.
Novikoff, A. B., Essner, E., Goldfischer, S., and Heus, M. (1963). *In* "The Interpretation of Ultrastructure" (R. J. C. Harris, ed.), p. 149. Academic Press, New York.
Palade, G. E., Siekevitz, P., and Caro, L. G. (1961). *In* "The Exocrine Pancreas" (A. V. S. de Reuck and M. P. Cameron, eds.), pp. 23-55. Little, Brown, Boston, Massachusetts.
Zeigel, R. F., and Dalton, A. J. (1962). *J. Cell Biol.* 15, 45-54.

Chapter VII

ACTIVE TRANSPORT

I. HISTORY

One of the first problems to be considered in studies of transport phenomena was the peculiarity most cells possessed of accumulating potassium ions in large quantities and maintaining a concentration gradient of potassium. In 1941, Boyle and Conway showed that freshly excised muscle concentrated potassium to a remarkable extent. They concluded that this process was due to an electrostatic attraction of the indiffusible anions. These anions were present inside the cell and were undoubtedly esters of organic acids, phosphoric acids, and various proteins. Potassium could enter the cell as the phosphate or the chloride and would be retained when the anions became indiffusible or when the potassium was bound to a protein or other nondiffusing anions.

During evolutionary processes, the indiffusible anions were, at one time, probably readily diffusible. Once they had entered the cell and as the process of cell growth continued, these anions became indiffusible, perhaps due to genetic alterations in membrane permeability (see Chapter II). Consequently, potassium must have accumulated automatically with the natural growth of the cell. The only special structure required was the characteristic membrane with its peculiar permeability properties. Alterations of temperature apparently had no appreciable effect on the potassium-accumulating ability of the muscle cell, suggesting to the early investigators that cellular accumulation of potassium per se did not require the expenditure of energy and probably was due to a type of double Donnan equilibrium phenomenon.[1] However, Boyle and Conway

---

[1] Whenever a colloidal electrolyte is present on one side of a membrane through which it cannot pass, it is found that other ionized substances to which the membrane is freely permeable, tend to become more concentrated on the opposite side of the membrane. This unequal distribution of ions at equilibrium is called the Donnan effect (1924).

did recognize that some cells behave differently from their muscle model. For example, potassium is concentrated in the cell sap of the one-celled marine plant *Valonia macrophysica* 42 times more than in the surrounding seawater, whereas the sodium concentration is only 0.18 times that of the external value (Osterhout, 1931). In addition, when the temperature is lowered, a signiffcant loss of potassium from the cell sap is noted with a concomitant gain of sodium. The problem of potassium concentration in *Valonia* cell sap prompted Boyle and Conway to postulate the role of an active secretion process in ion transport. They regarded the total process as occurring in two stages: (1) the entrance of potassium and chloride into the pellicle of protoplasm (Donnan equilibrium) and (2) the secretion of potassium chloride with constant energy expenditure. In the same year that Conway and Boyle published their classic paper, Dean reported the need for some sort of "pumping device" for the exclusion of sodium from the interior of cells (Dean, 1941).

One of the first individuals to suggest the concept of "active transport" was Huf (1935), who recognized the relationship between cellular metabolism and transport by stating a number of criteria essential to the definition of active transport:

1. A concentration and hydrostatic gradient should exist.
2. The process should exhibit a sensitivity to oxygen lack.
3. The process should exhibit a dependence on active metabolites.
4. There should be a sensitivity to enzyme poisons such as cyanide and bromoacetate.

These criteria are still applicable.

The third study which antedated modern active transport research was published in 1937 by Crow and Ussing, who found a particular specificity of a transport system for sodium.

The definition of active transport has been given (see Chapter II) and the reader is referred to numerous treatments of the mathematics involved in this process (Ussing, 1952; Rosenberg, 1948; Eisenman and Conti, 1965).

II. METHODS OF STUDY

A number of generally accepted methods are available for the analysis of active transport processes. We will briefly discuss two major procedures.

A. Isotopic Tracers Method

The use of labeled ions or labeled substances of any type for the study of active transport comprises the isotopic tracer method. The substance either is placed in the medium in which the tissue is being studied or is injected into the intact animal. After a suitable period of incubation (*in vitro* or *in vivo*, respectively), the tissue is extracted. The determination of the amount of flux that has taken place per unit time and per unit area by the use of inulin and mannitol (see Chapter II) allows for an estimation of intracellular and extracellular space and, hence, for the calculation of the quantity of the material that has accumulated in a particular tissue area. This procedure is fraught with many difficulties and artifacts, most of which are discussed in a review by Ussing (1952).

B. The Short-Circuit Method

This procedure was developed by Ussing and Zerahn (1950) and is diagrammatically demonstrated in Fig. 7.1. Essentially, a piece of tissue, usually a thin layer derived from the frog or other species, is employed. The so-called "skin potential" is short circuited so that both sides of the tissue, i.e., mucosal and serosal, possess the same potential. Therefore, no net passive transfer of ions can take place. However, active transport can occur; any current which runs through the short circuit would be a measure of this process (see Chapter II). If the current is equal to the sodium influx, for example, then the theory attributing the potential and current to the sodium influx would be justified. This is, in fact, borne out by the experiments of Ussing and Zerahn and others. The ionic compositions of the media bathing the mucosal and serosal sides of the tissue are identical; the presence of a sustained short circuit current (SCC) must, in fact, reflect active ion transport processes (Maffly and Edelman, 1963; Ussing and Zerahn, 1950; Eubank *et al.*, 1962). Continuous measurement of the short circuit current with calomel or silver chloride electrodes is made with only brief interruptions for measurements of the potential of the nonshorted skin. The skin, S, in Fig. 7.1, is placed as a membrane, separating the medium (in this case frog Ringer's solution) in the tube, C. Agar-Ringer bridges A and A′ open on either side in the immediate vicinity of the skin and the outer ends of A and A′ make contact with saturated KCl-calomel electrodes. The potential difference between the electrodes is read on the potentiometer, P. Another pair of agar-Ringer bridges, B and B′, open at either end of the tubes, as far as possible from the skin, and their outer ends dip into the beakers containing saturated KCl-AgCl solution. Spirals of stiff silver wire, immersed in these beakers, are used as electrodes through which an outer electromotive force can be applied. The voltage is applied and is adjusted so that the potential difference across the skin as read on the potentiometer

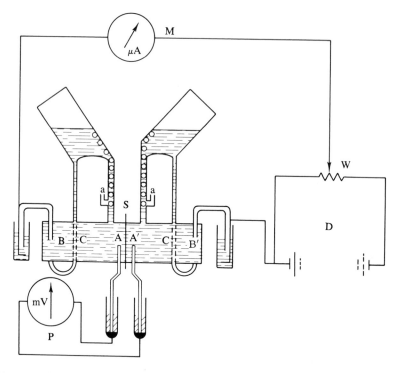

Fig. 7.1. Diagram of the short-circuit method. Redrawn from H. H. Ussing and K. Zerahn, *Acta Physiol. Scand.* **23**, 111 (1951).

is maintained at zero, thus producing a total short circuiting of the skin potential. The current passing through the tissue at this zero potential difference is read on the microammeter. Usually, readings are taken every few minutes, and the potential is carefully observed during the experiments so that any unexpected changes in current strength can be recorded. Appropriate devices may be constructed to maintain temperature, pressure, and gas content.

In actual practice, both methods A and B are used concomitantly. In fact, Ussing and Zerahn developed a short circuit method because they found that the transport rate of sodium ions across the frog skin was so low that a chemical demonstration, i.e., the use of ordinary procedures for ion determination such as flame photometry, of the current-transport relationship was impossible.

A double isotope labeling procedure is used (Levi and Ussing, 1949) wherein the sodium influx may be determined by the use of ^{22}Na, while simultaneously measuring the outflux with ^{24}Na. Samples are taken at various intervals from either side of the tissue. Data shown in Table 7.1 indicate a typical experiment in which both sodium influx and sodium outflux is listed along with the short circuit current. Note that the addition of a drug alters the outward flux and changes the relationship between net sodium flux (ΔNa) and electric current.

TABLE 7.1

Measurement of Short Circuit Current Using ^{22}Na (Influx) and ^{24}Na (Outflux) in Frog Skin[a]

	Microamperes			
Addition	In	Out	ΔNa$^+$	Current
Control	233	19.4	213.6	195
Epinephrine	208	110	98	156

[a]From H. H. Ussing, *Advan. Enzymol.* **13**, 21 (1952).

III. THEORIES

It should be pointed out that the term "active transport" is at best superficial and descriptive; as Solomon has stated (1962), it is analogous to a hypothetical description of an automobile climbing a hill, given by a member of a primitive society. Here, the individual could conceivably conclude that the car is somehow endowed with some type of independent energy or energy-producing system which moves it up the hill. Of course, we know that this is not true, that there is no occult energy system. The combustion of the gasoline-air mixture serves to activate the cylinder and drive shaft; and, consequently, the power which existed potentially in the gasoline is transmitted to the drive shaft which then moves the automobile. The molecular mechanism for movement of the automobile is fairly well known. The molecular mechanism for movement of ions and substances against their concentration gradient at the apparent expense of energy is not known. When we have become suitably enlightened, the appropriate terminology can be substituted for the rather general phrase, "active transport."

What is actually known of the mechanism of active movements of substances across membranous barriers? Two general theories have been promulgated during the past 20 years, the *"absorption" or "sorption" theory* and the *membrane theory*. The former was first described by Nasonov and Alexandrov in 1943 and further elaborated by Troshin (1961) and Ling (1949, 1952). Existing evidence does not completely favor this concept, although a number of aspects are of interest and are consistent with the observations.

A. Absorption Theory

The main contention is that the protoplasm, particularly the proteins located in the intracellular milieu, are invested with certain specific properties which can effect, under certain circumstances, a selective binding of ions and substances against concentration gradients. The main evidence was obtained by using certain artificial polyelectrolytes, particularly the nonselective cation exchanger KU-2, i.e., a sulfonated polystyrene resin. It is well known that proteins are polyelectrolytic in nature and, depending upon the predominant charge at a specific pH, may bind oppositely charged species. The asymmetric distribution of ions, particularly potassium, between the cell and the medium is explained on the basis of a selective type of sorption of the ion by certain proteins present in the sarcoplasm. The binding would not be covalent but rather ionic in nature. One of the main arguments against the theory is the existence of a potential difference of excitable tissues across the resting cell membrane. According to the Nernst equation

$$E_K = \frac{RT}{F} \ln \frac{[K_i^+]^*}{[K_o^+]}$$

the potential difference is of a magnitude which would be predicted if a potassium gradient existed. The "sorptionists" maintain that in a normal resting cell, i.e., the condition prior to depolarization, potassium is practically inactive; or in any case its activity does not exceed that of the external solution. In other words, there may not be any gradient. The insertion of a microelectrode into the cell is the method used to determine potential difference, resting potential, and ionic distribution; it involves the use of a microelectrode, developed originally by Osterhout (1931), by Hodgkin and Huxley (1939), and by Ling and Gerard (1944). It also produces considerable damage which may cause a local decomposition or denaturation of the protein-electrolyte complex. This breakdown or damage causes a considerable gradient of potassium ion activity. The potential difference measured by the aid of microelectrodes is due, therefore, to an intracellular alteration or "damage zone," resembling the denaturation of native proteins. The complex colloid structure of protoplasm, i.e., intracellular environment is disturbed, resulting in an increase in solubility of substances in the protoplasm. The altered cell proteins abruptly change their binding capacity for both inorganic and organic ions, either present in the cell or absorbed

---

* In cat heart, $K_i^+ = 151$ meq/liter; $K_o^+ = 4.8$ meq/liter; $\therefore E_K = 61.5 \log 31 = 92.6$ mV, which is close to the measured resting membrane potential.

by the proteins and leading, presumably, to a redistribution of substances between the cell and its external environment. According to the adherents of the absorption theory, this is the primary mechanism which brings about the progress of biochemical reactions characteristic for the excitation process. These particular reactions will produce energy available for cellular work, particularly for the recovery of "damaged" protoplasm from the state of excitation to the rest state.

Ungar has suggested that there is a correlation between ion transport and the conformation of certain cellular proteins. Stimulation of excitable structures, particularly the cerebral cortex, causes a significant increase in ionizable SH groups, including some type of structural rearrangement of certain protein molecules. This denaturization concept is consistent with both the sorption and the membrane theories, according to the *location* of the proteins which undergo the conformational changes. Ling, one of the strongest proponents of the sorption concept, maintains that solutes are concentrated within the cell by a mechanism not involving the interposition of the membrane. Although he has modified this suggestion somewhat during the past few years, he still maintains that ions and proteins can freely enter into the cell and that there is no "sodium pump." Absorption would be realized by bonding with protein molecules located within the cell. Preferential absorption of ions (otherwise known as "permeability") occurs as a result of differential sizes of hydrated ions. The average dielectric constant at the distance of closest approach between a sodium ion and, for example, a carboxyl group of an intracellular polyelectrolyte is, according to Ling, much higher than for potassium ions. Therefore, an electrostatic potential for potassium results, and it is greater than for sodium. The greater majority of anionic sites within the cell would then be expected to be associated with potassium ions rather than sodium ions, explaining the selective permeability for potassium in the resting cell. A number of criticisms of this hypothesis may be mentioned:

1. It is certainly true that the critical sizes in general differ for the various hydrated ions. For example, potassium ions appear to possess a size which is just at the threshold for the free passage across membranes. Sodium and lithium virtually would be excluded with respect to the ions; chloride appears to be at the critical level, while carbonate and sulfate are of the size which could be slowly transported across membranes. Ion size, however, must not be the sole determinate because cesium enters much less rapidly than potassium even though cesium and potassium have very similar hydrated nuclei.

2. Magnesium and calcium ions appear to enter the cell through the formation of nonionized complexes.

3. It has been shown clearly that sodium is excluded from muscle against its concentration gradient and with the expenditure of energy. In addition, potassium and sodium can exchange with one another by an exchange-diffusion process (see Chapter II).

Furthermore, if Ling's hypothesis were correct, various carboxyl exchange resins with fixed anionic sites should be excellent models for transport; but this, unfortunately, has not been the case. Ling's theory does not explain the Donnan relationship which exists with respect to potassium and chloride ions and offers no explanation for the adherence of the resting cell potential to the Nernst equation (see Chapter II).

While there are numerous fallacies in the sorption hypothesis, the importance of charges on intracellular proteins in biological transport function (Eisenman, 1961; Eisenman and Conti, 1965; Katchalsky, 1964) is not to be underrated. Eisenman (1961; Eisenman and Conti, 1965) has demonstrated the importance of *anionic field strength*, contributed by polyelectrolytes, as a prime factor in cation specificity and as an important key in the search for molecular structures of appropriate field strength for cation discrimination. It may point the way to a mechanism by which the field strength is controlled through the energetic interactions between the nearest neighbor to the cations (which is usually oxygen) as modulated by the next adjacent atom (usually carbon or phosphorus) and, in turn, by more distant atoms. The development of glass membrane electrodes, for studying transport, has enhanced the field strength concept. Variation in electrostatic field strength of the anion is sufficient by itself to produce a particular specificity pattern of eleven transition sequences which have a wide, natural occurrence and can, in fact, govern cation selectivity. The glass electrode, depending upon its compositions, approximates *fixed charges*. Consequently, this idea is termed the "fixed site mechanism" for transport (Eisenman and Conti, 1965). What is the origin of permeability of specific ions in this system? Eisenman has measured the sodium-to-potassium permeabilities of several different types of electrodes. The self-diffusion coefficients of sodium and potassium in the hydrated glass surface were measured by exposing the electrodes to sodium and potassium solutions, labeled with isotopic ^{24}Na and ^{42}K and by studying the uptake of tracer as a function of time and of solution composition. Both the mobility ratio and the ion exchange equilibrium constant contribute to the permeability ratio of potassium to sodium in different types of hydrated glass, with the peculiarity that the more permeable potassium ion is the less mobile (Table 7.2). Hydration of glass increases the mobility of sodium by about four orders of magnitude over its value in dry glass. Thus, *selectivity* or

TABLE 7.2
Glass Membranes and the Fixed-Site Hypothesis[a], [b]

	PK^+/PNa^+	$\mu K^+/\mu Na^+$	$^K Na^+ K^+$(calculated)
Type I glass	10.3 ± 0.1	0.30 ± 0.09	34 ± 10
Type II glass	8.5 ± 0.3	0.15 ± 0.05	55 ± 17

[a]Typical values of permeability ratio, mobility ratio, and ion exchange equilibrium constant for the two batches of K^+-selective glass electrodes are given.
[b]From G. Eisenman and F. Conti, *J. Gen. Physiol.* 48, 65 (1965).

permeability may be brought about by increasing or decreasing negative electrostatic field strengths due to hydration processes. Higher electrostatic field strengths apparently lead to hydrogen ion-selective systems which are indifferent to cations at physiological pH's. Lower electrostatic field strengths lead to alkali cation-selective systems. Low field strength systems are potassium selective while intermediate field strength systems are selective to sodium ions. The "fixed-site" concept does not discount the usefulness of the cell membrane. In fact, both carboxyl and phosphate groups present in the cell membrane possess field strengths suitable for cation discrimination.

We will return to the "field strength hypothesis" in our discussion of the "sodium-pump" enzyme system (see p. 344).

It is of interest to point out that whenever a controversy exists in biology, not infrequently the final solution embodies principles of both sides. As Eisenman has pointed out, "Whether one proposes sites, carriers, or pumps to explain the process of transport and whether the process be active or passive, some molecular structure must ultimately distinguish between these species through the operation of elementary atomic forces according to principles which have been well established in physics and chemistry."

B. Membrane Theory

Electrical forces alone cannot account for the separation of potassium and sodium in the living cell. Some type of chemical reaction must be involved. Osterhout (1935) was the first to suggest that electrolytes enter the cell by combining with one or more constituents of the protoplasm. Spiegelman and Reiner (1942) extended this concept and suggested the following:

(1) $\text{carrier} + X \longrightarrow \text{stable carrier-X complex}$

(2) $\text{carrier-X} + \text{enzyme system} \longrightarrow X + \text{carrier}$

According to this scheme, the formation and splitting of the carrier-X complex must be *spatially separated.*

The concept of carrier-mediated transport located within the membrane has been studied in great detail by numerous investigators, and notable among these are Ussing, Rosenberg, and Conway (see Conway, 1961). Two possible types of carriers can be postulated; mobile and fixed-site. Some investigators do not regard the fixed-site type as a carrier mechanism. However, since the ions or substances involved must attach themselves to a molecule or molecules whether or not the latter are mobile, we will use the term "carrier" in a general way. Let us now consider the ways in which membrane-mediated transport can occur. Following the summary list below, we will discuss in detail the important mechanisms:

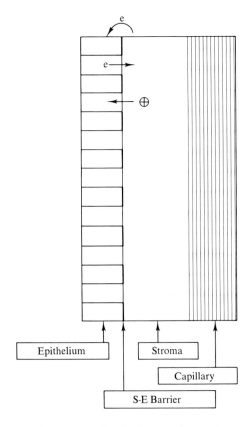

Fig. 7.2. A mechanism of secretion in the chorioid plexus. S-E Barrier = stromal-epithelial barrier. From R. D. Stiehler and L. B. Flexner, *J. Biol. Chem.* **126**, 603 (1938).

1. Coupled sodium-potassium pump, involving mobile carrier
2. Redox reactions and electron transport
3. ATP and ATPase systems
4. Reductive transphosphorylation
5. Chemi-osmosis
6. "Mechano-enzyme" concept

1. COUPLED SODIUM-POTASSIUM PUMP, INVOLVING MOBILE CARRIER

The movements of sodium and potassium ions are linked and mediated by a carrier that presumably is able to shuttle back and forth between the exterior and the interior of the membrane. There is no suggestion of any link between transport and electron transfer in this particular process. The sensitivity of transport to dinitrophenol (DNP) in many of these transport models indicates the involvement of energy-rich phosphate bonds in the process of transport. However, in systems which do not require aerobic metabolism, transport of ions and other substances still occurs against concentration gradients; the uncoupling agent, DNP, has no effect on this system. Therefore, it would seem that a more basic process is involved.

From the mass of data available it can be said with certainty that some type of membrane-located carrier system is involved in the active movement of substances. It is more prudent, perhaps, to refer to the carrier in a general sense, as a system, enzymic in nature, that somehow is able to selectively propel materials in appropriate directions. A number of possible candidates have already been discussed in detail (see Chapter II).

2. REDOX REACTIONS AND ELECTRON TRANSPORT

The redox theory originates from the work of Lund(1928) and Kenyon and Lundegardh (1939), who studied active transport of anions in plants and suggested that the process was linked directly with respiratory systems of the cell. The first definitive evidence, however, that electron transport was related to active transport of ions was delineated by Stiehler and Flexner (1938), who reasoned that "secretory work" was necessary for the formation of cerebrospinal fluid and that the chorioid plexus supplied the required secretory energy. It was shown that an electric current existed between the stroma and the epithelium of the ciliary body of the eye and that this current could account for transport mechanisms in this organ. A similar study was carried out using the chorioid plexus of fetal pigs. Three anatomical elements comprise the plexus: the epithelium, the stroma, and the barrier between the epithelium and the stroma as represented in Fig. 7.2. The passage of various acids and basic dyes across the stromal-epithelial barrier is a measure of transport. From the data listed in Table 7.3, it may be seen that the passage of the dyes from the *epithelial to the stromal layer* was not changed from normal conditions by anoxia, produced by the treatment with cyanide or with nitrogen. However, the passage of the dyes from the *stromal layer to the epithelial layer* was significantly depressed by the anaerobic conditions, indicating the *direction of active transport in this system*. The stromal-epithelial barrier is considerably more permeable when placed in an isotonic sodium chloride solution than when placed in a Ringer's solution. Thus, in isotonic sodium chloride, crystal violet dye slowly penetrates from epithelial layer to stromal layer but does not penetrate when the system is in a Ringer's solution. The explanation is founded upon the presence of calcium ions in the latter, since the permeability differences in salt solutions are resolved, upon the addition of traces of calcium to isotonic saline. This observation serves to demonstrate the importance of calcium in membrane permeability.

The early investigators suggested that an electric current exists between the epithelial and stromal layers of the system, the source of which may be the difference in *potential levels of these two tissues*. The stroma, with a lower potential, would tend to give up electrons to the epithelium (see Fig. 7.2) by means of a reversible oxidation-reduction system present in the stromal-epithelial barrier. In order to maintain electroneutrality, cations (represented by the basic dyes) must move from the stroma to the epithelium; anions (represented by the acid dyes) would move in the reverse direction. The stromal-epithelial barrier thus contains a reversible oxidation-reduction system, the first definite evidence of a redox role in active movement of substances across membrane barriers.

Conway (1953; Conway and Brady, 1948; Conway and Mullaney, 1961) has expanded this information into a general redox pump theory applicable to the active transport of inorganic cations across membranes. Very small concentrations of redox dyes exercise a considerable influence on the uptake of potassium ions and the excretion of sodium ions by yeast and other cells. Redox systems, therefore, can act as *carriers* and function as immediate suppliers of the energy necessary for the transport of substances. According to this theory, the essential mechanism for movement of substances against concentration gradients is *electron transfer* rather than the action of some high energy substance like ATP (Fig. 7.3). As we will see later, this view is not shared by all investigators, although part of the concept is useful.

If the short circuit current method of measurement, described on page 307, is used, frog skin shows great sensitivity to various oxidants and reductants. The epithelial side of the skin is much more responsive

TABLE 7.3

Penetration of Dyes Through the Stromal-Epithelial Barrier of Chorioid Plexus at Physiological pH[a]

Dye	Chemical type	Ionic type	Penetration normally		Penetration with asphyxia	
			Epithelium to stroma	Stroma to epithelium	Epithelium to stroma	Stroma to epithelium
Crystal violet	Triphenylmethane	Basic	0	+++	+	+
Cresyl violet	Oxazine	Basic	0	+++	+	+
Eosine	Fluoran	Acid	++	Trace	+++	+++
Bromophenol blue	Sulfonephthalein	Acid	++	Trace	++	++
Rhodamine B	Fluoran	Neutral	++	++		

[a]From R. D. Stiehler and L. B. Flexner, J. Biol. Chem. **126**, 603 (1938).

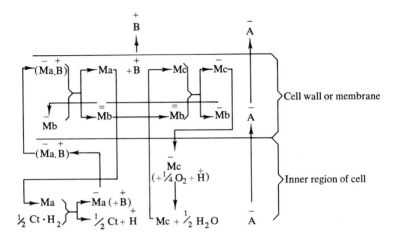

Fig. 7.3. The redox pump theory. The diagram indicates three oxidation-reduction cycles of the metal-containing catalyst systems, Ma, Mb, and Mc. Ma is the carrier system for the cation B^+ (the oxidized form Ma being reduced by a catalyst, CtH_2, of flavoprotein type, and binds the cation B^+). The catalyst system Mb oxidizes the reduced Ma and liberates B^+. The Mb cycle is completed in the membrane or outer cell region, transferring its electrons to the system Mc, which transfers them finally to the oxygen system within the cell. The anion A^- is carried out by the potential difference established by the active carriage of B^+. From E. J. Conway, *Science* **111**, 270 (1951).

than the dermal side. The reader will remember that the changes in current flow in this system are interpreted as a direct measure of active transport. The action of the redox reagents on the short circuit current can be correlated directly with actions on active transport systems. In frog skin, active sodium transport is dominant over the active transport of any other ion. Therefore, the short circuit current method measures, in this particular system, the active pumping of sodium. Oxidizing dyes, in general, inhibit sodium transport, while reducing dyes stimulate the transport. The net transport of sodium across the skin may be described in terms of a series of energy-linked ion exchange reactions in which potassium and hydrogen ions participate (Table 7.4). No net transport across the skin of potassium and hydrogen ions occurs. Exchange of sodium for potassium probably occurs near the inner boundary of the system while exchange of hydrogen for sodium occurs near the outer boundary of the epithelial cells. The free energy available in reaction 7 is sufficient to move 8 equivalents of sodium for each $\frac{1}{2}$ mole of oxygen. In other words, the Na^+ to O_2 ratio is 16. These reactions do not explain the mechanism by which total available free energy is utilized to move 8 equivalents of sodium, and requires the presence of some type of chemical carrier system, which will be discussed in some detail later.

An interesting and recent contribution to the redox mechanism for the transport of cations involves the use of respiratory-deficient mutants of yeast (Reilly, 1964). Figures 7.4, 7.5, and 7.6 show the effects of redox dyes on the uptake of potassium ions and the extrusion of sodium ions from normal and mutant yeast cells. The mutant yeast is not able to excrete sodium ions in exchange for potassium ions in the absence of glucose. Certain redox dyes significantly stimulate the uptake of potassium ions in normal cells but not in mutant yeast cells. Excretion of sodium ions is inhibited by increased potential in normal but not in mutant yeast.

The use of those mutant cells deficient in certain respiratory cofactors is a valuable adjunct to the study of active transport mechanisms. Genetic mutations in which the phenotypic manifestation is a malfunction of active transport of specific groups of substances in these cells can be identified.

3. ATP AND ATPASE SYSTEMS

The "ATPase" theory comes closest to explaining active transport on a molecular level. In contrast to the redox theory, ATP is specifically required for the pumping of ions against concentration gradients and across membranes. The main argument against the redox theory is that some cells which do not have any mechanism for oxidative phosphorylation still actively transport ions. The best example of this is found in the red blood cell, which depends upon glycolysis for energy. Dinitrophenol, a specific inhibitor of oxidative phosphorylation (see Chapter III), inhibits active ion transport only in those cells using the Krebs cycle. Therefore, some energy source *common* to all systems (other than

TABLE 7.4

Modified Redox Theory of Active Cation Transport[a]

(1)	Na^+ (cis) + HR $\rightarrow$ NaR + H^+
(2)	NaR + K^+ $\rightarrow$ KR + Na^+ (trans)
(3)	succinate + oxidized flavoprotein $\leftrightarrow$ fumarate + reduced flavoprotein
(4)	ATP + reduced flavoprotein + NAD^+ $\leftrightarrow$ ADP + P_i + NADH + oxidized flavoprotein + H^+
(5)	KR + H^+ $\rightarrow$ HR + K^+
(6)	ADP + P_i + NADH + H^+ + $\frac{1}{2}O_2$ $\leftrightarrow$ ATP + NAD^+ + H_2O
(7)	Na^+ (cis) + succinate + $\frac{1}{2}O_2$ $\rightarrow$ Na^+ (trans) + fumarate + H_2O

[a] From L. L. Eubank, E. G. Huf, A. D. Campbell, and B. B. Taylor, *J. Cellular Comp. Physiol.* **59**, 129 (1962).

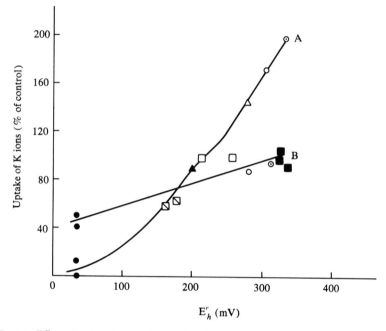

Fig. 7.4. Effect of redox dyes on the uptake of K^+ ions during the fermentation of normal (A) and mutant (B) yeasts. The yeasts were buffered by 0.04 M potassium hydrogen succinate at pH 4.5. The dyes employed were: □, none (control suspension, containing water in place of dyes); ●, Nile blue; ▲, Safranine T; ◨, Janus Green; ○, o-cresol-indo-2,6-dichlorophenol; △, phenol-indo-2,6-dichlorophenol; ■, o-chlorophenol-indo-2,6-dichlorophenol;◉, o-chlorophenol-indophenol. The dyes were used at a concentration of 0.1 mM. From C. Reilly, *Biochem. J.* **91**, 447 (1964).

a specific redox reaction) must be intimately concerned with the transport of substances.

In 1948 Libet found that the squid axon contained an active ATPase[2] which remained behind when the axoplasm was extruded. This implied that the ATPase activity was in fact in the membranes. Subsequent investigations defined a role for ATP in ion transport. The experiments of Caldwell *et al.* (1960) are particularly noteworthy. He found that potassium cyanide could effectively inhibit the active transport systems in squid axons. Microinjections of ATP through the membrane into the intracellular portion of the nerve restored active transport. ATP, incorporated into reconstituted ghosts[3] of red blood cells, also restored active

[2]Adenosinetriphosphatase: a reaction, enzymic in nature, which results in the hydrolysis of adenosine triphosphate to adenosine diphosphate and inorganic phosphate with the liberation of approximately 7000 calories per mole, at a pH of 7.0, under standard conditions.

[3]Erythrocyte "ghost," produced by hemolysis and washing procedure, consists of only an intact membrane (Dunham and Glynn, 1961).

sodium extrusion (Gardos, 1954). If an intact erythrocyte is poisoned with iodoacetic acid in glucose-free media, active transport is halted. The addition of glucose and a phosphorylating system can restore transport. It is obvious, then, that sodium and potassium transport in nerve muscle and erythrocyte depend somehow on ATP or on some energy rich phosphate compound (either ATP or creatine phosphate, or arginine phosphate).

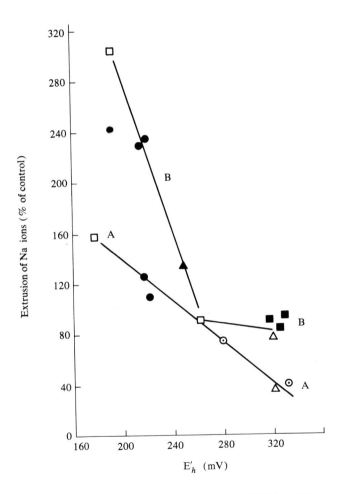

Fig. 7.5. Effect of redox dyes on the loss of Na^+ ions from Na^+ ion-rich normal (A) and mutant (B) yeasts during fermentation. Yeasts were fermented in 0.04 M sodium hydrogen succinate-buffered suspension at pH 4.5. The conditions and symbols are as given in Fig. 7.4. From C. Reilly, *Biochem. J.* **91**, 447 (1964).

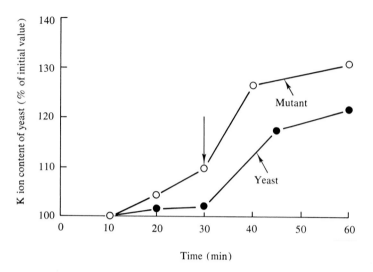

Time (min)

Fig. 7.6 Rates of K⁺ transport into normal and mutant yeast cells. The yeasts were suspended in 20 vol. of 0.1 M KCl and bubbled with nitrogen for 10 minutes. Oxygen-bubbling was substituted at 10 minutes, and at 30 minutes (arrow) glucose was added to give a final concentration of 5% (w/v). From C. Reilly, *Biochem. J.* 91, 447 (1964).

In 1953 a discovery was reported which started the probe for the molecular mechanism of active transport. Schatzmann found that cardiac glycosides specifically inhibited active cation transport in the red blood cell. Subsequently, it was shown that only those cardiac glycosides having a positive inotropic action on heart muscle were active inhibitors of the process (Fig. 7.7). The active glycosides consist of a cycloperhydropentanophenanthrene nucleus (which is similar to a steroid structure), a β-saturated lactone at C-17, and a special type of sugar or saccharide at C-3 in a glycosidic linkage. The first *in vitro* demonstration of an active transport system, enzymic in nature, was reported by Skou (1957), who isolated an ATP hydrolyzing enzyme from crab nerve membranes. This "ATPase" was markedly stimulated by Na⁺ plus K⁺ and cardiac glycosides reversed or prevented the stimulation.

a. PROPERTIES OF THE TRANSPORT SYSTEM. A number of characteristics of the active transport enzyme system have been delineated:

1. Active transport is specifically prevented by cardiac glycosides, i.e., digitalis drugs, at concentrations of from 10^{-8} M to 10^{-3} M. Cardiac glycosides in which the lactone ring is saturated or attached in an alpha configuration at C-17 are inactive, or much less active than other glycosides. The inhibition induced by the glycosides is prevented or reversed by increasing the potassium concentration of the system. Only those digi-

talis agents which produce a positive inotropic effect, i.e., increase force of contraction, on heart muscle act on this enzyme complex.

2. The enzyme system is spatially oriented in intact membrane systems such that internal sodium is "pumped out" when potassium is available on the outside to be "pumped in." Skou originally postulated (1960) the following sequence of reactions:

$$(1) \quad \text{enzyme} + \text{ATP} \xrightleftharpoons{\text{Mg}^{2+} + \text{Na}^{+}} \text{E} \cdot \text{ATP}$$

$$(2) \quad \text{E} \cdot \text{ATP} \rightleftharpoons \text{E} \sim \text{P} + \text{ATP}$$

$$(3) \quad \text{E} \sim \text{P} + \text{H}_2\text{O} \xrightarrow{\text{K}^{+}} \text{enzyme} + \text{P}_i$$

The sum total of (1) to (3) represents an "ATPase": $\text{ATP} \rightarrow \text{ADP} + \text{P}_i$. $\text{E} \sim \text{P}$ represents a phosphorylated intermediate. These reactions *suggest* an asymmetry of enzyme placement within the membrane such that one portion of the system is stimulated by sodium and the other portion stimulated by potassium.

The *erythrocyte membrane ghost* is an ideal system in this area of research since the "intra and extra membrane" environment can be easily altered and the consequences studied. The following are the important characteristics of this preparation:

a. The overall ATPase activity of ghosts consists of at least two measurable components: one requires the presence of magnesium ions and is insensitive to cardiac glycosides; the other requires the presence of magnesium, sodium, and potassium ions for maximal activity and is completely inhibited by active cardiac glycosides in concentrations that inhibit ion transport in intact red cells and other systems. Calcium ions inhibit the sodium and potassium stimulation in a manner similar to cardiac glycoside activity (Fig. 7.8).

b. The same molecular characteristics of cardiac glycosides are re-

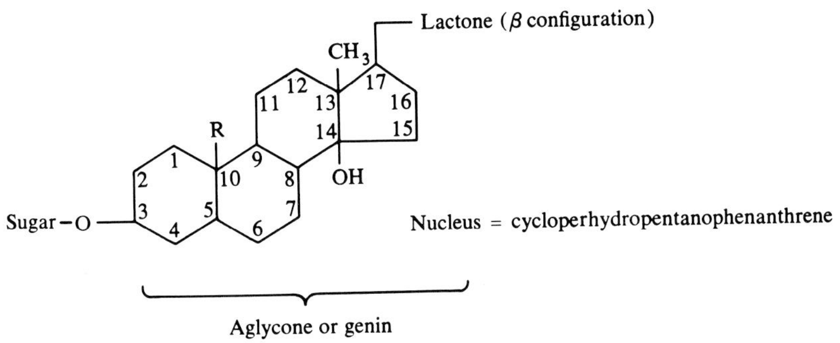

Fig. 7.7. General formula for active cardiac glycoside.

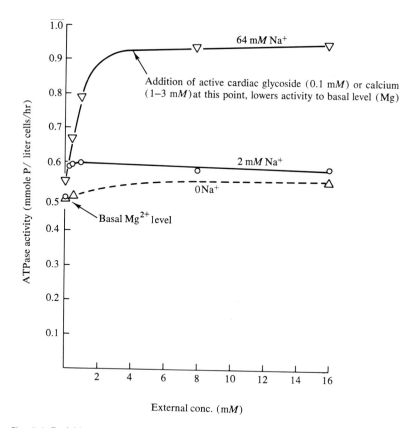

Fig. 7.8 Red blood cell ghost ATPase system. Temp. 37.1° C; pH 7.1; ATP 1.5 mM; Mg 1 mM; cysteine 1 mM; tris to make up 160 mM. Duration 1 hour. From E. T. Dunham and I. M. Glynn, *J. Physiol. (London)* **156**, 274 (1961).

quired for inhibition of the ATPase activity as for the inhibition of the ion pump in intact systems.

c. The ATPase inhibition by low concentrations of cardiac glycosides may be prevented or reversed by elevating the potassium concentration (Table 7.5). Interestingly enough, the potassium concentration must be raised on the external or outer side of the membrane. The K⁺-cardiac glycoside reaction, however, is more complicated than simple competitive inhibition kinetics would dictate (see p. 328).

d. Maximal stimulation of the ATPase enzyme system is accomplished by the addition of sodium ions *only on the inside of the membrane* plus the addition of potassium ions on the *external portion of the membrane*. The addition of sodium ions on the outside of the membrane system re-

TABLE 7.5

Effect of External K[+] on Cardiac Glycoside Inhibition of Erythrocyte Ghost ATPase[a,b]

K concn. (mM)	Total glycoside-sensitive activity (mmole P/ liter cells/hour)	Activity in presence of strophanthin (5 × 10[-8] gm/ml) (mmole P/liter cells/hour)	Inhibition (%)
0.25	0.188	0.027	86
0.5	0.364	0.134	63
1	0.619	0.290	53
2	0.778	0.498	36
4	0.855	0.686	20
8	0.926	0.782	15
16	1.02	0.823	19
32	1.04	0.964	7

[a]*Conditions of experiment:* Duration 1 hour; temp. 37°C; pH 7.2; ATP 1.5 mM; Mg 0.5 mM; Na 60 mM; cysteine 1 mM; tris to make up 160 mM. The tubes were placed in the water bath an hour before the addition of ATP. The total glycoside-sensitive activity was determined with 10^{-4} gm strophanthin/ml and at its highest accounted for 56% of the total ATPase activity.

[b]From E. T. Dunham and I. M. Glynn, *J. Physiol (London)* **156**, 274 (1961).

sults in an inhibition of activity (Fig. 7.9); addition of potassium ions on the inside of the membrane system also results in inhibition of activity.

e. The cardiac glycosides effect inhibition by a reaction with the *external portion* of the membrane.

b. THE Na[+]-K[+] ATPASE SYSTEM IN SUBCELLULAR COMPONENTS. The enzyme complex has been isolated from a variety of cells (see Appendix, this chapter). It is associated with endoplasmic reticulum and/ or plasma membranes in secretory cell types, and sarcoplasmic reticulum and/or plasma membranes and/or T-system in heart and skeletal muscle cells. The enzyme is also found in the smooth elements of reticula that are devoid of RNA but have esterase activity, cholesterol (when present in the cell), phosphoprotein, and phospholipid. The enzyme may be isolated by differential centrifugation techniques and may be partially purified by detergent and salt treatments and density gradient procedures, although the preparation from heart or skeletal muscle is complicated by the presence of contaminating ATPases from myofibrils, mitochondria, and other cytoplasmic constituents. A highly active, relatively uncontaminated transport ATPase can be isolated by a deoxycholate-sodium iodide, extractive procedure (Table 7.6A). This enzyme

TABLE 7.6A

Preparation and Purification of Na⁺, K⁺-ATPase[a]

Flow sheet diagram of the purification procedures for the Na⁺, K⁺-ATPase

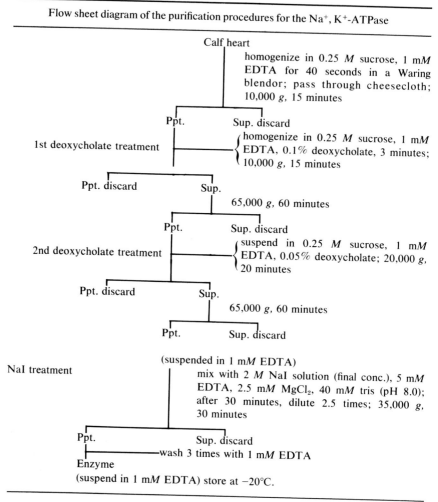

is markedly sensitive to Na^+ and K^+ ions and is almost completely inhibited by active cardiac glycosides (Table 7.6B). Typical ATPases from brain and parotid gland endoplasmic reticulum, i.e., microsomal fraction, are shown in Figs. 7.10, 7.11, and 7.12. Note that calcium inhibits the enzyme systems in a manner similar to the active cardiac glycosides and is of importance since the same type of reaction is observed in intact ion transport preparations. Furthermore, calcium appears to

TABLE 7.6A — *continued*

Purification data of Na$^+$, K$^+$-activated ATPase from calf heart[b]

Expt. Fraction	Specific activity (μmoles P$_i$/mg protein/hr)			Activity ratio	Percent recovery of	
	A + ouabain	B − ouabain	B − A	$\dfrac{B - A}{A}$	Protein	Activity
1 (a) Homogenate with deoxycholate	12.1	13.4	1.3	0.11	(100)	(100)
(b) 65,000 g ppt. after 1st deoxycholate treatment	9.5	19.6	10.1	1.06	3.3	24
(c) 65,000 g ppt. after 2nd deoxycholate treatment	13.0	24.0	11.0	0.85	1.7	14
(d) Enzyme after NaI treatment	1.7	24.9	23.2	13.7	0.47	7.8
2 (a) Homogenate with deoxycholate	10.2	12.0	1.8	0.18	(100)	(100)
(b) 65,000 g ppt. after 2nd deoxycholate treatment	20.2	30.5	10.3	0.50	1.2	6.6
(c) Enzyme after NaI treatment	1.1	28.1	27.0	24.6	0.26	4.0

[a] From H. Matsui and A. Schwartz, *Biochim. Biophys. Acta* 128, 380 (1966).
[b] The reaction system included 50 mM tris HCl (pH 7.4), 1 mM EDTA, 2 mM ATP, 5 mM MgCl$_2$, 100 mM NaCl and 20 mM KCl, in the presence or absence of 10^{-4} M ouabain, in a total volume of 1 ml. Incubation, 10 minutes at 37°C. Deoxycholate concentration, 0.1%, 1st treatment; 0.05%, 2nd treatment.

function like the cardiac glycosides on the intact heart, i.e., low concentrations produce an increased force of contraction; high concentrations result in toxicities and cardiac arrest. In addition, calcium produces a synergism with the cardiac glycosides, *in vivo* (an increase in activity not due to simple addition of effects) and on the membrane ATPase system. Consequently, it is of interest that many of the effects of an important drug and ion may be explained on an enzymic level.

Potassium ions, in concentrations beyond those which produce maximal stimulation of ATPase activity in the presence of physiological

TABLE 7.6B

A Highly Active Na$^+$–K$^+$ ATPase from Calf Heart[a,b]

	ATPase activity (μmoles P$_i$/mg protein/hour)	
Addition	With ouabain (10^{-4}M)	Without ouabain
Mg^{2+}	1.0	1.0
Mg^{2+} + Na$^+$	1.0	1.6
Mg^{2+} + K$^+$	0.9	0.9
Mg^{2+} + Na$^+$ + K$^+$	1.2	19
Na$^+$ + K$^+$	0	0

[a] The incubation medium consists of tris buffer, pH 7.4, 50 mM; EDTA, 1 mM and AT 2 mM. The ion concentrations are, MgCl$_2$, 5 mM; NaCl, 100 mM; KCl, 20 mM. The enzyme concentration was 60 μg protein/ml. Incubation was carried out at 37°C for 10 minute
[b] From H. Matsui and A. Schwartz, *Biochim. Biophys. Acta* 128, 380 (1966).

levels of Na$^+$, cause significant inhibition (Fig. 7.12B). This effect due to a competition with Na$^+$ (for a site which may lie on the inside the vesicle) and is analogous to similar actions observed on the red blood cell ghost ATPase (see p. 324). It is also consistent with effects on the intact "ion pump." It may be recalled that externally added K$^+$ to a ghost preparation appears to compete with cardiac glycoside for the same site, thereby reducing th inhibitory effect produced by the drug. The competition is also shown on a subcellular ATPase preparation (Fig 7.13A), but it is obvious that simple competitive kinetics do not describe the K$^+$-ouabain effects. Notice that both the slope and the maximum velocity of curve B are lower than A, suggesting that the drug and K$^+$ interact with different sites on the same Na$^+$,K$^+$-ATPase enzyme Detailed kinetic analysis (Fig. 7.13B) reveals that ouabain inhibition depends not only on K$^+$ but rather on a Na$^+$-K$^+$ ratio. In fact, the concentration of drug which is necessary to cause half-maximal inhibition may be expressed in the equation:

$$K_i = 3.5 \times 10^{-6} \times (Na^+ \div K^+)^{-0.82}$$

Information of this type along with recent binding experiments using tritiated digitalis preparations and rapid adenosine triphosphate-32 and inorganic phosphate-32 labeling studies, have led a number of investigators to conclude that the Na$^+$,K$^+$-ATPase system is a complex consisting of the following scheme:

$$\text{enzyme} + \text{ATP} \xrightarrow{\frac{\text{Na}^+}{k_1}} \text{``E} \sim \text{P''} \xrightarrow[\leftarrow]{\frac{\text{K}^+}{k_2}} \text{E} + \text{P}_i$$

$$\text{Digitalis} \updownarrow$$
$$\text{Dig} \cdot \text{E} \sim \text{P}$$
$$\downarrow$$
$$\text{Dig} \cdot \text{E}{-}\text{P}$$
$$(\text{stable})$$

Notice that sodium ions are required for phosphorylation of the enzyme protein while potassium ions stimulate the dephosphorylation (presumably two different rate constants, k_1 and k_2, operate). Digitalis inhibits activity by binding the phosphorylated enzyme producing a stabilization. The binding constant for digitalis is around 10^{-8} M which may be in the pharmacological range. Very recent evidence suggests that under certain circumstances, digitalis can actively bind to a nonphosphorylated form of the enzyme. This indicates the complexities involved in understanding intermediates in enzymes and implies the presence of numerous forms or types of intermediates (Schwartz et al., 1968).

If phosphorylation and dephosphorylation could occur sequentially at different sides of a membrane (inner and outer), the process of cation translocation might be explained by this enzyme system (see Fig. 7.22A). On the other hand, the processes of phosphorylation and dephosphorylation, catalyzed by specific cations, probably cause important conformational changes in the enzyme complex, which directly relate to cation transport. The various intermediate forms of the enzyme therefore [E $\sim$ P, E $-$ P, E $\cdot$ P, "E," etc.] may actually represent structural alterations which reflect in cation site orientation (e.g., inward directed or outward directed) and hence in cation affinity or "permeability." Agents such as digitalis, which interfere with transport may do so by stabilizing a particular enzyme configuration. Very small changes in protein structure, involving perhaps only a 2–3 Å displacement of a peptide chain, can result in marked functional alterations (Jardetzky, 1966; see Fig. 7.22B).

A number of important compounds exert site-specific inhibitory effects on the membrane ATPase system. These are summarized in Table 7.7 and in Fig. 7.13. The importance of these observations will become apparent in our discussion of the mechanism of action of this enzyme system (see p. 341).

Purification attempts, employing classical biochemical techniques (e.g., ammonium sulfate precipitation, detergent treatments, chromatography) have met with failure. This is due to the geometric requirement of the ATPase in the membrane or membrane fragment. Figures 7.14 and 7.15 clearly indicate the structural nature of the ATPase. Note

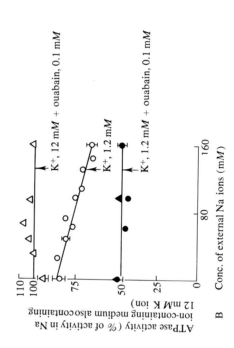

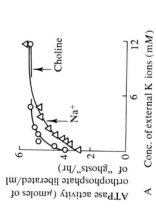

Fig. 7.9. Inhibitory effect of external Na⁺ of the K⁺-activation of RBC ghost ATPase. Stimulation of adenosinetriphosphatase activity by K⁺ ions in the presence and absence of external Na⁺ ions. The "ghosts" were suspended in medium containing K⁺ ions (0—11.5 mM) and ouabain (0.1 mM if present) and incubated for 1 hour at 37° C. From R. Whittam and M. E. ... Bio ... and M. E. ... ... 1.33 ...

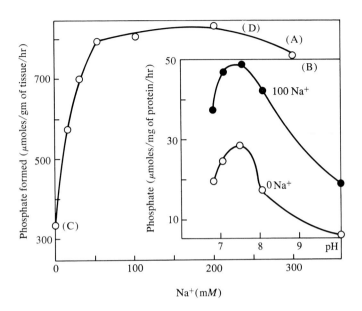

Fig. 7.10. Na⁺-K⁺ ATPase system in guinea pig brain. The assay medium consists of tris, 100 mM, pH 7.4; MgCl₂, 3 mM; ATP, 3 mM and KCl, 30 mM. Incubation was at 37°C for 15 minutes. A. ATPase activity dependency on Na⁺; B, pH optimum with or without Na⁺; C, basal level; D, addition of ouabain (10^{-4} M) at this point depresses activity to the basal level. From A. Schwartz, H. S. Bachelard, and H. McIlwain, *Biochem. J.* **84**, 626 (1962).

Primary Functions Obtained by Differential Centrifuging of Dispersions of
Cerebral Cortex[a]

Description of deposited material	Conditions of centrifuging			
	Material	g	Time (minutes)	Washing
N (nuclear)	Initial dispersion	600	10	1 (initial vol.)
Mt₁ (mitochondrial)	Supernatant and washings from N	10000	15	2 (0.5 initial vol.)
	Washings from Mt₁	10000	15	None
Ms (microsomal)[b]	Supernatants from Mt₁	20000	60	1 (initial vol.)
	Supernatants from Mt₂	20000	60	None
Mt₂	Ms	10000	15	None

[a]Tissue was dispersed in 0.32 M sucrose and centrifuged as in the figure. The deposits were thoroughly dispersed in a homogenizer with 0.32 M sucrose before centrifuging again. Final sediments were suspended in a small volume of 0.32 M sucrose.

[b]Almost all activity is found in this fraction.

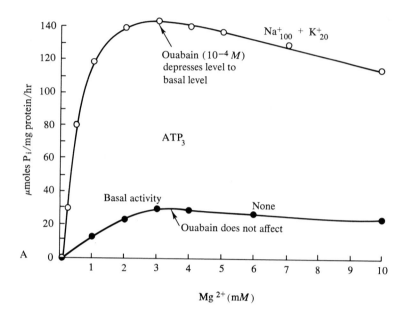

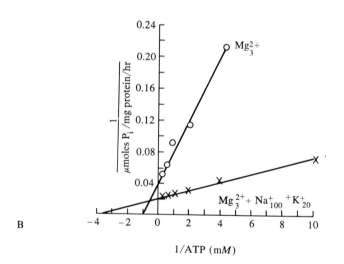

Fig. 7.11. A. Membrane ATPase of the parotid gland. B. A reciprocal (Lineweaver-Burk) plot of substrate and enzyme activity. The assay medium consists of tris, pH 6.8, 30 m*M* ATP, 3 m*M*, and the ion concentration (m*M*) as denoted by subscripts. Incubation was at 37°C for 30 minutes. From A. Schwartz, A. H. Laseter, and L. Kraintz, *J. Cellular Comp. Physiol.* **62**, 193 (1963).

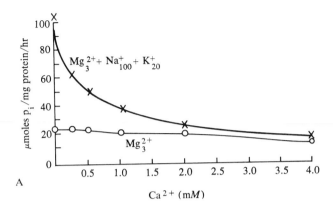

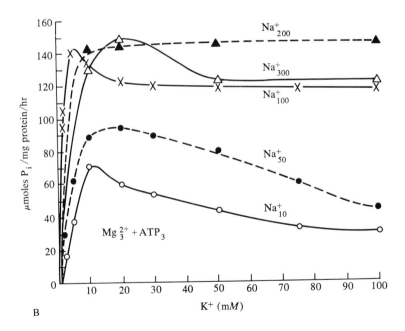

Fig. 7.12. A. Effect of Ca^{2+} concentration on parotid gland membrane ATPase. B. The effect of K^+ on parotid ATPase in the presence of varying concentrations of Na^+. The conditions were the same as indicated in Fig. 7.11. From A. Schwartz, A. H. Laseter, and L. Kraintz, *J. Cellular Comp. Physiol.* **62**, 193 (1963).

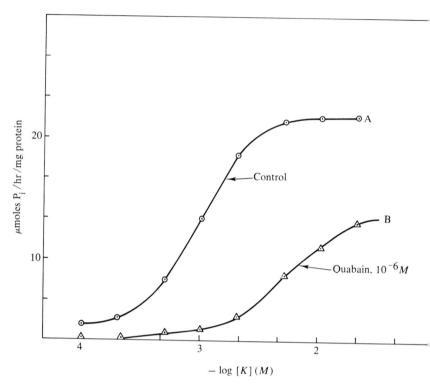

Fig. 7.13A. Nature of K⁺-cardiac glycoside "competition" in membrane ATPase of hea muscle. Enzyme isolated from calf heart; incubated at 37°C for 10 minutes in tris, 50 mℳ pH 7.4; MgCl₂, 5 mM; EDTA, 1 mM; NaCl, 100 mM; ATP, 2 mM. Enzyme protein, 6 μg/ml. From H. Matsui and A. Schwartz, unpublished observations (1967).

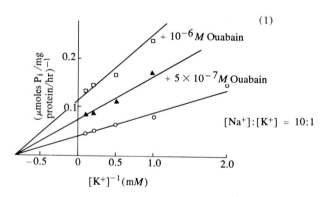

Fig. 7.13B. (1) See facing page for legend.

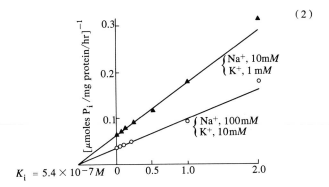

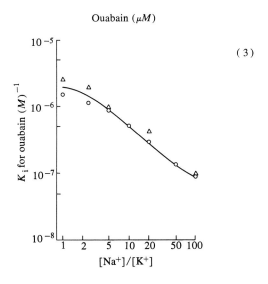

Fig. 7.13B. (1) Lineweaver-Burk analysis of ouabain inhibition at a constant Na^+/K^+ ratio. Na^+, K^+-ATPase activity was determined at various K^+ concentrations with a constant Na^+/K^+ ratio of 10 in the presence and absence of $5 \times 10^{-7}\ M$ and $10^{-6}\ M$ ouabain; $1/v$ was plotted against $1/K^+$ ($=1/Na^+ \times 10$). (2) Dixon plot of ouabain inhibition at a constant Na^+/K^+ ratio. (3) Relationship between K_i for ouabain and Na^+/K^+ ratio. K_i obtained was plotted against Na^+/K^+ ratios on a full logarithmic scale. $\bigcirc$, $\blacktriangle$ represent different enzyme preparations. From H. Matsui and A. Schwartz, *Biochem. Biophys. Res. Commun.* **25**, 147 (1966).

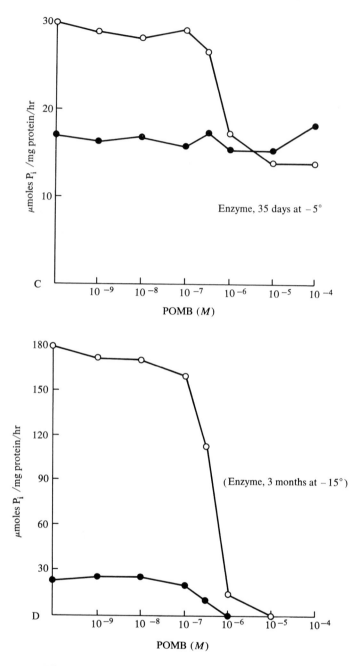

Fig. 7.13C, D. Effect of a sulfhydryl inhibitor [*p*-hydroxymercuribenzoic acid (POMB)] on membrane ATPases from heart and brain. C. The effect of various concentrations of POMB on cardiac microsomal ATPase. ○ : Mg^{2+} + Na^+ + K^+; ● Mg^{2+}. D. The effect of various concentrations of POMB on cerebral microsomal ATPase. ○ : Mg^{2+} + Na^+ + K^+; ● : Mg^{2+}. From A. Schwartz and A. H. Laseter, *Biochem. Pharmacol.* **13**, 337 (1964).

TABLE 7.7

Inhibitors of Membrane ATPase

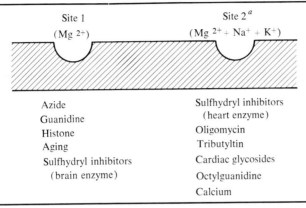

Site 1 (Mg $^{2+}$)	Site 2 ^{a} (Mg $^{2+}$ + Na$^+$ + K$^+$)
Azide	Sulfhydryl inhibitors
Guanidine	(heart enzyme)
Histone	Oligomycin
Aging	Tributyltin
Sulfhydryl inhibitors	Cardiac glycosides
(brain enzyme)	Octylguanidine
	Calcium

^{a} Site 2 presumably consists of a number of "subsites" characterized by sensitivity to some of the indicated reagents; there appears to be, e.g., a cardiac glycoside site which is distinct from the K$^+$ site and from a sulfhydryl-containing or oligomycin-sensitive site. It is emphasized that *independent* sites may not actually exist; rather the "sites" are expressions of different conformational forms of the same enzyme.

that both in conventionally prepared and negatively stained preparations (see Chapter I), the predominant features are enclosed saccules or vesicles. Any procedure which disrupts this configuration leads to a disorientation of the vectorially placed enzyme complex and, hence, to a loss of Na$^+$ + K$^+$-stimulating (and cardiac glycoside sensitivity) activity.

Thus, it is obvious that the ATPase enzyme system, located in membranes and membrane fragments, possesses the characteristics of the entire intact ion pump system, even to the inclusion of the system's vectoral character. The coupling of the chemical reaction to a particular directional flow of matter requires a specific, spatial arrangement of an enzyme system within the membrane. This ATPase enzyme system satisfies the requirements of a specifically localized pump enzyme. Whether or not the functional aspects of the system can be found in all tissues is still unanswered although presumptive evidence is strongly in favor of a membrane-bound ATPase as the basis for the movements of a wide variety of substances against their concentration gradients (see Appendix, this chapter).

The nature of the "E ~ P" intermediates are unknown although three possibilities have been suggested:

1. *The mobile type of phosphatidic acid cycle.* Alternate phosphorylation and dephosphorylation of phosphatidic acid was originally proposed as being responsible for the movements of sodium and potassium across

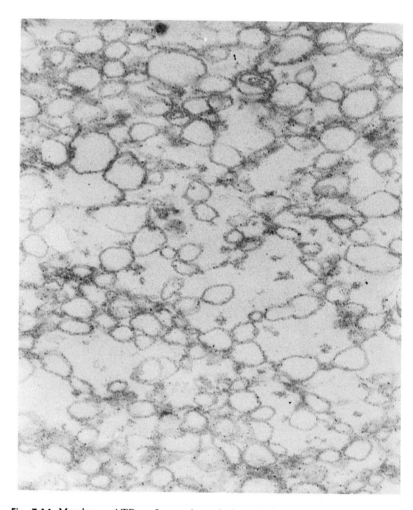

Fig. 7.14. Membrane ATPase from guinea pig heart (microsomal fraction); preparation was stored at 5°C for 1 month prior to assay. The suspension was centrifuged at 150,000 g for 2 hours; the resulting pellet was fixed in osmium-tetroxide, stained with KMnO₄ × 118,000. Note the enclosed vesicles representing fragmented sarcoplasmic reticulum. The black dots are probably glycogen. Courtesy of Dr. K. Smetana, Department of Pharmacology, Baylor University College of Medicine, Houston, Texas.

the membrane (Hokin and Hokin, 1965). The phosphorylation and dephosphorylation would occur at separate portions of the membrane according to this theory (Fig. 7.16). Unfortunately, the rates of phosphatidic acid turnover in various tissues are too slow to account for the rapid movements of ions and for the very active ATPase activities

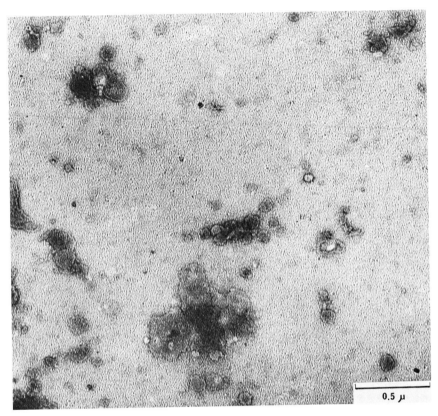

Fig. 7.15. Membrane ATPase from rat heart (microsomal fraction); preparation was stored at 5°C for 1 month prior to assay. The suspension was plated directly onto an aluminum grid and subjected to a negative staining procedure. Magnification as indicated. Courtesy of Dr. K. Smith, Department of Virology, Baylor University College of Medicine, Houston, Texas.

that have been found in these tissues. Accordingly, this hypothesis has been modified somewhat as shown in Fig. 7.17; it is suggested that it still may play some role in membrane function as part of the intermediate carrier system. Stimulation of the phosphatidyl inositol synthesis presumably occurs in the cytoplasm of the cell and hence may function in transporting protein (actively) in excitable tissues.

2. *The phosphoprotein hypothesis.* Interest in phosphoproteins developed from the work of Johnson and Albert (1952; Albert *et al.*, 1951), who demonstrated that trichloroacetic acid-insoluble proteins of most animal tissues rapidly incorporated $^{32}P_i$ (*in vivo*). Subsequently,

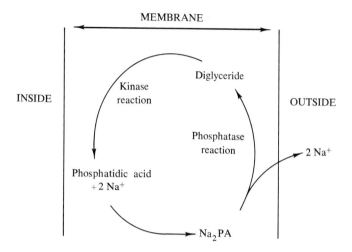

Fig. 7.16. Hokins' original theory. PA = phosphatidic acid. Taken from L. E. Hokin and M. R. Hokin, *Membrane Transport Metab., Proc. Symp. Prague, 1960* p. 205. Academic Press, New York.

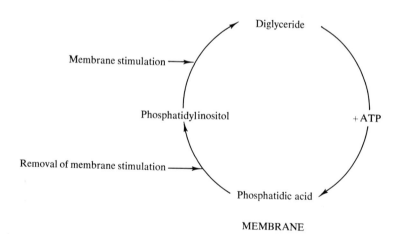

Fig. 7.17. Hokins' revised theory. From "The Chemistry of Cell Membranes" by L. E. Hokin and M. R. Hokin, Copyright © (1965) by Scientific American, Inc. All rights reserved.

it was shown that the $^{32}P_i$ originated from high energy compounds, chiefly ATP.

Electrical stimulation of isolated cerebral slices causes an increased ion flux and a significant, concomitant stimulation of $^{32}P_i$ incorporation into a "phosphoprotein" fraction (Heald, 1960). The phosphate is incorporated chiefly into phosphoserine, and the active particulate fraction appears to be associated with membrane elements (Trevor *et al.*, 1965). Therefore, phosphoproteins may function as intermediates in active cation transport as well as in the "pump" ATPase, as follows:

(1) $\text{ATP} + \text{protein} \xrightleftharpoons{\text{Na}^+ + \text{Mg}^{2+}} \text{ADP} + \text{P} \sim \text{protein}$

(2) $\text{P} \sim \text{protein} \xrightarrow[\text{Mg}^{2+}]{\text{K}^+} \text{protein} + \text{P}_i$

Reactions (1) and (2) would constitute the $\text{Na}^+ + \text{K}^+$-stimulated ATPase observed in tissues. It is vital that a proposed intermediate react at a rate consistent with the overall reaction, viz., that phosphorylation and dephosphorylation of the protein proceed at a rate equal to or faster than the cation-stimulated ATPase activity. Conclusive proof is not yet available, although recent evidence indicates that phosphorylation of serine, e.g., is too slow to account for the overall ATPase activity.

3. *Acyl phosphate.* The "E ~ P intermediates" might be represented by some type of acylated phosphate compound (Nagano *et al.*, 1965). Possible candidates are carbamyl phosphate (Izumi *et al.*, 1966) or an L-glutamyl-γ-phosphate peptide (Kahlenberg *et al.*, 1967). The reader is referred to several recent reviews discussing this important aspect (see General References, this chapter).

Studies on the molecular mechanism for the "pump"-ATPase have defined a specific stoichiometric geometry between Na and K transport and a high energy phosphate:

$$3\,\text{Na}^+ : 2\,\text{K}^+ : 1 \sim \text{P}$$

The inner surface of the membrane is coupled to the transport of 3 Na^+ outward and 2 K^+ inward at the expense of one energy rich phosphate bond. Presumably one H^+ is transported with each K^+ to maintain electrical neutrality (Fig. 7.18).

Regardless of the nature of the intermediary substances a molecular mechanism is necessary to explain how the hydrolysis of ATP is coupled to or how it effects the selective movements of cations. There are two possible explanations, one involving a fixed-site carrier and the other, a mobile carrier.

c. FIXED SITE. 1. *Direct ATPase participation.* The ATPase system is regarded as a complex consisting of the formation and breakdown of ATP, the two occurring at spatially separate sites; hydrolysis alters the charge on ATP and, therefore, the affinity of ATP for cations. The ATPase presumably has as its substrate: Mg·Na·K-ATP.

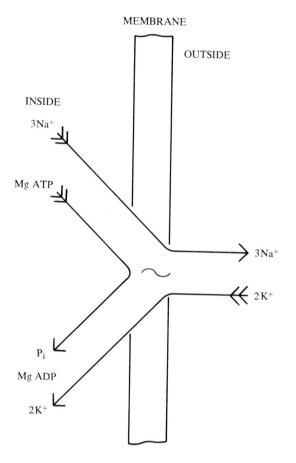

Fig. 7.18. Stoichiometry and geometry of the membrane ATPase system. From A. K. Sen and R. L. Post, *J. Biol. Chem.* **239**, 345 (1964).

The partial reactions occur as follows:

(1) $Mg^{2+} + K_2^+ + ATP + ATPase \longrightarrow Mg \cdot K_2 \cdot ATP\text{-}ATPase$ complex

(2) $Mg \cdot K_2 \cdot ATP\text{-}ATPase + Na^+ \longrightarrow Mg \cdot K \cdot Na \cdot ATP\text{-}ATPase + K^+$

(3) $Mg \cdot K \cdot Na \cdot ATP\text{-}ATPase \longrightarrow Mg \cdot K \cdot ADP\text{-}ATPase + P_i + Na^+$

Reaction (3) involves a separation of Na^+ from negatively charged products. Thus, an alternate attachment of Na^+ to acidic sites, e.g., gangliosides, and to ATP, spatially apart, would result in the transfer of the cation from the inside to the outside (Fig. 7.19). It is known that ADP-Na is much less stable than ATP-Na. Therefore, the hydrolysis of ATP results in a complex which readily yields the Na ions to negative sites. One might envision a type of "competition" control between the acidic sites of the fixed gangliosidic molecules and the acidic ATP-ADP system. The enzyme itself (ATPase) has a basic group which when freed from substrate and products after the hydrolysis of ATP, may recombine with an acidic group at a pore site, displacing Na$^+$.

Another possible ion control system might involve the interaction of polycationic compounds, i.e., histones and gangliosides, altering the cation-ganglioside-ATP complexes. Histones markedly affect ion transport across membranes associated with the endoplasmic and sarcoplasmic reticulum and with mitochondria (Schwartz, 1965a, Johnson *et*

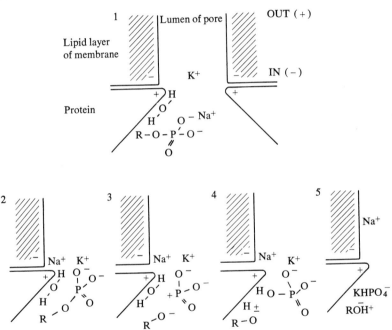

Fig. 7.19. Fixed site ATPase concept (McIlwain). Adenosinetriphosphatase-pore mechanism for active cation transport. ROH = adenosine diphosphate. 1. Acidic group at rim of pore (ganglioside?), forms part of the enzyme complex. In resting state, it is neutralized by a basic group at the active center of ATPase. Affinity for Na$^+$ is low. 2. Na$^+$ enters (depolarization) and forms the Na·K·ATP. Na$^+$ occupies the acidic group at base of pore, for which it has greater affinity. The phosphate anion becomes associated with the basic group of the ATPase. The Na-affinity of ATP and enzyme is high. 3. Hydrolysis of substrate occurs. 4. Negatively charged groups increase in number, repelling one another. 5. The basic group of enzyme is freed from substrate and products. It may recombine with the acidic group in the pore and displaces Na$^+$ toward the exterior. The enzyme may then return to stage 1. From H. McIlwain, "Chemical Exploration of the Brain." Elsevier, Amsterdam, 1963.

al., 1967). Polyanionic electrolytes are effective inhibitors of the histone induced changes.

2. *Electrostatic field strength-pore concept.* The membrane consists of pores along which are fixed, charged groups. The distribution of electrons changes with the hydrolysis of ATP. In Fig. 7.20 the ATP is attached to the inside of the membrane via an electronic bridge through Mg^{2+} and the NH_2 of adenine by hydrogen bonding. The electrons tend to move from the "a" site in the pore through the NH_2, adenine, P-O-P and Mg^{2+}, to site "b." The P-O-P segment of the ATP is an "electron

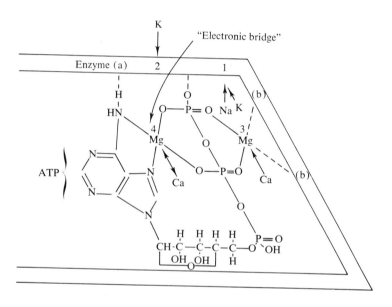

Fig. 7.20. Relationship between ATP, ions, and membrane ATPase. From J. C. Skou, *Biochim. Biophys. Acta* **42,** 6 (1960).

acceptor" while the adenine moiety serves as a "donor." The relative movement of electrons from the membrane pore sites effects changes in electrostatic field strength. It may be recalled that at low electrostatic field strength (EFS), the affinity for K^+ is much greater than for Na^+ (see Chapter II). Thus, alterations in EFS result in selective "replacements" of K ions for Na ions at specific sites. Hydrolysis of the ATP releases an electron which is transferred to the lowest empty orbital of a molecule at the site nearest the ATP, increasing the electrostatic field strength and resulting in a selective increase in affinity for Na^+ over that of K^+ (Fig. 7.21). Thus, sodium moves, stepwise, from the intracellular space toward the exterior of the membrane, and K^+ would move from the outside toward the inside.

ATPase activity can be a mechanism for the alteration of field strengths at specific membrane-located sites. The nature of the molecules donating negative sites is not explained by this concept although several candidates might include the gangliosides, phosphoproteins, or phospholipids. In partial support for this hypothesis, the pump ATPase system is sensitive to agents which affect electron transport (Table 7.7). This concept was originally proposed by Skou (1960, 1964, 1965).

3. *Conformation changes.* Phosphorylation and dephosphorylation of intermediates (or fixed sites) might result in significant structural alterations in the enzyme, which increase or decrease affinity of specific groups for cations (see Fig. 7.22B and Jardetzky, 1966 and p. 329).

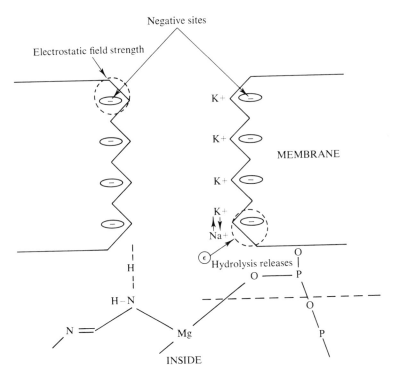

Fig. **7.21**. Fixed-site concept adapted from J. C. Skou, *Symp. 1st Intern. Pharmacol. Meeting, 1961*. Pergamon Press, Oxford, 1963.

d. MOBILE SITES. The difference between the mobile sites hypothesis and those previously described is essentially one of mobility. The carrier, instead of remaining fixed to a site, or actually representing *the* "site," is "free" to move or "shuttle" to and fro with the membrane, alternately transporting substances from outside to inside and vice versa. The phosphatidic acid and phosphoprotein "cycles" described above would define such a mechanism, although it is not known how these molecules alter their affinities for the particular cation requiring transport. Thus, as depicted in Fig. 7.22A, a carrier would first be phosphorylated, presumably at the inner surface, by a kinase reaction, at which point the affinity for Na$^+$ is maximum. Dephosphorylation of the complex occurs at the outer surface, where the Na ions are released and affinity for K$^+$ is maximum.

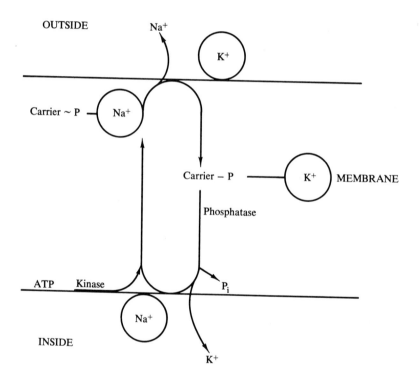

Fig. 7.22A. Mobile site concept.

The discovery and characterization of the membrane-bound, cardiac, glycoside-sensitive, ATPase enzyme system has facilitated detailed probes into the molecular mechanism for a basic and prime function of

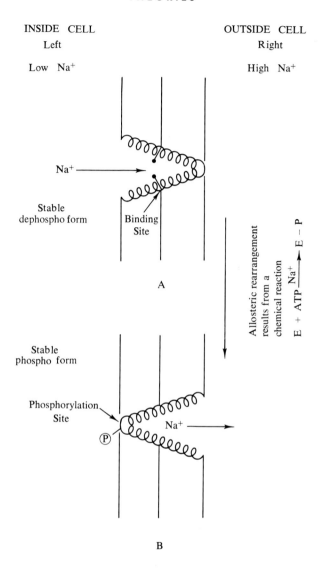

INSIDE CELL   OUTSIDE CELL
   Left   Right

   Low Na$^+$ High Na$^+$

Na$^+$ ———→

Stable
dephospho form   Binding
                          Site

                            A

                                        Allosteric rearrangement
                                        results from a
                                        chemical reaction

                                        $E + ATP \xrightarrow{Na^+} E - P$

Stable
phospho form

Phosphorylation
    Site   (P)   Na$^+$ ———→

                            B

Fig. 7.22B. Allosteric pump concept.

the membrane, active transport. Current investigators are hopeful of revealing the nature of the intermediate(s) involved in this process, realizing a fruitful beginning to an understanding of membrane transport function.

4. REDUCTIVE TRANSPHOSPHORYLATION

This concept is designed to move large numbers of ions efficiently, either by means of the energy liberated from the hydrolysis of small amounts of ATP or by some high energy intermediate generated during electron transport. Thus, the mechanism embodies principles of both the redox and ATPase ideas (see Chapter II).

The living cell uses a series of coupled chemical reactions to cause secretion of a large number of ions against a low electrochemical gradient. The mechanism, exhibited in Fig. 7.23, produces equal numbers of separated hydrogen and hydroxyl ions. The carrier X, representing another anion, possibly phosphatidic acid or phosphoprotein complex, is fixed to a rotating structure which is able to exchange sodium and hydrogen ions. Specificity is achieved by the difference in pore affinities that connect the rotating structure with the inside and outside of the membrane surfaces. The scheme may be driven by energy which is

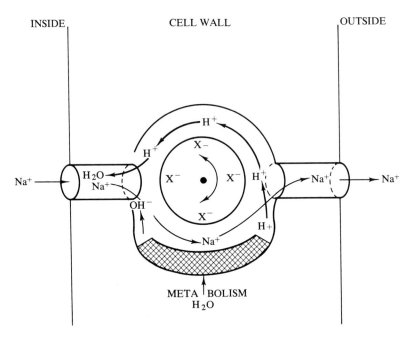

Fig. 7.23. Transport linked to proton movement: "a redox pump." Outline of a "pure" sodium pump driven by metabolically produced hydrogen and hydroxyl ions. The carrier, X^-, is an anion fixed to the rotating structure and can exchange sodium and hydrogen ions. From R. E. Davies, *Membrane Transport Metab., Proc. Symp. Prague, 1960*. Academic Press, New York 1961.

derived from the hydrolysis of ATP in a reductive transphosphorylation reaction (see Chapter III). A ferrous-ferric system, perhaps analogous to the iron-cytochrome system in the mitochondria, may be involved also. This system may be part of cytochrome b_5, known to occur in various subcellular membranes, particularly the endoplasmic reticulum (see Chapter V) and also in the plasma cell membrane itself. Hydrogen and hydroxyl ions are formed from the mutual serial oxidation and reduction of the ferrous-ferric system and the connecting carriers. An interesting candidate for the carrier role in this type of system is the quinone, coenzyme Q, which is thought to play a role in electron transport (see Chapter III). The only requirement is that there be some type of linked oxidation-reduction of the quinone and the iron. An effective source for the hydrogen ions necessary for this scheme is water. During the process, formed hydroxyl ions would balance the hydrogen ions made during the subsequent oxidation of the quinol by the cytochromes, vicinal to the quinol in the membrane. The overall process would work for a series of cytochrome b_5 molecules whose mutual oxidation-reductions are mediated by a series of quinol-like molecules, placed structurally in the labile protein membrane (Fig. 7.24). The hydrolysis of one mole of ATP could

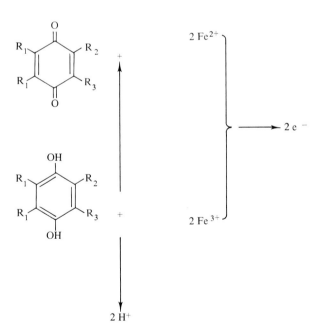

Fig. 7.24. Ubiquinone as a "H^+ carrier." From R. E. Davies, *Membrane Transport Metab., Proc. Symp. Prague, 1960,* p. 327. Academic Press, New York, 1961.

produce a gradient for one gram-ion of more than a million to one. Therefore, it is possible to account for the secretion of a very large number of ions for each oxygen molecule utilized to drive active transport. Coenzyme Q could function conceivably as the link between active transport and metabolism.

The energy derived from either the hydrolysis of ATP or the formation of a high energy intermediate via electron transport serves to form some type of intermediary complex which is intimately involved in the movement of ions across all membranes against concentration gradients.

5. CHEMI-OSMOSIS

In this hypothesis, emphasis is placed on the importance of cellular metabolic energy as well as the cell membrane and carriers. Metabolic energy is converted to osmotic work by the formation and opening of covalent links between "translocators" in the membrane and the carried molecules, exactly as in enzyme-catalyzed group-transfer reactions. Spatial and specific arrangement of the enzymes, coenzymes, and substrates may be of importance in transport processes. The translocation itself may be due to a thermal movement of the translocator which is catalyzed by a normal group-transferring enzyme called a "translocase"

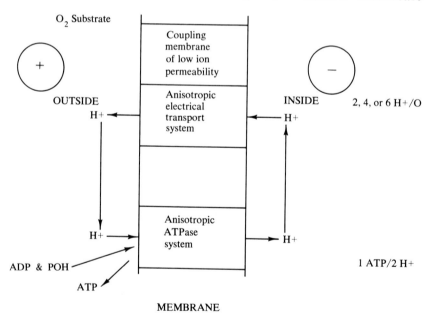

Fig. 7.25. The chemiosmotic coupling hypothesis of Mitchell (1961).

or a "carrier protein." Active transport of substances by this coupled system is caused by the movement of a substrate in the reverse direction to the solute being transported. According to Mitchell (1961; Mitchell and Moyle, 1965), the plasma membrane is regarded not simply as an osmotic barrier but as a chemical link allowing the exchange of one co-valent group for another, similar to the well-known group-transfer reactions. The covalent "translocator" intermediates located in the plasma membrane arise either from constituents of the external medium or from the protoplasm within the cell. The mitochondrial membranes, a type of membrane model system, have been used to study the chemi-osmotic theory. Extrusion of hydrogen ions by mitochondria has suggested that membranes, in general, possess a metabolically driven pump for the outward flux of H^+. The movements of other cations (sodium, potassium, calcium, strontium) would, therefore, occur secondarily to the proton translocation. There is also a metabolically driven pump for the uptake of hydrogen ions into the cell (Fig. 7.25). According to this hypothesis, phosphorylation is linked with the hydrogen pump, so that there are two protons for each ATP hydrolyzed. Therefore, the P/H quotient is $\frac{1}{2}$. Furthermore, as may be seen from the figure, this quotient refers to the movement of the protons from the *outside to the inside*. Movements of protons from the inside to the outside require an electron transport system.

This concept, while interesting, is still highly speculative and is not necessarily supported by existing information.

6. THE MECHANO-ENZYME CONCEPT

The mechano-enzyme hypothesis was originally suggested by the work of Danielli (1952, 1954) and Goldacre (1952), who referred to the mechanism as a "propelled carrier concept." The membrane presumably contains a contractile protein which can literally drag the carrier molecule through the membrane. Thus, penetration of substances which would normally diffuse too slowly for the needs of living cells might occur and would result in net transport. With the exception of the mitochondria (Chapter III), evidence for this concept as a general membrane phenomenon is not available, although there has been some suggestion that membranes of the erythrocyte and chloroplast may contain contractile proteins.

IV. AMINO ACID AND OTHER NONCATION TRANSPORT

It has been known for many years that amino acids can be transported, in various tissues, against concentration gradients (Noall et al., 1957) and that this transport may be under endocrine control. Also, it has been known that polysaccharides, pyrimidines, ethanol, phosphate, and bile acids are transported actively. The participation of sodium ions in these processes has been demonstrated, and it has been suggested that this ion plays a fundamental role in membrane transport processes of these substances. Ouabain, the cardiac glycoside which specifically inhibits active cation transport, also inhibits the transport of amino acids, glucose, and other substances. However, it should be emphasized that while the processes involved in the accumulation of noncation substances may have some similarities to electrolyte transport, important *qualitative* differences undoubtedly exist (Kennedy, 1967).

TABLE 7.8

Effect of Various Agents on Membrane Transport

Drug or agent	Summary of possible effect	References[a]
1. Cardiac glycosides	Inhibition of Na^+, K^+–ATPase system	1
2. Catecholamines	Inhibition; interaction with relaxing system	2
3. Quinidine	Inhibition; inhibits "Na^+–carrier"	3
4. Parathyroid hormone	Stimulation of Ca^{2+}, Mg^{2+}, K^+(?) pump in mitochondria	4
5. Aldosterone	Stimulation of RNA-polymerase to synthesize active transport protein	5
6. Histones and gangliosides	Control of active transport at membrane level	6
7. Neurohypophyseal hormone	Increase of membrane "pore size"	7
8. Acetylcholine	Inhibition of Na^+, K^+–ATPase system	8
9. Chlorpromazine	Mitochondrial membrane stabilizer; Inhibits QO_2 of electrically stimulated cerebral tissues	9
10. Thyrotropic-stimulating hormone (TSH)	Stimulation of Na^+, K^+–ATPase system	10

[a] *Key to references:*
1. Skou (1964, 1965).
2. Lee and Yu (1963); Stam and Honig (1962); Vales *et al.* (1964); Ussing (1952).
3. Kennedy and Nayler (1965).
4. DeLuca *et al.* (1962); Rasmussen (1965); Rasmussen and Ogata (1967).
5. Edelman *et al.* (1963).
6. McIlwain (1963); Schwartz *et al.* (1962); Schwartz (1965a,b); Johnson *et al.* (1967).
7. Ussing (1952).
8. Lee and Yu (1963).
9. McIlwain (1963); Spirtes and Guth (1963).
10. Turkington (1963); Wolff (1964).

TABLE 7.9

Effect of Neurohypophyseal Extract on Electromotive Force, E_{Na}, and Internal Resistance, R_{Na}, of Sodium Transporting Mechanism of Frog Skin[a]

Sample	E_{Na}(mV)	$R_{Na}(\Omega\ cm^{-2})$
Control	79.2	2300
Neurohypophyseal extract	75.9	1666
Control	98.2	4590
Neurohypophyseal extract	94.6	2640

[a] From H. H. Ussing, *Advan. Enzymol.* **13**, 21 (1952).

V. THE EFFECTS OF DRUGS AND HORMONES ON TRANSPORT

Ever since the early 1900's, when Overton postulated a mechanism of action for the anesthetic gases involving membranes, there has been a developing interest in the relationship between drug action and membrane permeability.

Otto Loewi once stated to Homer Smith, "Homer, it's all in the plasma membrane" (Smith, 1962), indicating a long-standing "pharmacological suspicion" that the activity of drugs depends upon membrane interactions. Table 7.8 sets forth a summary of some of the drugs and hormones that have been found to exert a relatively specific effect upon the primary function of membrane transport.

Ussing was one of the first to suggest that a hormone could act on membrane-linked transport. He found that neurohypophyseal extract lowered the resistance of the membrane to sodium ions, to about half its original value. This peculiar effect could best be explained as "an increase in pore size" (Table 7.9). He also found that epinephrine in low concentrations increased the short circuit current of frog skin (Table 7.10) and, in addition, significantly increased both influx and outflux of sodium ions. He suggested that epinephrine stimulation initiated a new source of electric current which appeared to be dormant in the nonstimulated skin. This effect arose mainly from an active transport of chloride ion outward, a process which was probably connected with a function of the skin glands. His model served to show the possible importance of the membrane in hormonal actions.

Iodide transport across the membrane of the thyroid gland depends directly upon the ability of the membrane to transport sodium and potassium ions and suggests the presence of an iodide carrier system. Thyroid-

TABLE 7.10

Influence of a Number of Agents on Sodium Flux and Total Current
Values as Obtained on Totally Short-Circuited Frog Skins[a,b]

Group A			Group B		
	Millicoulombs cm^{-2} $hour^{-1}$			Millicoulombs cm^{-2} $hour^{-1}$	
Expt. No.	Na	Short circuit current	Expt. No.	Na	Short circuit current
1. Control	99	86	1. Control	10.9	176
Cu²⁺ outside	91	87	Cu²⁺ outside	8.6	165
Cu²⁺ outside	99	89	Cu²⁺ outside	10.5	139
2. Control	57	49	2. Control	13.6	112
Epinephrine	87	76	Epinephrine	41.0	126
Epinephrine	111	88			
3. Control	47	92	3. Control	1.2	153
Epinephrine	129	140	Epinephrine	67.0	174
Epinephrine	115	101	Epinephrine	58.0	126
4. Control	105	100	4. Control	5.6	124
Neurohypophyseal extract 168		150	Neurohypophyseal extract	8.5	164
				9.5	164
5. Control	126	118	5. Control	1.6	77
Neurohypophyseal extract 246		232	Neurohypophyseal extract	3.1	129
				5.4	164
6. 5% CO_2 + 95% O_2	4.5	0	6. 5% CO_2 + 95% O_2	5.5	0
5% CO_2 + 95% O_2	3.8	0	5% CO_2 + 95% O_2	6.1	0
Atmospheric air	165	150	Atmospheric air	8.3	161
Atmospheric air	173	136	Atmospheric air	15.3	158

[a] Group A comprises results from influx experiments; group B results from outflux experiments. Each experiment is represented by a control period and one or two periods following the application of the agent in question. Duration of periods generally 1 hour. In experiments A3 and B3 the first period after application of epinephrine lasted only 30 minutes. These periods may not represent steady states.

The following dosages were used: $CuSO_4$ (aq.), 0.2 mg added to outside solution. Epinephrine, 50 µl 1% epinephrine hydrochloride, added to inside solution. Neurohypophyseal extract, the equivalent of 1 mg dry gland, added to inside solution.

[b] From H. H. Ussing. *Advan. Enzymol.* 13, 21 (1952).

stimulating hormone (TSH; thyrotropin) stimulates a Na^+, K^+-ATPase located in the membranes of the thyroid gland. This stimulation occurs only in the presence of sodium and potassium ions; high concentrations of potassium ions reduce this effect. Furthermore, ouabain inhibits the TSH stimulation of ATPase activity. Interestingly enough, the same concentration of hormone also stimulates oxygen consumption in thyroid slices *in vitro* and this, too, depends upon the presence of sodium and potassium ions and is inhibited by ouabain. These observations imply that an important effect of TSH resides in a membrane action on an enzyme system responsible for the movements of potassium and sodium ions. Thus, the "sodium pump" ATPase appears to be directly related to iodide transport in the thyroid gland and is significantly responsive to a hormone which affects transport.

It has been known for some years that aldosterone is a hormone which is involved in sodium transport in the kidney. It causes a very significant retention of sodium ions and an increase in circulating blood volume. Aldosterone has been implicated in hypertensive disease as well as in the physiological control of blood volume (Davis, 1963). Numerous *in vitro* experiments have shown that the hormone increases sodium transport, but only after a latent period of from 1 to $1\frac{1}{2}$ hours. Direct effects of this hormone on the membrane ATPase located in membranes have not been observed. Edelman *et al.* (1963) has now demonstrated, however, that aldosterone is preferentially localized in the nucleus of epithelial cells of the urinary bladder of the toad, a model membrane transport system. The hormone acts by promoting a specific DNA-dependent RNA synthesis which yields an increased rate of protein synthesis. This was shown by the use of actinomycin D and puromycin in a transport experiment depicted in Figs. 7.26 and 7.27. The antibiotic puromycin and actinomycin, inhibitors of protein synthesis and nuclear synthesis of RNA, respectively, both effectively block the aldosterone effects. Interestingly enough, similar results were found in cortisone-induced increases in liver tryptophan pyrrolase, tyrosine, transaminase, and glycogen deposition. The newly synthesized proteins may be enzymes involved in sodium transport, particularly the ouabain-sensitive ATPase. Here is the first indication that hormones can act on sodium transport via an indirect mechanism involved in protein synthesis of enzymes responsible for active transport.

Cardiac glycosides significantly increase the force of contraction of heart muscle. This is the basis for their therapeutic use in congestive heart failure. In fact, since the discovery of these glycoside principles in the middle of the eighteenth century, no new drug has been introduced for the specific treatment of this disease. Numerous attempts have been made to localize the mechanism of action of the glycoside drugs on either energy-liberating systems or energy-utilizing systems. The only specific effect, in very low concentrations, of these drugs has been, in fact, found on a membrane function, namely, transport. Cardiac glycosides specifically inhibit active cation transport. In addition, these drugs have been found to specifically inhibit the $Na^+ + K^+$-dependent ATPase system, suggesting a possible molecular explanation for mechanism.

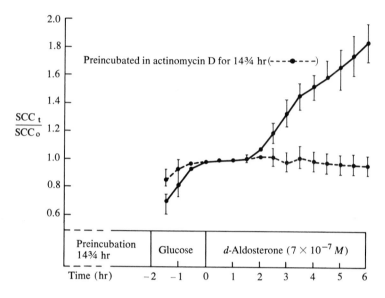

Fig. 7.26. Effect of actinomycin D on the action of aldosterone. The ratio SCC_t/SCC_o is defined in the legend of Fig. 7.27. The hemibladders of the toad pretreated with actinomycin D ($--\bullet--$) for $14\frac{3}{4}$ hours show no obvious response to aldosterone. The controls ($-\bullet-$) show the usual response. From E. S. Edelman, R. Bogoroch, and G. A. Porter, *Proc. Natl. Acad. Sci. U.S.* **50**, 1169 (1963).

Another drug which specifically affects cardiac function is quinidine. This drug is used to abort or prevent certain cardiac arrhythmias, particularly auricular fibrillation. It has been suggested that the mechanism of action might also involve an effect on a "sodium carrier" or perhaps the "sodium pump system."

Parathyroid hormone has been shown to specifically affect calcium metabolism in the organism. The classical concept of parathyroid hormone function may be stated as follows: a decrease in plasma calcium causes a release of hormone which elevates blood calcium by a mobilization of skeletal calcium. However, in recent years, the mechanism of action of this hormone has been ascribed to a membrane effect. It has been shown that in very low concentrations the hormone causes an increase in oxygen consumption of the mitochondria, a swelling of the mitochondrial membranes (Fig. 7.28) and an increase in "permeability" of the membrane to both potassium and magnesium ions. The effects of parathyroid hormone mimic the effects of small amounts of calcium ion on the mitochondrial membranes. Accordingly, this hormone might effect an increase in permeability to calcium ions through an action on a "potassium pump."

Attempts have been made to show that the catecholamines, epinephrine and norepinephrine, as well as the parasympathetic hormone, acetylcholine, both causing specific changes in heart muscle (see Chapter IX), exert their actions through specific effects on the membrane ATPase system of that tissue.

While these investigations are still in the embryonic stage, they do indicate that current interest in drug and hormone action is being directed toward a consideration of membrane transport.

VI. SOME CONCLUDING REMARKS ABOUT THE TRANSPORT FUNCTION OF MEMBRANES

It is hoped that this chapter has given the reader an insight into the complex, multiphasic nature of transport processes associated with membranes. The link between active transport and cellular metabolism is of such importance that both might be mutually regulatory processes in overall cellular activity. It is evident that a wide variety of both physiological and pharmacological agents produce specific changes in membrane

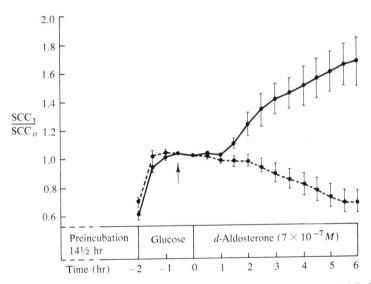

Fig. 7.27. Effect of puromycin on action of aldosterone on Na^+ transport. SCC_t denotes short circuit current recorded at time "t" after addition of aldosterone; SCC_o, short circuit current recorded at time of addition of aldosterone (time zero). Arrow indicates time of addition of puromycin (80 $\mu g/ml$) to media of one of each pair of toad hemibladders (— —●— —). The control group (— ●—) shows the usual response to aldosterone. Vertical bars represent ±1 SEM (standard error of the mean). From E. S. Edelman, R. Bogoroch, and G. A. Porter, *Proc. Natl. Acad. Sci. U. S.* **50**, 1169 (1963).

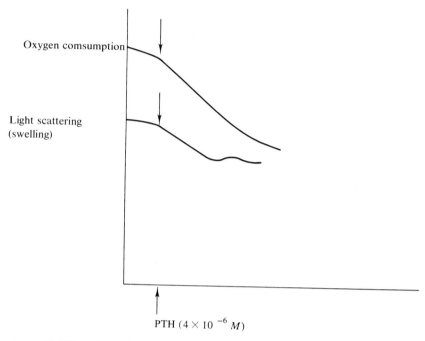

Fig. 7.28. Effect of parathyroid hormone on mitochondria using a polarographic recording system. From H. Rasmussen, *Symp. 5th Ann. Meeting Am. Soc. Cell Biol., Philadelphia, 1965.*

transport, which may represent the basis for mechanism of action. Poly-electrolytic compounds present within membranes may function as receptor sites for various endogenous substances. One such interaction has been described with respect to the action of serotonin (5-hydroxy-tryptamine). It is believed that part of the mechanism is derived from an interaction between serotonin and sialic acid present in the cell membrane, resulting in an increase in the transport of calcium ions (Wooley and Gommi, 1964).

The mechanism of muscle contraction appears to involve the operation of a "calcium pump" located in the smooth membrane portion of the sarcoplasmic reticulum of the cell. Activation of this calcium pump, via membrane depolarization, results in the release of calcium, which diffuses to sites adjoining the actin and myosin polymers, linking actin and myosin. Thus, the process of contraction ensues. Relaxation of muscle is due to a "pumping back" of calcium ions into membrane sites. The energy required for this process is derived from a specific type of ATPase system located in the membrane region.

The cell may be considered, therefore, as a complex multi-membraned system, compartmentalized into specific functional areas or "units." Each "unit" (or organelle) is bounded by a geometrically oriented membrane that is replete with spatially placed enzymes, whose chief function is the maintenance of a dynamically active transport system. The membrane-enzyme complex really comprises what we may call "active permeability," which is undoubtedly one of the most fundamental characteristics of the living cell.

APPENDIX

The Na^+, K^+-ATPase Enzyme System

Tissue	Fraction
Adrenal medulla	Mitochondria
Bacteria	Soluble protoplasm
Bird salt gland (gulls)	Homogenate
Bird renal tissue	Microsomes
Brain	Microsomes; cell membranes; homogenate; single neurons; axonal membrane; mitochondria
Cancer	Cell membrane
Crab nerve	Microsomes
Cerebrospinal fluid	Choiroid plexus homogenate
Erythrocyte	Intact cell; ghost; fragmented stromata
Eye	Epithelium of lens; Ciliary body; retina
Fish (electric eel; squid; sea urchin eggs; goldfish intestine)	Microsomes; axonal sheaths; Microsomes Cell membrane; mitochondria; Sarcoplasmic reticulum
Heart	
Intestine	Cell "debris" of mucosa
Kidney	Cell membrane; microsomes; mitochondria
Liver	Cell membrane; microsomes
Marrow	Ghost
Muscle, skeletal	Cell membrane; microsomes
Plant	Sediment obtained after 20,000 g
Salivary glands	Microsomes
Skin	Cell membrane
Toad bladder	Homogenate
Thymus gland	Homogenate
Thyroid	Cell membrane
Uterus	Homogenate

GENERAL REFERENCES

Albers, R. W. (1967). *Ann. Rev. Biochem.* **36**, 727.

Conway, E. J. (1957). *Phys. Rev.* **37**, 84.

Davies, R. E. (1961). *Membrane Transport Metab. Proc. Symp. Prague*, 1960 p. 320. Academic Press, New York.

Davies, R. E., and Ogston, A. G. (1950). *Biochem. J.* **46**, 324.

Heald, P. J. (1960). "Phosphorous Metabolism of Brain." Pergamon Press, Oxford.

Izumi, F., Nagai, K., and Yoshida, H. (1966). *J. Biochem. (Tokyo)* **60**, 533.

Katchalsky, A. (1964). *Biophys. J.* 4, Suppl. 2, 9.

Lehninger, A. L. (1964). "The Mitochondrion." Benjamin, New York.

Ling, G. N. (1952). *In* "Phosphorous Metabolism" (W. McElroy and B. Glass, eds.), Vol. 2, p. 748. Johns Hopkins Univ. Press, Baltimore, Maryland; (1965). *Ann. N. Y. Acad. Sci.* **125**, 401.

McIlwain, H. (1963). "Chemical Exploration of the Brain." Elsevier, Amsterdam.

Murphy, Q. R., ed. (1957). "Metabolic Aspects of Transport Across Cell Membranes." Univ. of Wisconsin Press, Madison, Wisconsin.

Nagano, K., Kanazawa, T., Mizumo, N., Tashima, Y., Nakao, T., and Nakao, M. (1965). *Biochem. Biophys. Res. Commun.* **19**, 759.

Reilly, C. (1964). *Biochem. J.* **91**, 447.

Rosenberg, T. (1948). *Acta Chem. Scand.* **2**, 14.

Sacks, J. (1948). *Chem. Rev.* **42**, 411.

Schwartz, A., and Matsui, H. (1967). *In* "Secretory Mechanisms of Salivary Glands," p. 75. Academic Press, New York.

Skou, J. C. (1963). *Symp. 1st Intern. Pharmacol. Meeting, 1961.* p. 41. Pergamon Press, Oxford.

Skou, J. C. (1964). *Progr. Biophys. Mol. Biol.* **14**, 131; (1965). *Physiol. Revs.* **45**, 596.

Solomon, A. K. (1962). *Sci. Am.* **207**, 100.

Troshin, A. S. (1961). *Membrane Transport Metab., Proc. Symp. Prague, 1960* p. 45. Academic Press, New York.

Ungar, G. (1963). "Excitation." Thomas, Springfield, Illinois.

SPECIFIC REFERENCES

Albert, S., Johnson, R. M., and Cohan, M. S. (1951). *Cancer Res.* **11**, 772.

Boyle, P. J., and Conway, E. J. (1941). *J. Physiol. (London)* **100**, 1.

Bungenberg de Jong, H. G. (1949). *Colloid Sci.* **2**, 335.

Caldwell, P. C., Hodgkin, A. L., Keynes, R. D., and Shaw, T. I. (1960). *J. Physiol. (London)* **152**, 561.

Conway, E. J. (1951). *Science* **111**, 270.

Conway, E. J. (1953). *Internat. Rev. Cytol.* **2**, 419.

Conway, E. J., and Brady, T. G. (1948). *Nature* **162**, 456.

Conway, E. J., and Mullaney, M. (1961). *Membrane Transport Metab., Proc. Symp. Prague, 1960* p. 117. Academic Press, New York.

Crow, A., and Ussing, H. H. (1937). *J. Exptl. Biol.* **14**, 35.

Danielli, J. F. (1952). *Symp. Soc. Exptl. Biol.* **6**, 1.

Danielli, J. F. (1954). *Symp. Soc. Exptl. Biol.* **8**, 502.

Davies, R. E., and Krebs, H. A. (1952). *Biochem. Soc. Symp. (Cambridge, Engl.)* **8**, 77.

Davis, J. O. (1963). *Yale J. Biol. Med.* **35**, 402.

Dean, R. B. (1941). *Biol. Symposia* **3**, 331.

DeLuca, H. F., Engstrom, G. W., and Rasmussen, H. (1962). *Proc. Natl. Acad. Sci. U.S.* **48**, 1604.

Dunham, E. T., and Glynn, I. M. (1961). *J. Physiol. (London)* **156**, 274.

Edelman, I. S., Bogoroch, R., and Porter, G. A. (1963). *Proc. Natl. Acad. Sci. U.S.* **50**, 1169.

Eisenman, G. (1961). *Membrane Transport Metab., Proc. Symp. Prague, 1960* p. 163. Academic Press, New York.

Eisenman, G., and Conti, F. (1965). *J. Gen. Physiol.* **48**, 65.

Eubank, L. L., Huf, E. G., Campbell, A. D., and Taylor, B. B. (1962). *J. Cellular Comp. Physiol.* 59, 129.

Gardos, G. (1954). *Acta Physiol. Acad. Sci. Hung.* 6, 191.

Glynn, I. M. (1962). *J. Physiol. (London)* 160, 18P.

Goldacre, R. J. (1952). *Intern. Rev. Cytol.* 1, 135.

Hokin, L. E., and Hokin, M. R. (1965). *Sci. Am.* 213, 78.

Huf, E. G. (1935). *Arch. Ges. Physiol.* 235, 655.

Jardetzky, O. (1966). *Nature* 211, 969.

Johnson, C. L., Mauritzen, C. M., Starbuck, W. C., and Schwartz, A. (1967). *Biochemistry* 6, 1121.

Johnson, R. M., and Albert, S. (1952). *Arch. Biochem. Biophys.* 35, 340.

Kahlenberg, A., Galsworthy, P. R., and Hokin, L. E. (1967) *Science* 157, 434.

Kennedy, E. P. (1967). Plenary Lecture. *7th Intern. Congr. Biochem., Tokyo.*

Kennedy, K. G., and Nayler, W. G. (1965). *Biochim. Biophys. Acta* 110, 174.

Kurella, G. A., (1961). *Membrane Transport Metab., Proc. Symp. Prague, 1960,* p. 54. Academic Press, New York.

Lee, K. S., and Yu, D. H. (1963). *Biochem. Pharmacol.* 12, 1253.

Levi, H., and Ussing, H. H. (1949). *Nature* 164, 928.

Libet, B. (1948). *Federation Proc.* 7, 72.

Lund, E. J. (1928). *J. Exptl. Zool.* 51, 265.

Lundegardh, H. (1939). *Nature* 143, 203.

McIlwain, H. (1963). "Chemical Exploration of the Brain." Elsevier, Amsterdam.

Maffly, R. H., and Edelman, I. S. (1963). *J. Gen. Physiol.* 46, 733.

Mitchell, P. (1961). *Membrane Transport Metab., Proc. Symp. Prague, 1960,* p. 22. Academic Press, New York.

Mitchell, P., and Moyle, J. (1965). *Nature* 208, 1205.

Noall, M. W., Riggs, T. R., Walker, L. M., and Christensen, H. N. (1957). *Science* 126, 1002.

Osterhout, W. J. V. and Hill, S. E. (1931). *J. Gen. Physiol.* 14, 385; (1935). *Proc. Natl. Acad. Sci. U.S.* 21, 125.

Rasmussen, H. (1965). *In* "Mechanisms in Hormone Action" (P. Karlson, ed.), p. 131. Academic Press, New York.

Rasmussen, H., and Ogata, E. (1966). *Biochemistry* 5, 733; Rasmussen, H., Shirasu, H., Ogata, E., and Hawker, C. (1967). *J. Biol. Chem.* 242, 4669.

Reilly, C. (1964). *Biochem. J.* 91, 447.

Schatzmann, H. J. (1953). *Helv. Physiol. Pharmacol. Acta* 1, 346.

Schwartz, A. (1964). *Cardiovascular Res. Center Bull.* 2, 73.

Schwartz, A. (1965a). *Biochim. Biophys. Acta* 100, 202.

Schwartz, A. (1965b). *J. Biol. Chem.* 240, 939.

Schwartz, A., Bachelard, H. S., and McIlwain, H. (1962). *Biochem. J.* 84, 626.

Schwartz, A., Matsui, H., and Laughter, A. H. (1968). *Science* in press.

Sen, A. K., and Post, R. L. (1964). *J. Biol. Chem.* 239, 345.

Skou, J. C. (1957). *Biochim. Biophys. Acta* 23, 394.

Skou, J. C. (1960). *Biochim. Biophys. Acta* 42, 6.

Smith, H. W. (1962). *Circulation* 26, 987.

Spiegelman, S., and Reiner, J. M. (1942). *Growth* 6, 367.

Spirtes, M. A., and Guth, P. S. (1963). *Biochem. Pharmacol.* 12, 47.

Stam, A. C., Jr., and Honig, C. R. (1962). *Biochim. Biophys. Acta* 58, 139.

Stiehler, R. D., and Flexner, L. B. (1938). *J. Biol. Chem.* 126, 603.

Teorell, T. (1935). *Proc. Natl. Acad. Sci. U.S.* 21, 152.

Trevor, A. J., Rodnight, R., and Schwartz, A. (1965). *Biochem. J.* 95, 883.
Turkington, R. W. (1963). *J. Biol. Chem.* 238, 3463.
Ussing, H. H. (1952). *Advan. Enzymol.* 13, 21.
Ussing, H. H., and Koefoed-Johnsen, V. (1958). *Acta Physiol. Scand.* 42, 298.
Ussing, H. H., and Zerahn, K. (1951). *Acta Physiol. Scand.* 23, 110.
Vates, T. S., Bonting, S. L., and Oppelt, W. W. (1964). *Am. J. Physiol.* 206, 1165.
Whittam, R., and Ager, M. E. (1964). *Biochem. J.* 93, 337.
Wolff, J. (1964). *Physiol. Rev.* 44, 45.
Woolley, D. W., and Gommi, B. W. (1964). *Nature* 202, 1074.

Chapter VIII

CELL LIFE CYCLE AND CELL DIVISION

I. INTRODUCTION

The individual cells apparently undergo an "aging" process which ultimately culminates in their death. Two factors are directly associated with death at the cellular level: (1) a limited capacity for growth, and (2) a finite life span of the cell. The former was effectively demonstrated by Hartmann in 1928 in *Amoeba proteus*. He succeeded in demonstrating an "experimental immortality" with this organism by periodically amputating bits of cytoplasm. As long as the nucleus was kept intact, the *Amoeba* would regenerate the lost cytoplasm. After completion of the regeneration process, he once again amputated the cytoplasm and observed another regeneration. The cycle could be repeated over and over again and Hartmann concluded that the process could probably have continued indefinitely. The corollary to this study is related to the ability of a cell to cast off its own cytoplasm and, in this way, achieve immortality. Unfortunately, most cells cannot throw off cytoplasm and require an additional mechanism for lengthening their potential life span. This process is *cell division* or *mitosis*.

When most cells reach a defined size, they will undergo mitosis, leading to the production of two daughter cells of composition almost identical to the parent. Mitosis must be very carefully controlled so that all parts of the cell are replicated and distributed equally. The distribution of materials takes place either by partition of free organelles, e.g., some mitochondria and ribosomes redistribute themselves between the daughter cells, or by some precise mechanism for redistribution. The latter is best exemplified by the events of nuclear division which are discussed later in this chapter.

Mitosis not only leads to perpetuation of cells but to their multiplication and to the expansion of the tissue mass. Multicellular organisms arise from a single fertilized cell, the *zygote*. By repeated multiplications, this zygote is transformed into a defined and organized tissue. Thus, mitosis is a necessary and distinct step in the normal cell cycle. It will be the purpose of this chapter to describe in some detail the morphological as well as the biochemical events occurring during the cell cycle which culminate in mitotic division. Unfortunately, the morphological description is the more complete and our understanding of the biochemical processes which characterize the stages during the life of the cell and the subsequent mitosis is in an embryonic state.

II. DESCRIPTION OF THE CELL LIFE CYCLE

Mitosis as first reported by W. Flemming in 1882 consists of a series of nuclear changes leading first to nuclear division or *karyokinesis*, then to cleavage of the cytoplasm, *cytokinesis*, and ultimately to the production of two identical daughter cells.

It has become apparent during the past few years that virtually all cells have a definite period in their life cycle during which DNA is synthesized and the genetic information is replicated. The cell life cycle as first suggested by Howard and Pelc (1953) is depicted in Fig. 8.1.

The entire cell cycle may be complete in a matter of minutes in bacteria, or in a matter of months in liver tissue; the duration of the phases of the cell life cycle and generation times for several mammalian cell lives are presented in Table 8.1. The G_1 phase consists of that period between the completion of cell division and the beginning of DNA synthesis. The G_1 phase is generally the longest and the most variable in duration and is followed by a period of replication of the genetic material, the S phase. The latter in most mammalian cells lasts approximately 7 hours, during which time the DNA content is doubled in preparation for mitosis. Autoradiographic studies have indicated that mitosis comprises only a small segment of the cell life cycle, i.e., less than 1 hour. The decision to divide generally occurs at the G_1 period since once the cell has passed through this interval and has entered the S phase, the cell will then proceed into mitosis without interruption. The latter is not without many exceptions, however.

The G_2 phase represents the lag period generally observed between the completion of DNA synthesis and the onset of the first recognizable signs of mitosis. Although no DNA synthesis occurs during G_1 and G_2, active RNA and protein syntheses are apparent.

III. INTERPHASE

The combined G_1, S, and G_2 phases are called the *interphase* or *"resting stage."* This terminology may be misleading and indeed is incorrect, since it connotes a period of inactivity. However, it is during interphase that the preparations for mitosis are made, and the bulk of the synthetic requirements are fulfilled. The interphase events include:

1. Chromosomal duplication
2. Increase in the mass of the cell, nucleus, and nucleolus
3. Energy liberation for mitosis
4. Increase in sulfhydryl compounds
5. Reproduction of the mitotic centers
6. Macromolecular syntheses for spindle formation

The importance of these activities will be discussed in subsequent sections. The reader is reminded that these categories are artificial designations and in biological systems, overlap to maintain a fluidity in cellular events.

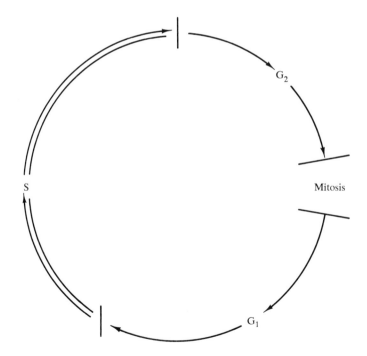

Fig. 8.1. Cell life cycle. S, synthetic period; G_1, presynthetic or "gap" period; G_2, postsynthetic period.

TABLE 8.1

Generation Time and Life Cycles of Various Cells

| Cell | Duration (hours) of | | | | |
	G_1	S	G_2	Mitosis	Generation time
HeLa	12	5	5	1	25
L-fibroblast	8	6 − 7	3 − 4	1	20
Mouse ear epidermal	10 − 20	6 − 8	1 − 4	1	30
Rabbit kidney (culture)	32 − 60	10	5	1	70

[a]From D. Mazia, in "The Cell" (J. Brachet and A. E. Mirsky, eds.), Vol 3, p. 77. Academic Press, New York, 1961.

Chromosomal Duplication. Implicit in any discussion of the mechanism underlying chromosomal duplication is a basic knowledge of the morphology of the chromosome. Although aspects of chromosomal structure have been briefly considered in Chapter IV, perhaps a more detailed discussion would better equip the reader for a proper understanding of mitotic events. The following description is based on the appearance of the chromosome during the *metaphase* (see p. 384).

Each chromosome is composed of two subunits, *chromatids,* joined at a constricted region which appears responsible for the regulation of the chromosomal movements, the *centromere* (or kinetochore). Morphologically (see Fig. 8.2), the metaphase chromosomes may be classified as *metacentric,* that is, possessing a median or nearly median centromere; *telocentric,* terminal centromere; *acrocentric,* centromere which is very near the end of the chromatid. The size of each chromosome may vary from 0.2 to 20 μ with a diameter of $0.2 - 2.0\mu$; in humans, the approximate length is $10 - 15$ μ. The number of chromosomes in cells varies from species to species. Among the organisms which possess the largest number is the protozoan of the group, *Agregata,* with over 300. In the human, the number of chromosomes is 46, comprising 23 pairs or a *diploid* set. Not only is the number of chromosomes constant for a particular species, but the shape, i.e., the number of metacentric, telocentric, and acrocentric chromosomes, is also well defined. The shapes of the human chromosomes have been classified and are depicted in pairs in Fig. 8.3A and B. Such a representation is referred to as a karyotype.

Autoradiography has proved an invaluable tool in the clarification of the events occurring during the replication of the chromosomes. A major breakthrough was achieved by Taylor (1957) by the use of tritiated thymidine (thymidine-^{3}H) as a precursor for DNA. Since tritium has a low beta energy of emission, excellent resolution of the particles may be

obtained on photographic film. In addition, precursors with extremely
high specific activities may be prepared. Taylor and other workers,
using this technique, were able to answer the fundamental question of
how the parental material is distributed between the daughter chromosomes. Several possibilities are available and duplication of the genetic
material may be accomplished by one of the following methods.

1. There may be a *semiconservative mechanism*, in which each
daughter chromosome is composed of 50% newly synthesized material,
i.e., from the medium, and 50% of the conserved parental chromosome.
In this mechanism, the daughter chromosomes must be equally labeled
with thymidine-³H after the first division. After a second duplication in
the presence of unlabeled thymidine, the daughter chromosomes are
unequally labeled.

2. A *conservative mechanism* may exist, in which the chromosomal
unit retains its integrity and in some manner is able to reproduce a
completely new chromosome which is then passed to the daughter cells.

3. A *dispersive mechanism*, in which the chromosomal strands are
disrupted, destroyed, and new structures resynthesized for the daughter
cells. The evidence upon which this theory is based is not the most sound.

Evidence indicating the correct mechanism for chromosomal replication came from Taylor's laboratory from studies with chromosomes of

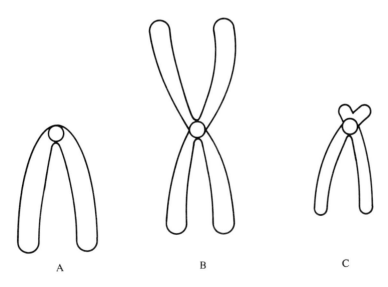

A   B   C

Fig. 8.2. Various positions of the centromere. (A), telocentric; (B), metacentric; (C),
acrocentric.

the roots of *Vicia faba*. Their results are depicted in Figs. 8.4 and 8.5. The chromosomal material was duplicated in the presence of the labeled compound for one generation, i.e., until both chromatids were labeled, then was removed from the thymidine-^{3}H and allowed to double for one generation. The results were clearly in accord with a semiconservative mechanism for the duplication of chromosomes.

The evidence substantiating this mechanism on a molecular level was supplied by Levinthal in 1956 who studied the distribution of ^{32}P-labeled DNA in the bacteriophage T$_2$. Additional and definitive proof was provided by the ingenious experiments of Meselson and Stahl (1958) and Meselson *et al.* (1957), who grew cells of *Escherichia coli* for a number

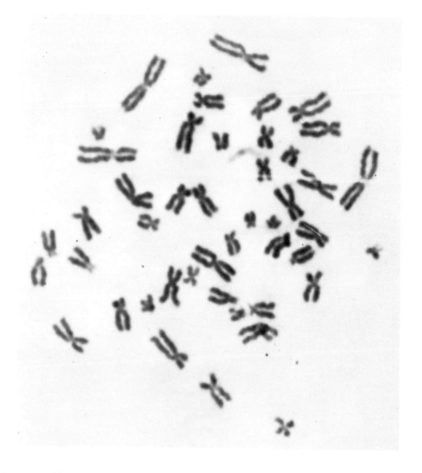

Fig. 8.3A. Normal human male idiogram. Courtesy of A. Sinha and H. Rosenberg.

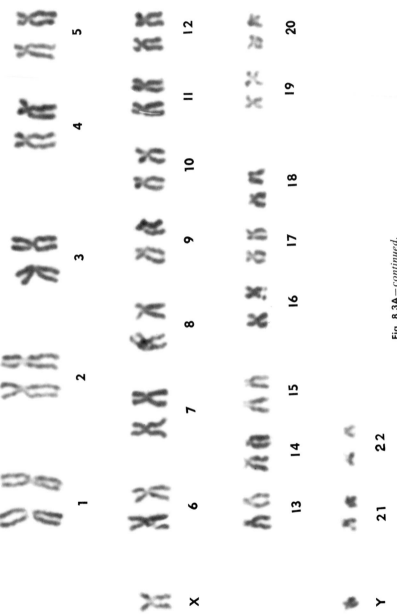

Fig. 8.3A—continued.

of generations in an ^{15}N-containing medium. At timed intervals after the addition of an excess of ^{14}N substrate, the *E. coli* was sampled and lysates were prepared. The DNA which now contained varying amounts of ^{14}N and ^{15}N was then subjected to density gradient centrifugation in cesium chloride. Since the DNA containing the ^{15}N is heavier than its counterpart with ^{14}N, the two substances separate under the conditions of the centrifugation (Fig. 8.6). In the first ultraviolet tracing of Fig. 8.6, the peak is entirely due to DNA-^{15}N. Their results clearly demonstrate that after one generation of growth in ^{14}N, the DNA consisted of a hybrid of ^{15}N and ^{14}N which banded in a different region than the original DNA-^{15}N. The uniqueness of the hybrid DNA is indicated by banding

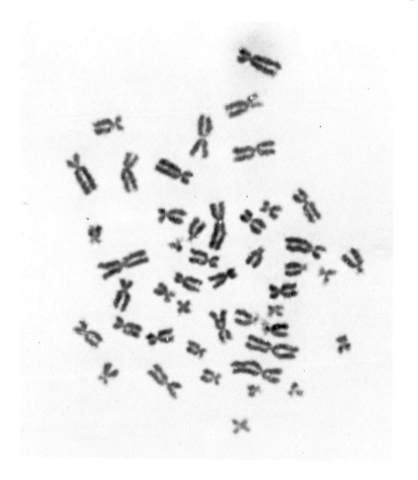

Fig. 8.3B. Normal human female idiogram. Courtesy of A. Sinha and H. Rosenberg.

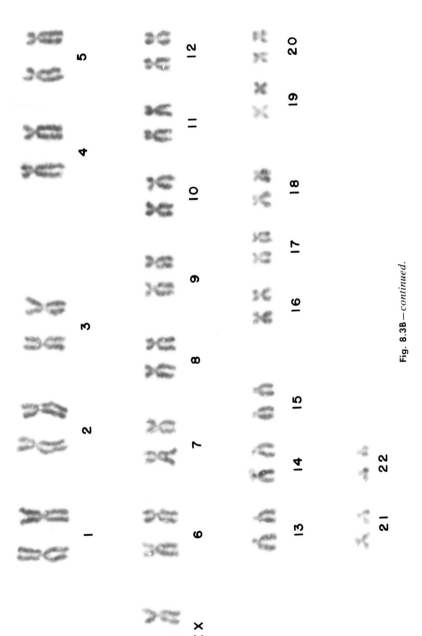

Fig. 8.3B—*continued.*

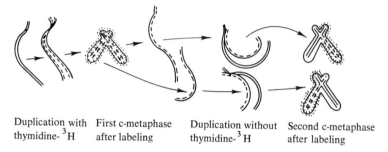

Duplication with   First c-metaphase   Duplication without   Second c-metaphase
thymidine-³H   after labeling   thymidine-³H   after labeling

Fig. 8.4. Diagrammatic representation of the mode of chromosome duplication as revealed by thymidine-³H. The labeled DNA subunits are shown as dashed lines, the unlabeled ones as solid lines. From J. H. Taylor, P. S. Woods, and W. J. Hughes, *Proc. Natl. Acad. Sci. U. S.* **43**, 122-128 (1957).

the original DNA-¹⁵N with the material obtained after growth of the organisms for 1.9 generations in the presence of ¹⁴N (mixed 0 and 1.9 in the figure). We see here three distinct peaks, representing from left to right ¹⁴N, ¹⁴N + ¹⁵N, and DNA-¹⁵N. After a second period of replication in the presence of ¹⁴N, DNA-¹⁴N made its appearance (1.9 generations in the figure); the amount of DNA-¹⁴N increased with subsequent replications. The last tracing shows the presence of only original DNA-¹⁵N and the DNA obtained after growth of the organisms in ¹⁴N for several generations, DNA-¹⁴N. Very little, if any, hybrid material is apparent. These observations at the gross and molecular levels established that the DNA of the bacterial chromosome is duplicated by a semiconservative mechanism. What is the structural arrangement of the DNA in the chronally along the chromosome meets with serious physical difficulties. The total DNA of the nucleus *would be more than a meter in length!* Obviously, the DNA must exist in some intricately coiled structure within the chromosome. In fact, a coiled structure was observed as early as 1880 by Baranetzky and subsequently was called a *chromonema*. The major problem is not whether coiled DNA molecules exist within the framework of the chromosome but in what manner this coiling is effected.

A model has been proposed by Freese (1958) in which the DNA molecules are linearly linked together by blocks of proteins (see Fig. 8.7). Such an arrangement allows for expansion into a linear structure, or formation of a tight helix as demonstrated in Fig. 8.7. Taylor (1963) has modified this model to account for the variations in structure as seen during the replication cycle. Before or during replication, the chromosomal structure may be stabilized by the presence of a "linker" molecule, the H-linker, which eventually forms a chromatid with its attached DNA chain. In addition to the H-linker, Taylor has postulated the existence of

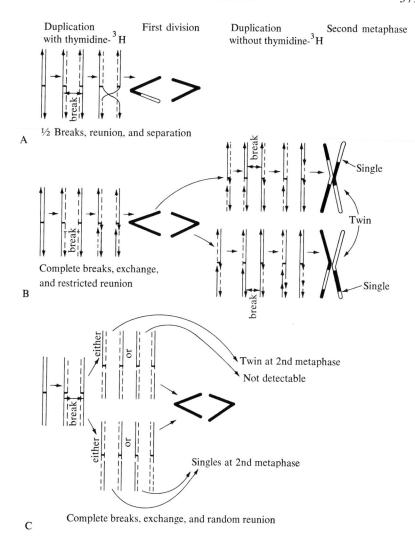

Duplication with thymidine-³H   First division   Duplication without thymidine-³H   Second metaphase

A   ½ Breaks, reunion, and separation

B   Complete breaks, exchange, and restricted reunion

Single

Twin

Single

C   Complete breaks, exchange, and random reunion

Twin at 2nd metaphase
Not detectable

Singles at 2nd metaphase

Fig. 8.5. Diagrams showing the predicted results of chromosome reproduction and sister chromatid exchange in which DNA was labeled during one replication with thymidine-³H; the dashed lines represent labeled subunits and the chromosomes shown in solid black represent regions showing grains in autoradiograms. The unbroken lines represent regions without grains in the autoradiograms as shown in Fig. 2 in Taylor (1963). A. The predicted results of exchanges involving only one subunit in each chromatid (these were undetected in the experiments and perhaps do not occur). B. The predicted results of exchanges when the two subunits are restricted in their reunion by a difference in sense analogous to the two chains of a DNA double helix. From J. H. Taylor (ed.), *in* "Molecular Genetics," Part I, pp. 65-111. Academic Press, New York, 1963.

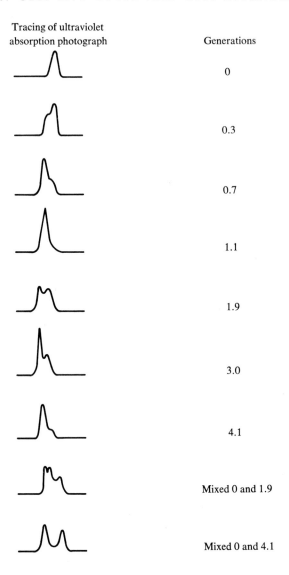

Fig. **8.6**. Density gradient centrifugation of DNA from *E. coli*. *E. coli* was grown in a medium containing ^{15}N for a number of generations, resulting in the formation of DNA-^{15}N. An excess of ^{14}N-substrate was then added and at various times thereafter, samples of *E. coli* were removed and lysates were prepared. The lysates were subjected to density gradient centrifugation and the ultraviolet absorption of the DNA bands was determined. A tracing of the latter is presented in the figure with the time of sampling. Note the formation of the "hybrid" DNA, i.e., mixed $^{15}N - ^{14}N$-DNA shortly after the introduction of the ^{14}N substrate. From M. Meselson and F. W. Stahl, *Proc. Natl. Acad. Sci. U. S.* **44**, 671-682 (1958).

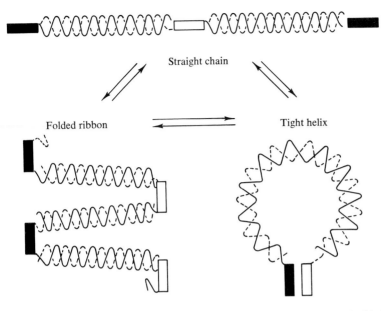

Fig. 8.7. Freese's model of basic chromosome structure. DNA shown as a double helix, protein links shown solid or in outline. From E. Freese, *Cold Spring Harbor Symp. Quant. Biol.* 23, 13-18 (1958).

two other substances, the 3'-linker which joins the 3'-hydroxyl groups of adjacent members of the DNA-helix, and the 5'-linkers which unite the 5'-groups (Fig. 8.8).

During the duplication of the chromosomes, the 3'-linkers are enzymatically cleaved, with the production of two subunits, each consisting of alternate 5'-linkers. Only hydrogen bonds hold together the H-linkers with the attached DNA chains. With the insertion of a new 5'-linker, replication of the two chains of adjacent helices may begin (Fig. 8.9). The progression of these events is depicted in Fig. 8.10. As replication ensues, a new 3'-linker which is resistant to enzymic degradation, i.e., 3'-R linker, is inserted, and at the completion of chromosomal duplication, another resistant 3'-linker is introduced at the growing 3'-ends of the strands.

Although the identity of the "H-linker" is unknown, a peptide structure has been suggested. The binding portions of 3' or 3'-resistant linkers may consist of two phosphoserine residues coupled by an esteratic bond to the terminal nucleoside of DNA, in a manner analogous to the amino acyl-transfer RNA linkage.

The replication of the DNA in the chromosomes does not occur simultaneously among all the chromosomes or concurrently along the length

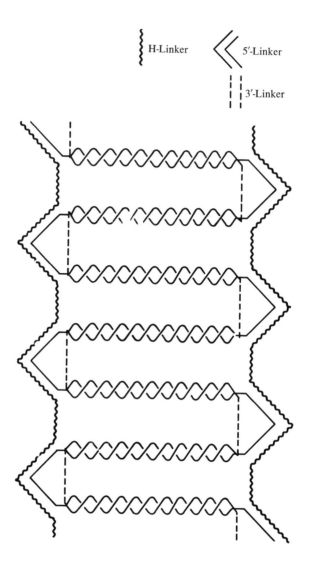

Fig. 8.8. Before or during replication the chromosome is assumed to be stabilized by the addition of H-linkers to form a double axis, each of which with its attached DNA chains will become a chromatid by the time replication is complete. The stabilizing material is represented as a polymer attached to alternate 5'-linkers. Although shown as a continuous chain for simplicity in illustration, it probably would consist of many small subunits attached in some way as yet unknown. From J. H. Taylor (ed.), *in* "Molecular Genetics," Part I, pp. 65-111. Academic Press, New York, 1963.

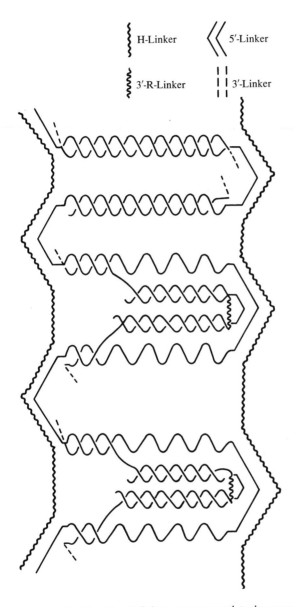

Fig. 8.9. During replication the 3'-linkers are assumed to be opened by appropriate enzymes. The two subunits of the chromosome consisting of the alternate 5'-linkers and the H-linkers with the attached DNA chains are then held together only by the hydrogen bonds in the DNA. Two chains of adjacent molecules are assumed to begin replication by the insertion of a new 5'-linker. From J. H. Taylor (ed.), *in* "Molecular Genetics," Part I, pp. 65-111. Academic Press, New York, 1963.

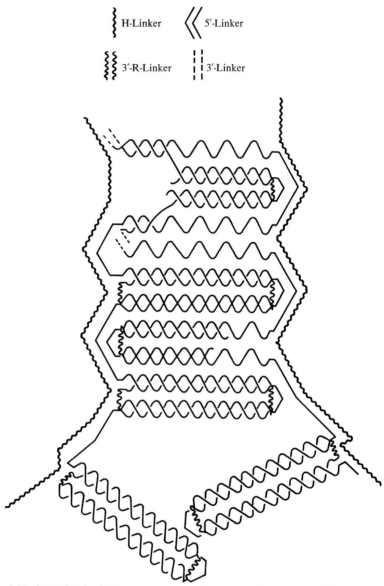

Fig. 8.10. As each pair of chains begins replication, a new 3′R (resistant)-linker is assumed to be inserted which is resistant to the attack of the enzyme that opens the regular 3′-linkers. When the pairs of chains finish replication their growing 3′-ends are assumed to be closed by the addition of another 3′R-linker. The two new chromatids would consist of these pairs of DNA molecules attached to a single axial element. Another axial element (half-chromatid) would be added to each during the succeeding late interphase or prophase. From J. H. Taylor (ed.), *in* "Molecular Genetics," Part I, pp. 65-111. Academic Press, New York, 1963.

of a single chromosome. The kinetics of autoradiography have established that the DNA replicates sequentially along the chromosome toward the centromere with the latter structure labeled last.

IV. MITOSIS

Mitosis, as we have seen, occupies only a minute portion in the life of the cell, with an average duration between 30 minutes and 3 hours. The organism with the shortest mitosis is the *Drosophila* egg, 9 – 10 minutes. The major events of cell division are the reapportionment of the nuclear material into the two daughter nuclei by *karyokinesis* and the cleavage of the cell cytoplasm into two sister cells by *cytokinesis*.

During karyokinesis, several events take place leading to (1) alterations in the morphology of the chromosomes which are prerequisite to the replication of the chromosomal material, and (2) the appearance of the *mitotic apparatus*, a highly organized network that assists in the separation of the sister chromatids. These events will be considered in detail subsequently.

Based upon the histological appearance (see Fig. 8.11), karyokinesis may be divided into several stages: prophase, prometaphase, metaphase, anaphase, and telophase. The duration of each phase varies from cell to cell although anaphase is considered the briefest stage. Some examples of the duration of the mitotic phases in various cells have been tabulated in Table 8.2. Reference will be made to the figure describing these mitotic phases, Fig. 8.11, during the course of this chapter.

A. Prophase

Hypertrophy of the cell, nucleus, and nucleolus are principal occurrences during the prophase. Increases in the refractivity, turgidity, and viscosity of the cytoplasm are noticeable during this stage. These physicochemical changes appear directly related to the movement of chromosomes and to the formation of the mitotic apparatus.

During prophase, the chromosomes begin their movement, the most conspicuous event of which is the coiling and subsequent shortening of these structures, suggesting a condensation of chromosomal constituents (see Fig. 8.11). Each chromosome is composed of two special filaments, the chromatids, which though not united are closely associated. It is these chromatids which undergo condensation and thickening during progression of the prophase. With the shortening and increased coiling of the chromatids, the centromere region becomes more accentuated as the chromosomes move toward the nuclear membrane.

The ultimate explanation for the increase in the chromosomal coiling is not precisely understood although several theories have been suggested

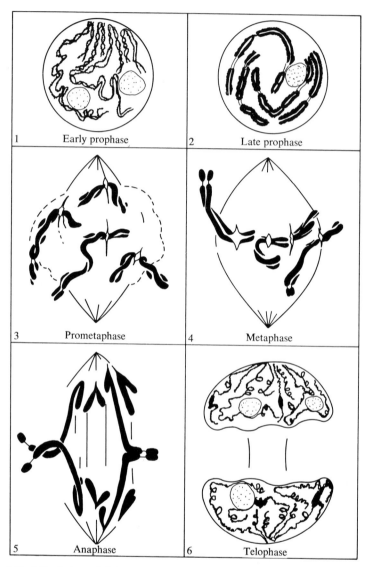

Fig. 8.11. Mitosis in a diploid cell. Four chromosomes are illustrated; two of these are small metacentrics with a subterminal heterochromatic segment. The other two are large acrocentrics with a nucleolar constriction in the long arm. From K. R. Lewis and B. John, "Chromosome Marker." Little, Brown, Boston, Massachusetts, 1963.

attributing the coiling to: (1) some property of the internal chromosomal matrix containing the DNA molecule, or (2) polycationic compounds, e.g., histones, that may induce chromosomal coiling. The latter theory is presently in favor.

TABLE 8.2

Some Examples of Data on the Duration of the Mitotic Phases, Representing
Direct Observations of Living Dividing Records[a]

Cell	Duration (minutes) of			
	Prophase	Metaphase	Anaphase	Telophase
Yoshida sarcoma (35°C)	14	31	4	21
MTK-sarcoma I (35°C)	10	44	5	18
Mouse spleen in culture	20 − 35	6 − 15	8 − 14	9 − 26
Triton liver fibroblast (26°C)	18 or more	17 − 38	14 − 26	28
Grasshopper neuroblast (38°C)	102	13	9	57
Pea endosperm	40	20	12	110
Iris endosperm	40 − 65	10 − 30	12 − 22	40 − 75
Desmid	60	21 − 24	6 − 12	3 − 45

[a]From D. Mazia, *in* "The Cell" (J. Bracket and A. E. Mirsky, eds.), Vol. 3, p. 156.
Academic Press, New York, 1961.

1. Mitotic Apparatus

The mitotic apparatus represents the ensemble of structures comprising the chromatic and achromatic figures, the spindles, asters, centrioles,
nuclei, and chromosomal structures (see Fig. 8.12). Accordingly, it occupies a large proportion of the cellular volume. The fundamental function of the mitotic apparatus is to aid in the polarization of the chromosomes and hence in the fulfillment of cell division. The mitotic apparatus
is formed as a gel from preexisting macromolecular material (see later
section).

The *central spindle* is a basic structure of the mitotic apparatus and is
responsible for the pole-to-pole connections and for the chromosome-to-
pole attachments. The spindle formation begins in the vicinity of the
centriole or cell center at the side of the nuclear membrane. At the beginning of cell division, the centriole splits into two parts, from which
emanate the *astral rays* or *asters*. A bundle of delicate filaments − *the
spindle* − is located amid the two asters.

The movement of the centrioles with their asters toward the opposite
poles of the nucleus constitutes much of the prophase activity. We are indebted to the pioneering work of Mazia and Dan (1952) (see p. 384) for
much of the knowledge of the chemistry of the mitotic apparatus.

2. Nucleolus

The disintegration of the nucleolus is one of the most characteristic
events of prophase. The nucleolar substance may be transported by the

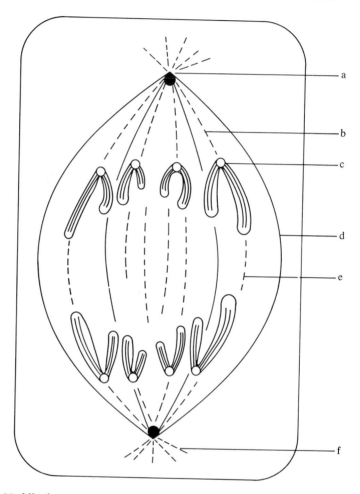

Fig. 8.12. Mitotic apparatus at metaphase. a, centriole; b. chromosomal fiber (chromosome-to-pole connection); c, centromere; d, pole-to-pole connection; e, interzonal connection (chromosome-to-chromosome connection); f, aster.

chromosomal threads into the daughter nuclei or may commute between the nucleus and cytoplasm during mitosis. In most nuclei, certain secondary constrictions on the *chromatin* may be observed that are intimately associated with the nucleolus. These structures represent the *nucleolar-associated chromatin*. During prophase, the nucleolus prior to dissolution becomes disengaged from the latter. Perhaps, this condition is responsible for the instability of the nucleolus and its consequent dispersion.

3. NUCLEAR MEMBRANE

The nuclear membrane generally has disintegrated by the end of prophase, although in several instances the progression of mitosis has been reported with an intact nuclear membrane. The underlying mechanism is not understood. An electron micrograph of a tumor cell in prophase is presented in Fig. 8.13. The dissolution of the nuclear membrane has just begun; the nucleolus is still intact although more diffuse.

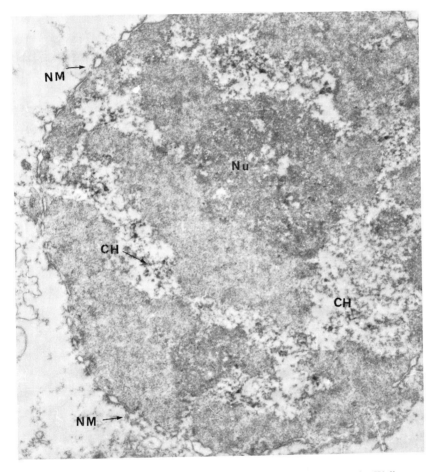

Fig. 8.13. Isolated nucleus in prophase. The nuclei were isolated from the Walker carcinosarcoma by the Chauveau method (Chapter IV); fixed in glutaraldehyde, stained in OsO₄ and poststained in uranyl acetate. Note the dispersing nuclear membrane (NM), the intact nucleolus (Nu) and the compactness of the chromatin (CH). × 31,000. Courtesy of H. Busch and K. Shankar.

B. Prometaphase and Metaphase

The dissolution of the nuclear membrane and the concomitant increase in fluidity observable at the center of the nucleus signal the advent of *prometaphase*. The chromosomes begin their movements toward the equatorial plane of the cell, a process termed *metakinesis* (see Fig. 8.11). This phase may be extremely ephemeral, lasting only a matter of minutes.

With the arrival of the chromosomes at the equatorial plane, the cell passes into *metaphase*. The end result of metaphase is the partial separation of the chromosomes. Toward this end, the chromosomes arrange themselves at the periphery of the spindle. Prior to the separation, the mitotic apparatus must engage the chromosomes and it is the centromere which serves as the point of attachment and which appears responsible for the subsequent movement of the chromosomes. The chromosomal arms, on the other hand, act merely as passengers, a point aptly described by Mazia (1961):

> Indeed, the role in mitosis of the chromosome arms, which carry most of the genetic material, may be compared with that of a corpse at a funeral: they provide the reason for the proceedings but do not take an active part in them.

Each of the chromosomes are attached to the spindle fibers by their respective centromeres which now divide the chromatids. The adjacent daughter centromeres move apart, separating the daughter chromatids and causing the migration toward the opposite poles until the anaphase position is reached.

1. BIOCHEMISTRY OF THE MITOTIC APPARATUS

Much of the mystery of the previously described cytological events may be correlated with a gap in our knowledge of the composition of the mitotic apparatus. The mitotic apparatus is composed of tightly arranged vesicular and tubular elements and is sensitive to high temperatures and high pressure.

The successful mass isolation of the mitotic apparatus from sea urchin eggs by Mazia and Dan (1952) was the initiating point in our understanding of the activity of this structure. Previous efforts had been hampered by the extreme lability of the mitotic apparatus. However, Mazia and Dan succeeded in stabilizing the dividing sea urchin eggs by exposure to hydrogen peroxide and ethanol at $-10°$ C, dispersing the cytoplasm with detergents, e.g., digitonin, or with ATP, and collecting the structures that morphologically retained most of the characteristics of the mitotic apparatus (see Fig. 8.14A and B).

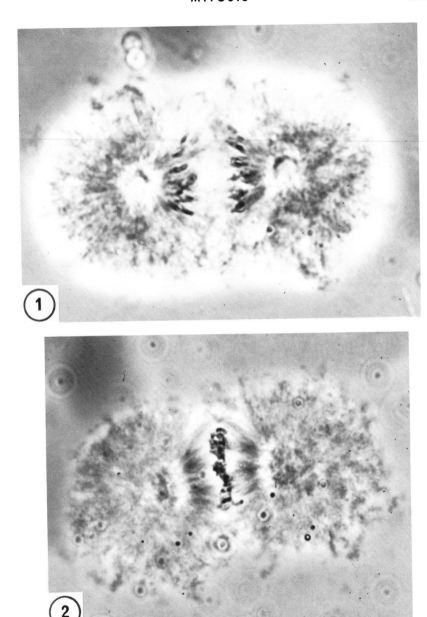

Fig. 8.14A. Isolated mitotic apparatus. Mitotic apparatus isolated by the alcohol-digitonin method from the sea urchin, *Strongylocentrus purpuratus*. 1, the apparatus at metaphase; 2, at anaphase. Courtesy of A. Zimmerman.

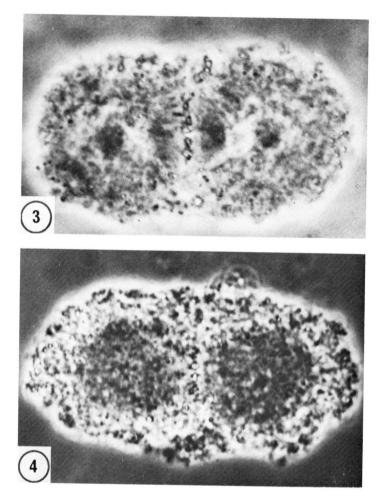

Fig. 8.14B. Isolated mitotic apparatus. 3, the mitotic apparatus isolated from the sea urchin, *Arbacia puntulata*, by the alcohol-digitonin method at metaphase; 4, at anaphase. Courtesy of A. Zimmerman.

Mazia reasoned that the gel properties of the mitotic apparatus were due to the presence of numerous disulfide bonds and the inability to isolate the mitotic apparatus was the result of cleavage of these bonds by reducing agents. Consequently, the stabilization of this structure could be effected with the use of disulfide-protecting agents. The agent of choice, however, must be able to permeate the sea urchin eggs, be relatively nontoxic, and must be soluble in aqueous media. The compound possessing the most suitable properties for this task was dithiodiglycol (DTDG). With DTDG, these structures were isolated in greater yield and greater purity than had previously been possible.

The isolated mitotic apparatus was composed of more than 90% protein, with an average molecular weight of 34,700, and an amino acid composition similar to that of the contractile protein of muscle, actin. The protein(s) was acidic with an isoelectric point of 4.5. Electrophoretic and ultracentrifugal data indicated the presence of one major protein and another minor constituent. A more sensitive immunological technique has demonstrated the existence of two other minor components and has confirmed the presence of the major structural protein. In addition to protein, the mitotic apparatus contains some RNA, some unidentified polysaccharides, and a lipid present perhaps in the form of lipoprotein, i.e., related to the filamentous and vesicular structure of the various membranes of the mitotic apparatus.

The investigations of Rapkine in 1931 (see Mazia, 1961) suggested the existence of sulfhydryl groups in the mitotic apparatus. He observed a rhythmic fluctuation in the content of a soluble sulfhydryl-containing compound, which he presumed to be reduced glutathione. The concentration of the sulfhydryl-containing compound gradually declined until the onset of metaphase. Thereafter, an increase in its concentration was observed. Rapkine, accordingly, referred to these events as the "glutathione cycle" and suggested that a reversible denaturation of proteins favored the formation of fibers and the production of a gelatinous structure.

A number of other investigators have confirmed the existence of sulfhydryl groups in the mitotic apparatus, although with the utilization of more sensitive and selective techniques the fluctuations in glutathione content as proposed by Rapkine have not been substantiated. However, a cyclic variation has been observed in a sulfhydryl-containing polypeptide or protein that was soluble in trichloroacetic acid and that could not be precipitated by ammonium sulfate. Mazia (1955) has proposed the creation of the spindle fibers by a polymerization of these small protein fibrils as depicted in Fig. 8.15. The lateral orientation of the sulfhydryl-linked protein chains was accomplished by hydrogen bonding.

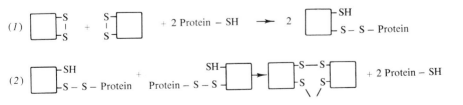

Fig. 8.15. Spindle formation from sulfhydryl groups.

2. PROTEIN SYNTHESIS DURING METAPHASE

Macromolecular synthesis is at a very low level in the metaphase cells (Salb and Marcus, 1965). In particular, the following effects have been observed: a marked decrease in the incorporation of amino acids into mitotic cells; a diminution in the numbers of polyribosomes in the metaphase-arrested cell. Salb and Marcus (1965) have shown that the ability of the metaphase-arrested HeLa cell to synthesize protein *in vitro* can be regained by treatment of the cells with the proteolytic enzyme, trypsin. Presumably, the ribosomes (or polyribosomes) are prevented from incorporating amino acids into protein during metaphase by the presence of a trypsin-sensitive substance, perhaps a histone. Apparently during mitosis, a *"translational inhibitor"* is elaborated which is removed during the interphase period. These findings are analogous to the events which transpire in the unfertilized amphibian egg (see Chapter V). Only upon fertilization can protein synthesis proceed, although the machinery appears intact. Trypsin treatment of preparations from unfertilized eggs allows for protein synthesis.

We may summarize the events preceding the metaphase by:

1. Construction of preformed macromolecular components, including one major protein, into a gel structure by sulfhydryl linkages.
2. The execution of pole-to-pole and chromosome-to-pole connections by a "central spindle" and "chromosomal spindle," respectively.
3. The ordering and orientation of the mitotic apparatus by hydrogen bonding with a coincident "focusing" of the spindle fibers and asters.
4. The centromere-directed-movement of the chromosomes toward opposite poles.

C. Anaphase

During anaphase, the chromatids shorten longitudinally and the centromeres are pushed apart by some repulsive force. The repulsive force represents the summation of two movements, an elongation of the central spindle tending to push the poles apart, and the second, a movement of the chromosomes to the poles. The anaphase movement is begun simultaneously at a uniform speed, and in a linear direction by all the chromosomes.

As the sister chromosomes separate, the *"interzonal"* region now becomes prominent. The properties of the interzonal region differ from those of the polar regions. The gel of the former is neither as strong nor as stable as the rest of the mitotic apparatus. By the end of anaphase, a marked accumulation of RNA has occurred in the interzonal region, the origin of which may have been the chromosomes themselves. These alterations of RNA content are related to the "chromosomal RNA cycle" wherein the chromosomes accumulate RNA during prophase and perhaps may discard the "excess" at anaphase. Electron microscopy has revealed the proliferation of the endoplasmic reticulum at the poles in the extra-spindle material before the onset of anaphase. During anaphase, however, the endoplasmic reticulum of the spindle increases and perhaps it is this proliferation which is responsible for the interzonal RNA.

D. Telophase

Telophase is marked by a despiralization or swelling of the chromosomal structure, i.e., unwinding with the ultimate aim of returning the nucleus to the original interphase condition. Other conspicuous events in telophase include: the re-formation of the nucleolus at specific regions on the chromosomes, the *nucleolar organizer sites*; the breakdown of the spindle and the asters; and the reappearance of the nuclear membrane, a process that apparently involves the interaction of the endoplasmic reticulum with the chromosomes. In grasshopper spermatocytes, vesicular endoplasmic reticula of approximately 0.5 μ in length collect around the chromosomes during despiralization or swelling, orient themselves along the chromosomal surface, and weld together in a membranous structure forming the nuclear membrane (Barer *et al.*, 1959). The furrowing of the cell surface and the eventual cleavage into two cells, *cytokinesis*, signals the end of telophase.

V. CYTOKINESIS

Cytokinesis is regulated by the mitotic apparatus. The plane of division of the cell is correlated with the equatorial plane, the position of the mitotic apparatus. The plane of cytokinesis is usually perpendicular to the center region of the spindle.

Cytokinesis is independent of karyokinesis and indeed, in some cells, e.g., striated muscle, nuclear division is not followed by a cleavage of the cell. In this manner, the multinucleated cells may arise.

The separation and segmentation of the cytoplasm are problems that have interested many investigators as judged by the number of reviews dealing with this subject matter (Wolpert, 1963; Brown and Danielli, 1964). The problems manifest in cytokinesis are: (1) the exact definition of the plane of cleavage; (2) the factors responsible for the changes in cellular form; (3) the mechanism for the evolution of the new cellular surface. Among the many theories that have been promulgated, the *astral relaxation theory* comes closest to explaining all the experimental observations.

The *astral relaxation theory* was proposed by Wolpert in 1960 (see Wolpert, 1963). The "rounding up" of the cell during anaphase produces a uniform tension along the surface of the cell membrane. Cleavage is initiated by a fall in the tensive force at a particular "differentiated" region, allowing the furrow site to actively contract and eventually to constrict the cell into two parts. The plane of cleavage is determined by asters which "differentiate" the specific regions of the cell surface so that they will relax. The exact mechanism for the formation of the new membrane, however, is not known.

Implicit in most of the theories on cytokinesis is a contraction or expansion of the cell membrane. Some fundamental similarity therefore is apparent in *contraction* and *cytokinesis*. The similarity may be more than a casual one, since in the former, the hydrolysis of ATP is required. One also finds in the cell membrane, actomyosin-threadlike proteins (see Chapter III) which may require the energy from ATP for cytokinesis. However, much more evidence must become available before this relationship is firmly established.

VI. MEIOSIS

As early as 1883, Van Beneden observed the parental contribution of equal numbers of chromosomes during *syngamy*, or fertilization, to the fertilized cell, the zygote. Thus, the zygote possessed a diploid number of chromosomes. Since each daughter cell is the recipient of identical sets of chromosomes during a mitotic division, they too are diploid in nature. In Metazoa, the majority of the cells are in the diploid state while only the sexual cells, the oogonial and spermatogonial cells, are haploid, i.e., possess only a single set of chromosomes. Obviously, some mechanism must exist within the organism that allows for the formation of a cell with a haploid number of chromosomes from a diploid parent, i.e., reduction division.

In 1887, Weismann postulated that the reduction in the number of chromosomes in the zygote from the diploid to the haploid state occurred without a longitudinal division. This far-reaching prediction was subsequently verified. The process by which the reduction division takes place has been termed *meiosis*. The haploid state is brought about through two divisions and is an obligate prerequisite for fertilization. In addition, an assortment and distribution of the parental genetic potentialities into the daughter cells, recombination, is accomplished.

Several of the events of mitosis (see p. 379) are also seen during meiosis: karyokinesis, cytokinesis, the dissolution and re-formation of the nucleolus and nuclear membrane, spindle formation and breakdown, and chromosomal spiralization. All of these events, in meiosis as well as in mitosis, occur in an orderly fashion and appear coordinated with respect to each other. The stimulus responsible for the initiation of either meiosis or mitosis is not understood although it is known that both processes are under genetic control since mutant organisms have been obtained with specific defects of one of these processes.

Meiosis consists of several phases figuratively depicted in Fig. 8.16. The first division is accompanied by a very lengthy prophase which has been subdivided into the leptonema, zygonema, pachynema, diplonema, diakinesis; metaphase; anaphase; and telophase. The second division is initiated by an interphase, and followed by a second metaphase, anaphase, and telophase. The end result of the two divisions is the elaboration of four genetically different gametes.

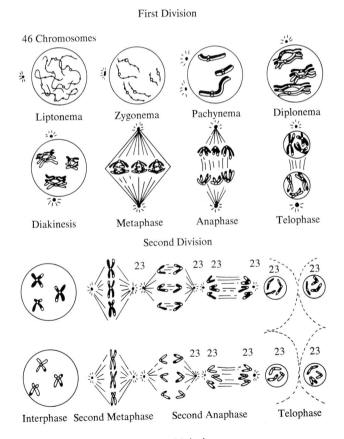

First Division

46 Chromosomes

Liptonema   Zygonema   Pachynema   Diplonema

Diakinesis   Metaphase   Anaphase   Telophase

Second Division

Interphase   Second Metaphase   Second Anaphase   Telophase

Fig. 8.16. Meiosis.

A. Leptonema

The onset of leptonema is characterized by an increase in the nuclear volume and in the gradual alteration in the nuclear structure. During the preleptonema stage, the chromosomes unwind and gradually extend to a final state resembling uncoiled, single threads. At early leptonema, the structures become maximally extended and dense regions representing more tightly coiled sections are observed at fixed irregular intervals along the threadlike chromosome. These areas are referred to as *chromomeres* while the thread adjoining two chromomeres represents the *interchromomeric region*. The free ends of the chromosome are attracted to the side of the nuclear membrane closest to the centromere which itself is usually polarized to one side of the nuclear membrane. Although microscopically the chromosomes appear undivided, as opposed to the mitotic prophase, this view is probably misleading. Each chromosome must actually consist of two chromatids in close juxtaposition resembling a single structure.

The size of the nucleolus gradually increases during the leptonema stage and may double in both volume and RNA content. During leptonema, the nucleolar organizers remain attached to the nucleolus. An enhancement in the protein biosynthetic activity is observable concurrent with the increase in volume and RNA content.

B. Zygonema

By the end of the leptonema, the chromatids begin to shorten and thicken in preparation for the pairing of the homologous chromosomes, i.e., the corresponding chromosomes from the maternal and parental cells.

The next stage of meiosis, zygonema, is introduced with these events. During mitosis, pairing between homologous chromosomes is only rarely seen, while in zygonema, the homologous single chromosomes pair in a defined manner, usually near their ends or close to the centromere, with each chromomere facing its sister structure and held in apposition by some unknown force. Once pairing has begun at a specific region, the process unfolds in a zipperlike manner. In diploid cells, the pairing occurs between homologous chromosomes, one originating from the egg and the other contributed by the sperm. While pairing continues, the chromosomes twist around each other, shorten and thicken. By the end of the zygonema, the pairing and twisting of the chromosomes has been completed.

C. Pachynema

At the pachynema, the chromosomes consist of paired homologs of chromatids, i.e., bivalents. The bivalents continue to shorten and thicken by progressively increasing the diameter of the coils until the length of the pachytene chromosomes has been reduced to one fourth to one sixth the leptotene length. Pachynema is a stable condition that may be prolonged indefinitely. The decision to proceed into the next stage of meiosis, diplonema, is dependent upon four changes which take place:

1. The partner chromosomes may begin to fall apart.
2. The two centromeres of each bivalent begin moving apart, forming two paired chromatids.
3. The paired chromatids criss-cross over one another and exchange partners at these crossover points, *chiasmata*. The chiasmata hold together the chromatid members and restrict the complete separation of the parental chromosomal apparatus. The term crossover is perhaps a little misleading since the chromatids do not actually cross over one another but the arrangements of genetic material suggest a breakage and reunion with the other strand (see Fig. 8.17).
4. The crossing-over, chiasmata formation, and reduction in the relational coiling of the chromosomes occur.

When these events have been completed, the meiotic cell passes directly into the diplonema stage. During the preceding stages, certain chromosomal regions may be observed to stain intensely with basic dyes. Consequently these areas are referred to as *heterochromatin*. The heterochromatin, first observed in 1928 by the biologist Heitz, is a region which remains in a condensed state during interphase. In contrast to the heterochromatin material, the rest of the chromosome, the *euchromatin*, appears uncoiled, in a swollen state, and stains only lightly with basic dyes. The two chromatin materials differ not only in their staining properties but also in their respective activities and physiological functions. A further discussion of the nature of the heterochromatin and euchromatin will be presented later in this chapter.

D. Diplonema

The end result of meiotic diplonema is the longitudinal separation by some undetermined force of the paired chromosomes. As the chromatids separate, and open up, loops and nodes which are characteristic of the diplonema stage may form. The final diplonema shape of the chromosomes is determined by the number of chiasmata. The chiasmata help

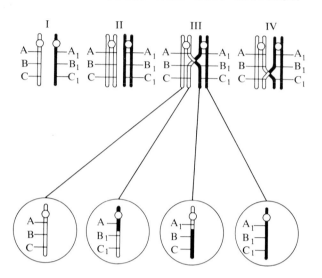

Fig. 8.17. Crossing-over. I. A pair of homologous chromosomes heterozygous for three pairs of loci, AA, BB, CC. II. 2-strand (tetrad) stage. III. Crossing-over between two strands only in region AA, BB. IV. Same in regions BB, CC.   From D. G. Harnden, *in* "Chromosomes in Medicine," Heineman, London, 1962.

preserve the bivalent structure of the chromosomes by preventing the complete separation. The bivalent with one nonterminal chiasmata would assume a crosslike shape whereas the bivalent with one terminal chiasmata would appear rod shaped; bivalents with terminal chiasmata in each arm assume a ring shape. The appearance of chiasmata at the nodal regions is suggestive of an exchange of partners between the two pairs of chromatids.

The chromosomes are also undergoing an active shortening process and an increase in coiling during this stage. In addition, a diminution in the size of the nucleolus may be noted.

E. Diakinesis

Diakinesis represents the last stage of the meiotic prophase. The chromosomes continue to contract and the nucleolus gradually disappears or becomes detached from the nucleolar organizer site on the chromosome. By late diakinesis, when the last remnant of the nucleolus has disappeared, the chromosomes appear now in a very contracted heteropycnotic, i.e., deeply staining, form. During the diplonema and diakinesis, the decrease in the nucleolar size coincides chronologically with the rapid increase in the chromosomal basophilia, suggesting a reincorporation of the nucleolar material into the chromosomes. Substantial biochemical proof for this attractive hypothesis is not presently available.

The bivalents migrate to the periphery of the nucleus during diakinesis where although separated from one another they lie in close apposition to the nuclear membrane. Other events include the dissolution of the nuclear membrane and the formation of the spindle. With the complete disappearance of the nuclear membrane, diakinesis and prophase come to an end.

F. First Metaphase

After the disappearance of the nuclear membrane, the formation of the bipolar spindle begins. The lapse of time between the disintegration of a nuclear membrane and the initiation of an orientation of the bivalents is considered the prometaphase. At full metaphase, the two homologous centromeres lie along the longitudinal axis of the spindle on opposite sides of the equatorial plate and at equilibrium, are equidistant from the spindle plate. The shape and appearance of the bivalent during the first metaphase is determined by the number and position of the chiasmata and by the relative lengths of the arms of the bivalent. The centromere is apparently the site of production of the chromosomal fibers which connect the centric regions to the nearest pole. These fibers are responsible for the movement of the chromosomes toward the poles at anaphase. In mitosis, the metaphase chromosomes possess the same number of centromeres as the diploid number of chromosomes while in meiosis each bivalent has two centromeres, i.e., the number of centromeres per cell has been halved. An active repulsion related to the movement of the chromosomes occurs between the homologous centromeres.

G. Anaphase I

The onset of anaphase is associated with the separation of each bivalent into two pairs by the movement of the centric regions toward opposite poles. As this poleward movement continues, the chiasmata loosen and slip toward the end of the chromosome, a process called *terminalization*. The chiasmata gradually slip off the free ends of the chromosome, releasing the chromatids. The homologous chromosomes have separated and reassortment of the genetic complement has been completed. The overall effect is the production of a haploid number of chromosomes instead of diploid structures.

H. Telephase I and Interphase I

With the arrival of the chromosomes at the spindle poles, the nuclear membrane reappears and the cell enters into the telophase. The cell then passes into a period of interphase followed by a second division, often called a meiotic mitosis. During the process of telophase, the chromosomes elongate by loosening of the coils and once more assume the diffused state characteristic of the interphase. The extremely short duration of the interphase may not permit the full extension of the chromosomes or the re-formation of the nucleolus as may be observed during mitotic interphase.

I. Prophase II

The second prophase is similar to a mitotic prophase. The two chromatids of each chromosome are joined by a common centromere and appear X-shaped. The four arms become widely separated and are without relational coiling. The length of the chromosomes is extended and is several-fold that of the telophasic structure. The chromatid arms progressively shorten and consequently stain more intensely with basic dyes than at the later prophasic stages.

J. Metaphase II and Anaphase II

With the disappearance of the nuclear membrane prior to the second metaphase, spindle formation will proceed. The centromere once again functionally divides and the chromosomal fibers arising from the centromeres pull the chromatids toward the opposite poles during the anaphase. It is the centromere which again is instrumental in the separation of the chromosomes. A reduction in the chromosomal number and a reconstitution is achieved during these stages.

The important consequence of the two divisions is the reduction in the number of the chromosomes to the haploid number. However, the original makeup of the chromosomes during the meiotic progression has been altered. A recombination of parts of each of the parental chromosomes has occurred during the crossing-over. Meiosis thus gives rise to four nuclei, all of which are different in regard to parental origin of each member of its haploid set.

VII. BIOSYNTHETIC ACTIVITIES DURING MEIOSIS

Autoradiographic evidence has clearly demonstrated the completion of DNA synthesis during the premeiotic interphase. In this respect, the situation is similar to the events occurring in the mitotic cell. As has been shown for mitosis, the distribution of thymidine-^{3}H in the chromatids of spermatocytes of the grasshopper conforms to a semiconservative mechanism, i.e., all the chromatids are labeled at the premeiotic interphase. It is at this time, preceding prophase, that the chromosomes also double.

The demonstration of DNA synthesis during the premeiotic interphase does not explain the subsequent processes of pairing and crossing-over. Recently, Stern and his colleagues (Hotta et al., 1966) have demonstrated with intact flower buds of Lilium and Trillium, synthesis of DNA during the meiotic prophase, i.e., zygonema and pachynema. This time period would overlap with the duration of chromosome pairing and crossing-over. The product was distinctly different from the interphase DNA in its physiochemical properties. Several possibilities are available for the explanation of these findings: (1) The crossing-over may require DNA synthesis as a break-repair mechanism. (2) The late DNA synthesis may represent the expression of a chromosomal segment which is not replicated during the premeiotic interphase and the crossing-over may be closely associated with this delayed replication. In mitosis, where essentially all DNA synthesis is complete in interphase, neither crossing-over nor delayed replication is observed. Stern's group favor the latter explanation.

Virtually all of the RNA synthesized during meiosis in the locust spermatocyte is chromosomal in origin with little RNA labeling occurring in the nucleoli. In the mitotic nucleus, the bulk of the RNA is synthesized within the nucleolus. In addition, the RNA is synthesized during the latter stages of meiotic prophase, in grasshopper spermatocytes or plant cells, stages at which the nucleoli are no longer present, at least as discrete entities. RNA synthesis may also be noted during the leptotene, zygotene, pachytene, and diplotene stages but not during metaphase or anaphase. The synthesis of RNA ceases along with the dissolution of the nuclear membrane before metaphase.

Some correlation exists between the extent of coiling of the chromosome and the degree of RNA synthetic activity. When the coiling of the chromosome is somewhat relaxed, as is apparent in the euchromatin, the capacity for RNA synthesis is greater. A number of investigators had previously observed the relative inactivity of the genes in the condensed state, i.e., heterochromatin. Protein synthesis continues through most of the meiotic cycle although the rate of synthesis decreases markedly with the increased condensation of the chromosome, i.e., metaphase.

The results of a number of studies have indicated that the most active periods of RNA and protein synthesis are associated with the leptonema and pachynema. The further development of meiosis is retarded when inhibitors of RNA or protein synthesis are applied at these stages of meiotic development. The synthesis of both RNA and protein is required for the maintenance of the condensed state of the late prophase chromosomes, the initial separation of the paired homologous chromosomes, and the orderly function of the spindle.

VIII. GENETICS AND MUTATION

Occasionally, a gene may undergo a sudden alteration with the resultant production of a new and hereditable character. The gene is said to have undergone a *mutation*. Although spontaneous occurrence is infrequent, e.g., one per 10^6 cellular divisions, the frequency may be significantly enhanced by various physical and chemical agents.

The concept of mutation has played an important role in development of our knowledge of evolution and indeed, in the very understanding of the genotype. At the turn of the nineteenth century, the prevailing scientific community, represented by Lamarck, held that a new species gradually evolved through the crossing of old existing species and the inheritance of acquired traits. Darwin, however, proposed the origin of new forms as a result of natural selection, arising from some chance occurrence. It has taken a number of years to definitively establish the latter point.

A further analysis of the transmission of these hereditable characteristics was made by the Austrian monk, Gregor Mendel, in 1865. This culminated in the postulation of the laws of heredity in which Mendel claimed that the various phenotypic events were controlled by pairs of factors which we now know to be the genotype. One of the factors was derived from the male parent, the other from the female. These factors, which occupy specific genetic loci on the homologous chromosomes and which are responsible for the same phenotypic expression, are referred to as *alleles.*

Mendel selected the pea as his experimental tool and was able to obtain a variety of strains differing in such characteristics as the shape of the seed, its color, the length of the stem, etc. After assuring himself that each type of pea would breed true, i.e., the progeny would elicit the identical phenotype as the parents, Mendel made genetic crosses between parent peas differing in a single characteristic. He observed that the progeny, the F_1 *hybrid,* exhibited the appearance of *one* parent. The parental phenotype that appeared in the F_1 hybrid was termed "*dominant*"; the nonappearing characteristic, "*recessive.*" However, one gene pair is not always clearly dominant or recessive, i.e., the phenotype may be intermediate.

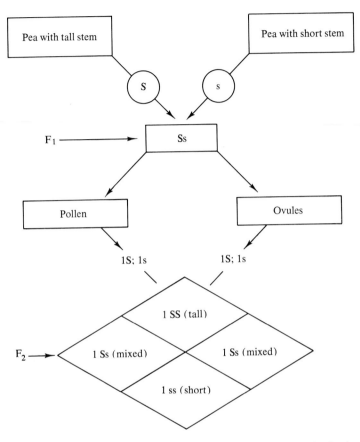

Fig. 8.18. Genetic crosses between peas with short and tall stems. S, the dominant characteristic; s, the recessive. The ratio of dominant to recessive in the F_2 offspring is 3:1.

Mendel now allowed the F_1 hybrids to cross by self-fertilization and observed the number of progeny, F_2 offspring, that bore the dominant or recessive trait. For each phenotypic expression examined, he noted the same ratio of dominant to recessive, 3:1 (see Fig. 8.18). Since the recessive allele may reappear in the F_2 offspring, his experiments indicate that the genes are not altered in any fashion but that both alleles, recessive and dominant, are transmitted and segregate during the formation of the sperm and ovum.

In Fig. 8.18, it is apparent that three types of F_2 offspring may arise: SS, which bears the identical allele for the tallness characteristic; hybrid Ss, bearing mixed alleles; ss, possessing the identical allele for the shortness characteristic. The former and latter progeny are said to be *homo-*

zygous, i.e., the allele on the homologous chromosomes is identical, and the progeny of a genetic cross between homozygous cells will always breed true. The mixed offspring, *Ss,* is said to be *heterozygous,* i.e., the alleles on the parental chromosomes differ and the resultant of a cross between heterozygotes will be of mixed constitution.

Mendel's observations are founded upon the independent behavior during the process of meiosis of the chromosomes bearing the different alleles. As we have seen, the mechanism of crossing-over first described by the Danish cytologist, Janssens, is responsible for the exchange of genes on homologous chromosomes. Consequently, those genes which lie close to one another on the chromosome would assort with each other more regularly during meiosis than genes located at a greater distance from each other. One can take advantage of this probability of recombination in mapping the relative location of one gene to another along a chromosome, i.e., a genetic map. Mutant alleles produced by a variety of techniques have been most useful in this area of research.

Chromosomal abberrations most frequently arise during the crossing-over stage of the meiotic prophase. Five such processes or types of changes have been recognized: inversion, deletion, duplication, translocation, and isochromosome formation (see Figs. 8.19 and 8.20).

During crossing-over, a region on the chromosome may break. During recombination, the segments may become inverted so that some of the genes now lie in an entirely different order (see Fig. 8.19). In this chromosomal alteration, *inversion,* the same genetic material is present although the expression of genetic information has been affected. Inversion has been observed in the giant salivary gland of *Drosophila* as a reversal in the order of the heteropycnotic regions. If the centromere is involved in this modification, the chromosomal appearance may be changed, e.g., from acrocentric to metacentric (see Fig. 8.19).

In *deletion,* an incorrect joining of the broken ends of the chromosome lying in a looped position may lead to the elimination of a chromosomal segment (Fig. 8.19). In addition to the production of the major chromosomal structure, we may also observe a minor looped or ring chromosome.

In *duplication,* the same sequence of the genes may appear twice in the same chromosome. The abnormality actually represents a special

---

Fig. 8.19. Chromosome structural change (1) paracentric inversion, (2) pericentric inversion, (3) deletion, (4) ring chromosome, (5) duplication, (6) shift, (7) translocation, (8) dicentric chromosome. From D. G. Harnden, *in* "Chromosomes in Medicine." Heineman, London, 1962.

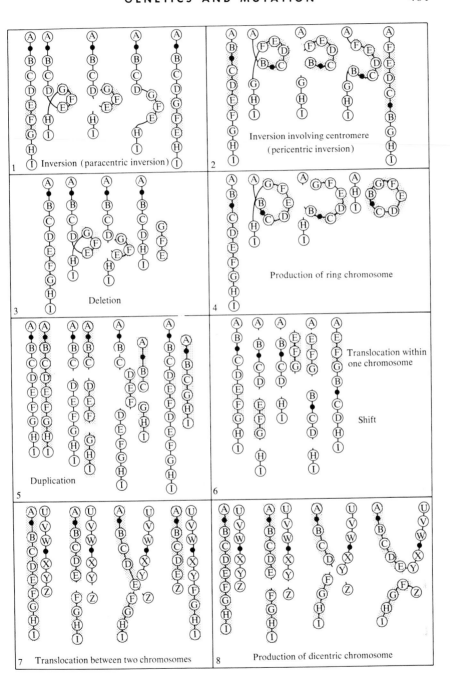

1 Inversion (paracentric inversion)

2 Inversion involving centromere (pericentric inversion)

3 Deletion

4 Production of ring chromosome

5 Duplication

6 Translocation within one chromosome

Shift

7 Translocation between two chromosomes

8 Production of dicentric chromosome

NORMAL SPLITTING

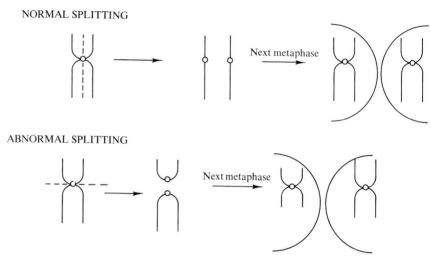

Next metaphase

ABNORMAL SPLITTING

Next metaphase

Fig. 8.20. Chromosomal structural abnormalities in isochromosome formation.

case of *translocation* involving an exchange between homologous chromosomes (see below). During *translocation*, a segment of one chromosome may be transferred to a completely different site on the same chromosome, e.g., a shift, or even to a site on a different chromosome. The broken ends may rejoin in various ways with the formation of chromosomes possessing two centromeres, *dicentric chromosomes;* a piece of chromatin which possesses no centromere, an *acentric fragment.* The latter cannot take part in the formation of the spindle and is generally lost to the daughter cells.

An altered chromosomal structure may also be produced in *isochromosome formation* in which a perfectly metacentric chromosome with two completely homologous arms may be united at the centromere (Fig. 8.20). Isochromosomes may arise at anaphase by a horizontal instead of a longitudinal splitting of the chromosome.

Another situation may arise which, although it does not give rise to a chromosomal structural abnormality, does produce a cell with a chromosomal lesion. In *nondisjunction*, the two members of the chromosome pair fail to separate at meiotic anaphase, perhaps as a consequence of an unsuccessful pairing during meiotic prophase, e.g., the two chromosomes do not orient at metaphase. In this case, both chromosomes may migrate to the same pole of the nucleus.

Several of the abnormalities in chromosomal structure or in the chromosomal separation may arise during or before the process of crossing-over and may result in the production of a *mutant* cell. It is in this area of molecular genetics where progress has very recently been made. The molecular definition of some of the inborn errors of metabolism originally postulated by Garrod in 1909 and of the hemoglobinopathies has proved intellectually enlightening. Unfortunately, research in this area is in a perinatal stage and the future prospects must include a more intensive study at the molecular level of underlying genetic organization and of procedures for the correction and prevention of these chromosomal abnormalities.

IX. CHROMOSOMES IN GENETICS

The existence of several unusual chromosomes has proved of enormous importance to the field of genetics in providing valuable information about the structure of the genetic material. These include (1) the *polytene chromosomes*, (2) the *lampbrush chromosomes*, and (3) the *sex chromosomes*. A brief discussion of these structures will follow.

A. Polytene Chromosomes

The polytene chromosomes are found in certain larval tissues and especially in the salivary gland nucleus of dipterous flies (see Fig. 8.21). Although they were first observed in 1881 by Balbiani, little attention was accorded them for the next 50 years. The homologous pairs of the polytene chromosomes are closely associated and are considered to be in permanent prophase. Their appearance resembles a giant tape which may total 2000 μ in length as compared to 7.5 μ for the somatic cell chromosomes. Bands rich in DNA and resembling discs of varying sizes traverse the polytenes while the interband regions are relatively devoid of nucleic acids (Plaut and Nash, 1965).

The interbands, on the other hand, are fibrillar in nature. The discs are constant in number along a chromosome and are located and distributed in homologous chromosomes in defined regions. Detailed maps of the linear structure of each of the chromosomes have been constructed.

In *Drosophila*, the polytene chromosomes occur as long strands (see Fig. 8.22) which, along with a nucleolus, are attached to a central region called a *chromocenter*. The polytene chromosomes are not uniform in diameter and, periodically, puffs or bulbs arise at defined positions along the chromosomal length. In addition to these local swellings, one may also observe structures that are larger than the puffs but of similar nature, the Balbiani rings. A correlation exists between the position of these rings and some defined stage of development. The synthetic activity of Balbiani rings is extremely pronounced and the puffing that is the principal feature of a Balbiani ring is generally accompanied by a burst in the syntheses of RNA and protein. A localized increase in the incorporation of labeled thymidine into DNA may also be noted at these regions (see Fig. 8.22). The completion of the synthetic activity is accompanied by a gradual deflation of the puffed region.

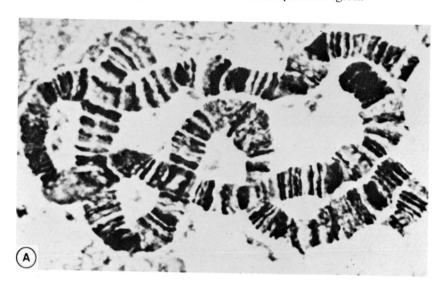

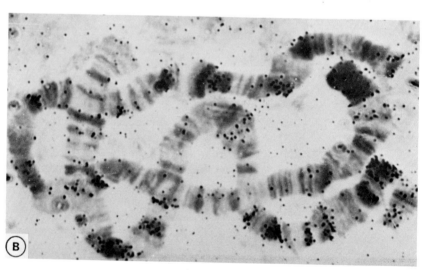

Fig. 8.21. Polytene chromosomes from *Drosophila* salivary gland. A. Phase contrast photograph of two chromosomes of squashed *Drosophila* salivary gland nucleus after staining with light acetoorcein. × 1900. B. Same field as above. Incubated for 5 – 8 minutes with thymidine-³H (15 μc/ml, 10 mc/μmole). AR-10 autoradiographic emulsion was applied and was exposed to the chromosomes for 17 days. Courtesy of W. Plaut.

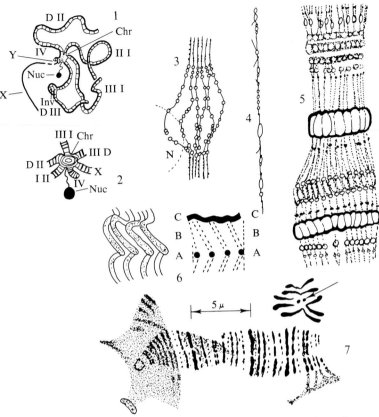

Fig. 8.22. Structure of the polytene chromosomes. 1. General schematic aspect of the chromosomes of the salivary gland of a male of *Drosophila melanogaster* after they have been spread out by crushing the nucleus. The paternal chromosome (in white) and the maternal one (in black) are paired. Chr, chromocenter; D II and II I, right and left arms of chromosome II; D III and III I, right and left arms of the third chromosome; *IV*, the fourth chromosome; Inv, an inversion in the right arm of the third chromosome; Nuc, nucleus; X and Y indicate respectively the sex chromosomes. 2. The chromocenter (Chr) formed by the union of the heterochromatic parts of all the chromosomes in a female of *D. melano-gaster.* (The other symbols are the same as for 1.) 3. A heterochromatic region of the X chromosome of *D. pseudoobscura,* showing its relations with the nucleolus (N) and the filamentous (chromonemic) constitution of the chromosome. 4. Detail of a component chromonema of the polytene chromosome in which the different chromomeres are seen. 5. Schematic structure of the chromosome of *Simulium virgatum,* showing the organization of the chromonemata, chromomeres, and vesicles, which together give the appearance of the bands. The segment drawn corresponds to a euchromatic zone. 6. Diagram to illustrate the interpretation of the helicoidal chromonema and the false chromomeres produced by the turns of the spiral. A zone (B) with four chromonemata is shown between two consecu-tive bands (at the left). To the right is the aspect of the same region when observed in a different focusing plane. *A* has a granular aspect, which stimulates chromomeres. *C* appears as a continuous solid line. From E. D. P. DeRobertis, W. W. Nowinski, and F. A. Saez, "Cell Biology." Saunders, Philadelphia, Pennsylvania.

B. Lampbrush Chromosomes

The lampbrush chromosomes were discovered in 1892 by Ruckert and occur as extremely long pairs held together at only a few points. They are found in oocytes during the leptonema phase of meiosis and are even longer than the polytene chromosomes. The total length of the chromosomal pair may approach 5900 μ, i.e., three times the expanse of the polytene chromosomes. Chromosomes of ordinary size are also present in oocytes but only the giant lampbrush chromosomes occur at the time of augmented growth (Fig. 8.23); the periods of growth are closely correlated with the presence of lampbrush chromosomes. The chromosomes are held together at only a few points, the *chiasmata*, and each chromosome exists as a row of *chromomeres*, approximately 1 μ

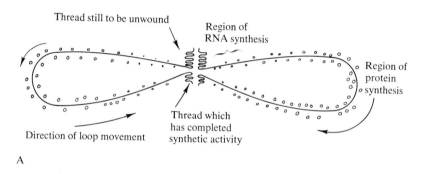

Thread still to be unwound

Region of
RNA synthesis

Region of
protein
synthesis

Direction of loop movement

Thread which
has completed
synthetic activity

A

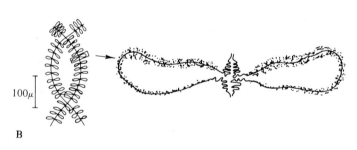

100μ

B

Fig. 8.23. A. Diagram of a loop of a lampbrush chromosome. Based on labeling with RNA precursors (uridine-^{3}H), it is postulated that first RNA synthesis and then protein synthesis take place along the loop. B. Diagram of the lampbrush chromosomes of the oocyte of *Triturus*. Left, low magnification. Right, higher magnification, showing the lateral expansions in the form of a handle and the spiralization of the chromonemata. From E. D. P. De-Robertis, W. W. Nowinski, and F. A. Saez, "Cell Biology." Saunders, Philadelphia, Pennsylvania.

in diameter, and along with pairs of loops of varying dimensions are laterally extended. The central axis of the lampbrush chromosomes consists of at least four chromatids to which the loops are projected out laterally. Each of the loops is composed of DNA surrounded by protein and RNA-like material. The loops are very similar to the Balbiani ring-like structures previously discussed in that at both these regions bio-synthetic activity is more marked. However, unlike the polytenes, the loops of the lampbrush chromosomes are present at almost all the chromomeres while puffs occur at only 2% of the bands of the polytenes. The lampbrush chromosomes are also programmed for replication after fertilization while the polytene chromosomes will not divide and indeed are removed at pupation.

The existence of the loops is directly dependent upon the continuing function of the chromosome and especially upon the functional capacity for RNA biosynthesis (Allfrey and Mirsky, 1962). The addition of an inhibitor of DNA-dependent RNA polymerase, actinomycin D, to either the intact oocyte or to the lampbrush chromosome itself abolishes RNA synthesis at the loops and also leads to their disintegration. Agents which do not block the synthesis of RNA have little effect upon the presence or structure of the loops.

C. Sex Chromosomes

In 1891, Henking published his observations on the behavior of the chromosomes during the formation of sperm. Of the 23 chromosomes in the cell of the organism he was studying, 11 were paired, i.e., both members exhibited the same morphological appearance. The extra chromosome had no mate and was called by Henking, X. During the first meiotic division, the 22 chromosomes synapsed into pairs which later formed bivalents. The X chromosome, however, did not pair with any other chromosome. As a result of the meiotic divisions, four progeny cells of two cell types arise which possess (1) 11 regular chromosomes or *autosomes*, or (2) the autosomes and the X chromosome or accessory chromosome. The X chromosome is unique in its affinity for basic dyes, its movement to the poles of the spindle, and in the lack of any mate.

The establishment of the accessory chromosome as a determinant of the sex of the cell was made by McClung in 1901 and further elaborated by the American cytologist, Wilson. Wilson reported the presence in some species, such as the human, of another sex chromosome found only in cells from males, the *Y chromosome.*

Two types of chromosomal behavior have been delineated, the XO-XX type and the XY-XX type. In the former, a single X chromosome is found in the male, i.e., XO, while the female possesses two X chromosomes, XX. The number of autosomes is identical in both sexes. During meiosis, two types of sperm are elaborated which contain as their chromosomal complement (1) only autosomes or (2) autosomes and a single X chromosome. In the ova, on the other hand, are found both the autosomes and a single X chromosome. During fertilization, or *syngamy*, two combinations are possible, (1) XO and autosomes characteristic for a male, and (2) XX and autosomes, a female.

In the XY-XX type, both the male and the female possess sex chromosomes; this is the predominant type in man (see Figs. 8.3A, B). In the female, the pair is identical and is symbolized as XX. The male, on the other hand, has only one X chromosome and the other morphologically distinct structure, the Y chromosome. Although the latter is structurally unique, synapsis (pairing) does take place between the X and Y, indicating the existence of some common features. After a mitotic division, the newly formed ova would possess autosomes and one X chromosome while the sperm may exist in two forms, one in which are located the autosomes and one X chromosome and in the other, the autosomes and one Y chromosome. Syngamy of an ovum with an X-bearing sperm would yield a zygote with the diploid set of autosomes and XX, a female. Syngamy of an ovum with a Y-bearing sperm would yield a zygote with the diploid set of autosomes and XY, a male. One may conclude from these studies that (1) only at fertilization is sex determined, (2) sex determination is the responsibility of the sex chromosomes, (3) the two sexes should be found in approximately equal numbers, (4) *the Y chromosome is the sex determinant in man.*

Sex Chromatin. In 1949, Barr and Bertram made an important discovery—a small chromatin-containing body present in the nerve cells of the female but absent from the male cat. Similar observations were then made in other tissues from other species, including the human. The chromatin-containing body, the *sex chromatin* or Barr body (see Fig. 8.24), is found in the nucleus of cells.

The sex chromatin stains vary intensely with basic dyes, yields a positive reaction for DNA, and its position within the cell is uniformly constant in each tissue and species. It may be found attached to the nucleolus, attached to the nuclear membrane, in the nucleoplasm or as an extension to the nucleus. The study of the sex chromatin has proved of great importance in the field of medical genetics and makes possible the correlation of a number of congenital diseases to abnormalities of chromosome structure or to presence of abnormal Barr bodies.

The sex chromatin originates from only one of the two X chromosomes. The number of Barr bodies generally equals one less than the total

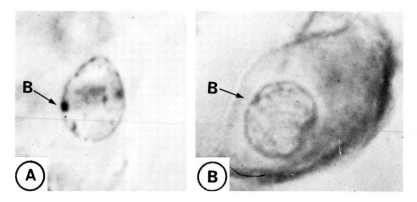

Fig. 8.24. Barr bodies (sex chromatin). A. The Barr body (B) is present in the nucleus of a cell derived from normal human skin. The cell had been stained with hematoxylin and eosin (approximately × 1800). B. The Barr body (B) is evident in the nucleus of a cell derived from normal human buccal mucosal membrane (stained with Feulgen's reagent; approximately × 2000). Courtesy of G. Clayton.

number of X chromosomes, a relationship which is particularly important in conditions wherein an abnormal number of sex chromosomes is present.

The X chromosome which has given rise to the sex chromatin is generally considered to be genetically inactive and the genes carried on the heterochromatic Barr body are considered to be unexpressed. The DNA of the latter replicates at a later time than the other chromosomal segments and appears also to be much less active in RNA synthesis. Autoradiographic studies with cultured cells from normal mammalian females indicate that only one of the X chromosomes will complete its duplication along with the autosomes while replication continues in the remaining X chromosome. The late-replicating X chromosome is the one that becomes genetically inactivated during the development of the embryo.

A number of other phenotypic events are associated with the sex chromosomes, expressions of the so-called *sex-linked genes*. Three types of sex-linked genetic expression may appear: X-linked by genes localized in the nonhomologous section of the X-chromosome; Y-linked; XY-linked, localized in one chromosomal segment that is homologous in both X and Y.

In *Drosophila,* the gene responsible for the red coloration of the eye is carried on the X chromosome. In man, the genes that determine red-green color blindness and hemophilia are carried on the X chromosome. The latter, a defect in a blood clotting mechanism, is inherited as a sex-linked recessive gene and is rarely found in a female. In *Drosophila,* approximately 150 genes are carried on the Y chromosome, although in man, only a few have been discovered. Certain skin diseases have been linked to the human Y chromosome. Several defects in the human are also associated with the SY chromosomal segment, for example, total color blindness, night blindness, several skin diseases, and a disease in which the predominate feature is bleeding from the nose or other mucosae.

X. CONCLUSION

In this discussion, we have not touched upon several questions of crucial importance. What is the "triggering force" which incites the cell to undergo mitosis or meiosis? This question exposes a veritable chasm in our knowledge of cellular activities and is part of the overall mystery of the cell. Other basic questions are: What causes the cell to differentiate during normal embryonic development? What is the "triggering" force which dedifferentiates a normal into a neoplastic cell? Akin to these enigmas is another point worthy of research endeavours but one which is beyond the scope of this text—what are the events that transpire during syngamy?

With the advent of more sensitive autoradiographic techniques, the science of cytogenetics has been extended. Hopefully in the future, a more complete picture of chromosomal activities *in situ* may result. During the next few years, perhaps the mechanism for chiasmata formation will be uncovered, the reasons for the late replication time of the sex chromosomes will be elucidated, and inroads to the sophisticated science of molecular genetics will be broadened so that we may better understand not only the processes of growth and development but also events of disease.

GENERAL REFERENCES

Darlington, C. D. (1939). "The Evolution of Genetic Systems." Basic Books, New York.
DeRobertis, E. D. P., Nowinski, W. W., and Saez, F. A. (1965). "Cell Biology." Saunders, Philadelphia, Pennsylvania.
Haggis, G. H., Michie, D., Muir, A. R., Roberts, K. B., and Walker, P. M. B. (1965). "Introduction to Molecular Biology." Wiley, New York.
Harnden, D. G. (1962). "Chromosomes in Medicine." Heineman, London.
King, R. C. (1962). "Genetics." Oxford Univ. Press, New York.
Lewis, K. R., and John, B. (1963). "Chromosome Marker." Little, Brown, Boston, Massachusetts.
Moore, J. A. (1963). "Heredity and Development." Oxford Univ. Press, London and New York.
Swanson, C. P. (1957). "Cytology and Cytogenetics." Prentice-Hall, Englewood Cliffs, New Jersey.

Watson, J. D. (1965). "Molecular Biology of the Gene." Benjamin, New York.
White, M. J. D. (1961). "The Chromosomes." Wiley, New York.
Winchester, A. M. (1958). "Genetics." Houghton, Boston, Massachusetts.

SPECIFIC REFERENCES

MITOSIS AND CELL LIFE CYCLE

Barer, R., Joseph, S., and Meek, G. A. (1959). *Exptl. Cell Res.* 18, 179-184.
Brown, F., and Danielli, J. F. (1964). *In* "Cytology and Cell Physiology" (G. H. Bourne, ed.), pp. 239-310. Academic Press, New York.
Howard, A., and Pelc, S. R. (1953). *Heredity* 6, Suppl., 261-273.
Hughes, H. (1952). "The Mitotic Cycle." Academic Press, New York.
Luykx, P. (1965). *Exptl. Cell Res.* 39, 643-657.
Mazia, D. (1961). *In* "The Cell" (J. Brachet and A. E. Mirsky, eds.), Vol. 3, pp. 77-412. Academic Press, New York.
Mazia, D., and Dan, K. (1952). *Proc. Natl. Acad. Sci. U.S.* 38, 826-838.
Monesi, V. (1964). *J. Cell Biol.* 22, 521-532.
Pritchett, R. H. (1960). *Symp. Soc. Gen. Microbiol.* 10, 000.
Salb, J. M., and Marcus, P. I. (1965). *Proc. Natl. Acad. Sci. U.S.* 54, 1353-1358.
Wolpert, L. (1963) *In* "Cell Growth and Cell Division" (R. J. C. Harris, ed.), pp. 277-298. Academic Press, New York.

CHROMOSOMAL REPLICATION

Freese, E. (1958) *Cold Spring Harbor Symp. Quant. Biol.* 23, 13-18.
Levinthal, C. (1956). *Proc. Natl. Acad. Sci. U. S.* 42, 394-404.
Meselson, M., and Stahl, F. W. (1958). *Proc. Natl. Acad. Sci. U. S.* 44, 671-682.
Meselson, M., Stahl, F. W., and Vinograd, J. (1957). *Proc. Natl. Acad. Sci. U. S.* 43, 581-588.
Taylor, J. H. (1957). *Am. Naturalist* 91, 209-221.
Taylor, J. H. (1962). *Intern. Rev. Cytol.* 13, 39-73.
Taylor, J. H. (1963). *Mol. Genet.* Part I, pp. 65-111.
Taylor, J. H., Woods, P. S., and Hughes, W. J. (1957). *Proc. Natl. Acad. Sci. U. S.* 43, 122-128.

MEIOSIS

Das, C. C., Kaufman, B. P., and Gay, H. (1964). *Exptl. Cell Res.* 35, 507-514.
Das, N. K., Siegel, E. P., and Alfert, M. (1965). *J. Cell Biol.* 25, 387-395.
Hotta, Y., and Stern, H. (1963). *J. Cell Biol.* 19, 45-49.
Hotta, Y., Ito, M., and Stern, H. (1966). *Proc. Natl. Acad. Sci. U. S.* 56, 1184-1191.
Kemp, C. L. (1964). *Chromosoma* 15, 652-665.
Lima de Faria, A. (1962). *Progr. Biophys. Biophys. Chem.* 12, 282-294.
Rhodes, M. M. (1961). *In* "The Cell" (J. Brachet and A. E. Mirsky, eds.), Vol. 3, pp. 1-79, Academic Press, New York.
Taylor, J. H. (1965). *J. Cell Biol.* 25, 57-67.

POLYTENE, LAMPBRUSH AND SEX CHROMOSOMES; HETEROCHROMATIN

Allfrey, V. G., and Mirsky, A. E. (1962). *Proc. Natl. Acad. Sci. U. S.* 48, 1590-1596.
Barr, M. L., and Bertram, E. G. (1949). *Nature* 163, 676-677.
Frenster, J. H., Allfrey, V. G., and Mirsky, A. E. (1963). *Proc. Natl. Acad. Sci. U. S.* 50, 1026-1033.

Hamerton, J. L. (1961). *Intern. Rev. Cytol.* 12, 1-68.

Izawa, M., Allfrey, V. G., and Mirsky, A. E. (1963). *Proc. Natl. Acad. Sci. U. S.* 49, 544-551.

Izawa, M., Allfrey, V. G., and Mirsky, A. E. (1963b). *Proc. Natl. Acad. Sci. U.S.* 50, 811-817.

Lima de Faria, A. (1959). *J. Biophys. Biochem. Cytol.* 6, 457-466.

Mukerji, B. B., and Sinha, A. K. (1964). *Proc. Natl. Acad. Sci. U. S.* 51, 252-259.

Plaut, W., and Nash, D. (1965). *In* "The Role of Chromosomes in Development" (M. Locke, ed.), pp. 113-135. Academic Press, New York.

Chapter IX

SPECIAL CELLS

While basic physiological and biochemical aspects of the general cell type have been discussed in detail in previous chapters, a number of specific cells having specialized functions, are of sufficient interest to warrant separate consideration. These include the muscle cell and the cerebral cortical cell.

I. THE MUSCLE CELL

There are three distinct types of muscles in the mammalian organism: skeletal or striated, smooth, and modified-striated or heart muscle. In general, the three are qualitatively similar in that they all perform the basic function, contraction. The differences are generally noted in the amount and distribution of the sarcoplasmic reticulum, mitochondria, and in the arrangement of the contractile fibers, depending upon whether or not the muscle is rapid acting, slow acting, or sustained acting.

Skeletal muscle may be divided into two general types, "red" and "white," with the former being the slower but more sustained and the latter being in general, more rapid acting. We will consider the cardiac muscle cell in depth since it represents a type of modified skeletal muscle, probably analogous to the red variety. Since the basic contractile apparatus appears to be similar in all three general types of muscle, basic aspects of cardiac muscle probably will apply to the others. However, whenever differences are known, they will be discussed.

THE HEART

A. Introduction

In approximately one human life span, about 70 years, the heart beats over 2,700,000,000 times and pumps over 380,000,000 quarts of blood.

Each minute this muscle performs at least 45 foot-pounds of work, which is enough work every hour to lift a compact car a foot off the ground at a remarkable efficiency of over 40%. Of all body tissue, muscle, and particularly heart muscle, is unique in that the conversion of *chemical energy* into kinetic energy (contraction) is as obvious as the transformation of gasoline into automobile movements.

Muscle biochemistry began in 1859 with the discovery of myosin, by Kühne. He showed that skeletal muscle gels *in vitro* upon the addition of certain salt solutions. In 1907 Fletcher and Hopkins observed the accumulation of lactic acid in contracting frog muscle. This led Meyerhof, in 1930, to the discovery that the lactic acid was derived from glycogen · through a glycolytic pathway; he suggested that lactic acid was involved in muscle contraction. However, in the same year Lundsgaard clearly showed that lactic acid was not necessary for contraction; he blocked the glycolytic pathway with iodoacetic acid, which prevented the formation of lactic acid, and was able to observe contractions. The simultaneous discovery of ATP and creatine phosphate by Löhmann and by Fiske and SubbaRow in 1930 opened the door to a modern biochemical consideration of muscle contraction. Szent-Györgyi and his colleagues made the initial contributions which established the beginnings of molecular muscle biology. These included the development of muscle models, the characterization of the muscle proteins, myosin and actin; and the construction of theories on the mechanism of action of muscle contraction on a biochemical and biophysical level.

The first evidence that the contraction of muscle is an enzymic process was presented by Engelhardt and Ljubimova in 1939 who showed that myosin was, in fact, an ATPase. The fundamental thermodynamic aspects of muscle are credited to A. V. Hill and his colleagues, who studied and developed such phenomena as *birefringence, latency, tension-length concepts, the "active state," force-velocity concept,* and energetics involved in the various stages of contraction and relaxation. During the decade 1950 to 1960 the electron microscope brought forward a new era in muscle research. Through the use of this instrument as well as the X-ray defraction, polarizing and interference microscopes (see Chapter I), Huxley and others have "dissected" the contractile apparatus to a degree of sophistication which has culminated in the development of a rational hypothesis of muscle contraction.

Modern muscle physiologists and biochemists are now probing into the proteins involved in the contractile event, studying amino acid content, active sites, etc. The relaxation event is now considered to be an active process, requiring energy. The involvement of calcium ions, sodium and potassium, is receiving major attention. The interrelationship among the cellular organelles and the contractile apparatus in the muscle cell is being examined in detail. Thus, it may be seen that during the past 50 years many disciplines have contributed toward our basic understanding of muscle function.

B. Ultrastructure

The use of the electron microscope has greatly extended our knowledge of the basic structural aspects of the muscle cell. The reader is referred to the electron micrographs and the diagrammatical representation of the cardiac muscle cell (Figs. 9.1–9.4) for orientation during the discussion of morphology.

1. GENERAL ORIENTATION

The muscular nature of the heart was first recognized by Stensen in 1664. With the advent of the light microscope, the resemblance of the heart to skeletal muscle became evident, with the main difference residing in the syncytial arrangement of the fibers of the heart and the presence of a characteristic structure, the *intercalated disc.* Like skeletal muscle, cardiac muscle is striated, due to the orderly arrangement of the myofibrillar elements (this will be discussed below). The position of nuclei of the cardiac muscle cell resembles that found in smooth muscle, i.e., they are centrally placed. The branching of the cells led early workers to postulate an *anatomical syncytium concept,* meaning that the heart is, essentially, one cell. Since the muscles appear to function in an all or nothing manner, i.e., the rhythmic beats appear to include all the fibers of the muscle at once, the concept of an anatomical syncytium was logical. However, as we will discuss later, this idea is no longer tenable.

The sarcoplasm of the cardiac cell is much more granular than skeletal or smooth muscle, and the transverse striation of the fibers appears less distinct.

2. THE SARCOLEMMA

Schwann and Bowman were the first to describe this as a "thin, structureless membrane investing a striped muscle". The sarcolemma surrounds the longitudinal surfaces of the cardiac muscle cell and consists of two membranes, the outer one being referred to as the *basement membrane* and appearing somewhat thicker than the underlying *plasma membrane.* Both membranes, however, are essentially similar in structure. The basement membrane may be up to 500 Å thick and may have a polysaccharide coating, perhaps similar to the cell wall of plants. The plasma membrane as well as the basement membrane seem to adhere to the unit

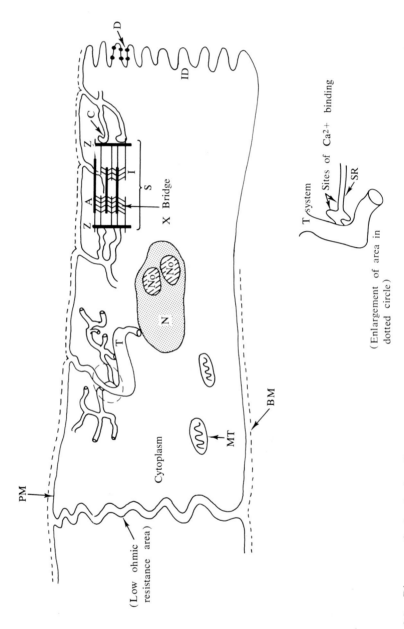

Fig. 9.1.   Diagrammatic sketch of cardiac muscle cell, PM, plasma membrane; BM, basement membrane; T, T-system; C, cisterna of SR; MT, mitochondrion; S, sarcomere of myofibril; I, I band of sarcomere (actin); A, A band of sarcomere (myosin) (note cross bridges between A and I); N, nucleus; No, nucleolus; ID, intercalated disc; D, desmosome; SR, sarcoplasmic reticulum; Z, Z line (borders one sarcomere).

membrane structure as defined by Robertson and discussed in Chapter II. In some portions of the plasma membrane, vesicles called *caveolae intracellulares* are evident. These may be pinocytotic vesicles or part of the sarcoplasmic reticulum. The dimensions of the plasma membrane are

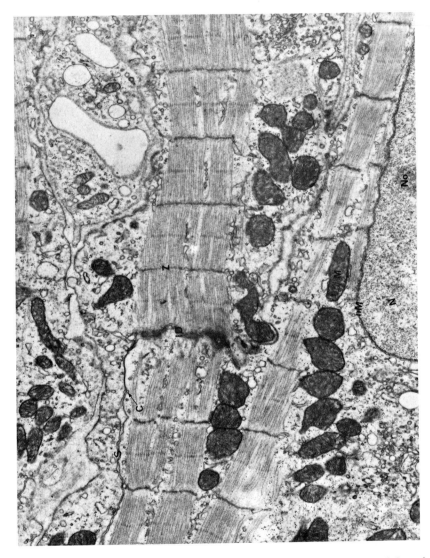

Fig. 9.2.   Ultrastructure of the human heart cell. Fixed in osmium tetroxide and glutaraldehyde. S, sarcolemma; ID, intercalated disc; M, mitochondrion; Z, Z line; N, nucleus; No, nucleolus; Nm, nuclear membrane; C, caveola intracellulares. × 16,500. Courtesy of Dr. E. Rabin, Department of Pathology, Baylor University, College of Medicine.

quite similar to those described for all membranes. In some sections or areas of the plasma membrane, invaginations occur. This inversion of the membrane becomes the *intercalated disc*, which is an intercellular boundary (Figs. 9.2 and 9.4). The *perimembrane* or *basement membrane* is not a part of the intercalated disc. Consequently, there appears to be no

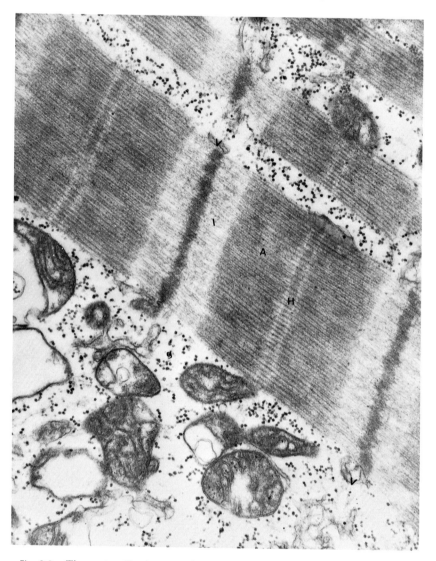

Fig. 9.3. The contractile elements of a human heart cell. Fixation and preparation same as Fig. 9.2. G, glycogen; I, I band (actin); A, A band (myosin); V, vesicular element of the T-system; H, H zone. × 48,750. Courtesy of Dr. E. Rabin.

direct communication between the intercellular and the extracellular space at the intercalated disc area. In other portions of the surface membrane the entire sarcolemma invaginates to become part of the *transverse sarcoplasmic reticulum* or T-system (see below). These latter invaginations occur characteristically at the Z line.

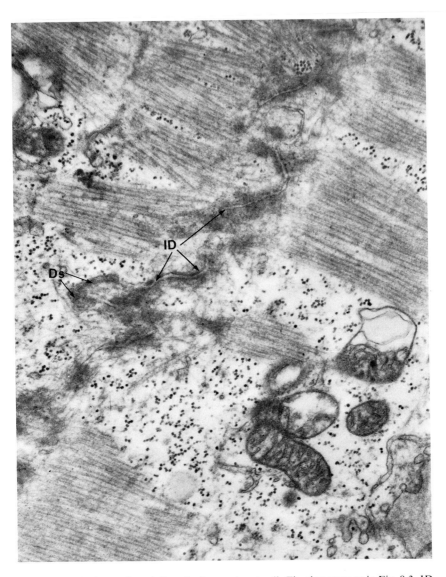

Fig. 9.4. The intercalcated disc of a human heart cell. Fixation same as in Fig. 9.3. ID, intercalated disc; Ds, desmosome areas. × 42,000. (Courtesy of Dr. E. Rabin.)

3. THE INTERCALATED DISC

As may be seen from the electron micrographs (Figs. 9.2 and 9.4) and the diagrammatic sketch (Fig. 9.1), the intercalated disc is the end-to-end separation boundary of cardiac muscle cells. Through the efforts of Sjöstrand, Anderson, Poche, Lindner, and Fawcett, it is now generally accepted that the intercalated disc separates cell territories in a transverse direction. The original concept of an anatomical syncytium, therefore, is no longer true.

At the edge of the intercalated disc the basement membrane bridges the gap between the opaque layers of the plasma membranes without showing any deviation from its course. In the intersarcoplasmic regions of the intercalated disc, one finds the very dense osmiophilic "S-region." The function of the S-region, known also as *desmosome*, or *desmosomic region* (Fig. 9.4), is still obscure, but the fact that this type of structure is most prevalent in the more rapidly conducting portions of the heart muscle suggests that these are areas of high enzymic activity. The desmosomic type of intercalated disc is usually found in pairs, in apposition, on the cytoplasmic aspects of the limiting cell membranes. The desmosomes are located along the lateral surfaces of the cardiac cells at the level of the Z line and are thought by some to be precursors of the intercalated disc structure. It is of interest that the intercalated disc does not appear, or at least is not clearly evident, during embryogenesis or early fetal life. After birth, they increase in number and complexity. However, the failure to observe intercalated discs in embryonic tissue may be explained by the minute dimensions of the embryonic discs.

It has been suggested that the intercalated disc is the site of growth of myofibrils. The structure may also secure or anchor myofibrils to the plasma membrane at the level of the Z line (see Figs. 9.1 and 9.7). The junction between cells, therefore, is very firm.

The possible functions of the intercalated discs may be summarized as follows:

1. Represent areas of low electrical resistance, to facilitate conduction from cell to cell.
2. Anchor and secure the myofilaments.
3. Represent the site of protein synthesis of the myofibrils.
4. Contains a high concentration of enzymes, which may serve as "boosters" of the contraction wave, from cell to cell.

4. THE NUCLEUS

This organelle in the heart has been the least investigated. The main difficulty has been techniques involved in the isolation of uncontaminated and intact nuclei. Recent methods developed in this laboratory have produced relatively purified nuclei (see Fig. 9.5). These are pale, usually centrally located, and ellipsoidal in shape (Fig. 9.6). At the end of each nucleus there usually appears to be a cone-shaped accumulation of sarcoplasm, free from myofibrils, containing small yellowish-brown pigment granules which appear to increase with age of the heart muscle. The number of nuclei found in the embryonic heart is far greater than that found in the adult heart.

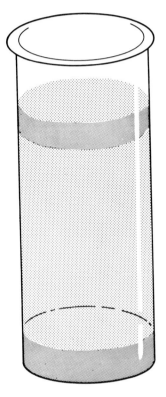

Fig. 9.5 Preparation of heart muscle nuclei from adult guinea pig. The tissue is homogenized in 0.32 M sucrose + 10 mM CaCl$_2$ so that a 5% (w/v) homogenate is obtained. The homogenate is centrifuged at 1900 rpm (about 600 g) for 10 minutes. The supernatant liquid from this centrifugation is decanted and the pellet resuspended in 10 ml of the homogenizing medium, using a motor-driven, loose-fitting rubber pestle in a polyethylene centrifuge tube. The suspension is centrifuged at 1900 rpm for 6 minutes. This process was repeated five times. The pellet is resuspended in the 2.0 M sucrose solution containing 10mM CaCl$_2$ and is centrifuged at 25,000 rpm for 1 hour in a Spinco Model L refrigerated ultracentrifuge (Rotor No. SW-25.1). Appearance of tube after centrifugation for one hour at 100,000 g in the SW-25.1 rotor. Top layer, mostly myofibrils and cell debris; few nuclei. Central clear area, little cellular material present. Bottom layer, fluffy layer of purified nuclei. From G. Ferguson, K. Smetana, A. H. Laseter, and A. Schwartz, *Cardiovascular Res. Center Bull.* **4,** 13 (1965).

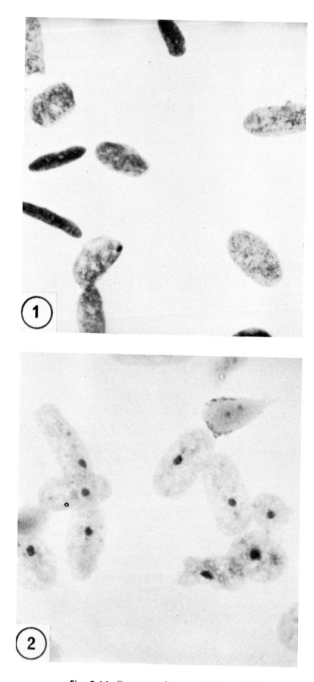

Fig. 9.6A. See opposite page for legend.

The micrographs indicate that the nucleus of the heart muscle cell is similar in structure to nuclei from other cell types, possessing a perinuclear membrane, one or more nucleoli, nucleolar-associated chromatin, and a somewhat granular matrix. The nucleoli (Figs. 9.2 and 9.6) possess characteristic light central areas, which are indicative of a low rate of RNA

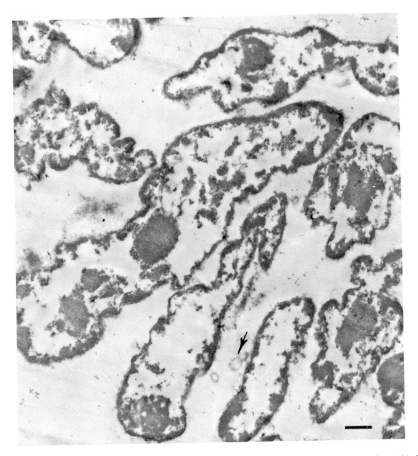

Fig. 9.6B. Microscopic appearance of heart muscle nuclei from adult guinea pigs. (A) A suspension of purified nuclei. Stained with azure C. × 600. (B) Electron microscopy. The nuclear pellet was fixed for 2 hours in 2% osmium tetroxide at pH 7.4 and postfixed for 1 hour in neutral formalin at 4°C. The absolute ethanol used in the dehydration procedure was diluted with 1% uranyl acetate. The specimens, embedded with Epon-Araldite, were cut with an LKB ultramicrotome and were observed with a Philips Norelco 200 electron microscope. An electron micrograph of an ultrathin section of the purified nuclear pellet. Note minimal contamination with cytoplasmic structures as represented by the arrow. × 6,700. From G. Ferguson, K. Smetana, A. H. Laseter, and A. Schwartz, *Cardiovascular Res. Center Bull.* **4**, 13 (1965).

synthesis. Staining procedures have delineated the presence of both DNA and RNA. While the rate of nucleic acid turnover is very low in the normal adult heart, RNA content and protein synthesis are markedly increased in the hypertrophied muscle (Table 9.1).

5. MITOCHONDRIA

More than 50% of the dry mass of heart muscle consists of mitochondria. The cristae of heart mitochondria are much more abundant than those derived from other cell types, and the dimensions of the cardiac mitochondrion are larger than other mitochondria. These differences would be expected since the mitochondria in heart are more metabolically active than in other tissues. Morphological and biochemical aspects of mitochondria are discussed in Chapter III.

6. THE SARCOPLASMIC RETICULUM

The importance and structure of the sarcoplasmic reticulum was first recognized and discussed in detail by Veratti in the early 1900's. The structure, however, was neglected for the next 50 years until the development of the electron microscope, whereupon a resurgence of interest developed, particularly through the efforts of Bennett, Porter, Palade, Fawcett, and Sjöstrand (see review "The SR," 1961).

There appear to be two types of sarcoplasmic reticulum (SR): one, communicating in a *transverse* manner throughout the cell and consisting of a double membrane (T-system); the other, traversing in a *longitudinal* direction and consisting of a single membrane (LSR or, simply, SR). The longitudinal SR generally is confined to a single *sarcomere,* the unit of contraction (see Fig. 9.1). The SR appears to arise from or is confluent with the T-system in the form of dilated transverse channels, forming a complex, called *terminal cisternae* (Fig. 9.1). The complex, oriented at the Z band of the sarcomere, consists of two terminal cisternae plus an intermediate row of small vesicles or short tubules, called a "triad."

The system of sarcoplasmic tubules has been interpreted to be a special form of endoplasmic reticulum, essentially lacking in RNA and ribosomes (Chapter V). It has a very precise orientation to the cross-banded structure of the myofibrils and, therefore, may have a special function in muscle contraction (see below). It has been shown clearly that the T-system arises from or is an extension of the plasma membrane and digresses downward from the membrane precisely in the area of the Z band (Figs. 9.1 and 9.3). *Elements* of the T-system may also communicate directly with the membranes of the other subcellular organelles, such as the nucleus, the mitochondrion, and the Golgi bodies, although this is not clear.

There appears to be a correlation between contractile rate and the

TABLE 9.1

Nucleic Acids and Protein Metabolism in Cardiac Hypertrophy and Failure[a]

Animal	Specific radioactivity[b] of protein (cpm)		RNA(μg/mg)		DNA(μg/mg)		RNA/DNA Ratio
	LV	RV	LV	RV	LV	RV	
Control or sham	1515	1373	22	22	14	15	1.57
Hypertrophy (1–2 days)	5882	4862	80	40	24	20	3.30
Chronic failure (6 months or more)	2019	2340					

[a]Data taken from S. Gudbjarnason, M. Telerman, and R. J. Bing, *Am. J. Physiol.* **206**, 294 (1964), and L. Gluck, N. S. Talner, H. Stern, T. Gardner, and M. V. Kulovich, *Science* **144**, 1244 (1964).

[b]LV = left ventricle; RV = right ventricle.

Glycine –^{14}C administered iv 4 hours prior to sacrifice of the animal. Total trichloroacetic acid-precipitable protein was measured.

amount of SR in a particular muscle cell; for example, the SR is exceptionally well-developed in the toadfish swim bladder, which has a peak contraction of only 5 – 8 milliseconds. Mammalian cardiac muscle, on the other hand, is generally a slow muscle and, consequently, has a less well-developed SR; the rat or mouse heart, maintaining a faster rate than the human heart, exhibits a more abundant SR system. The LSR represents a sparse network of single membrane-lined tubules, closely applied to the surface of the myofibrils, thus affording longitudinal continuity within the length of one sarcomere but rarely with the adjacent sarcomeres of the same fibril.

Recent evidence indicates that the T-system is made up of channels which are filled with extracellular fluid which extends through and in close association with the Z lines of the myofibrils. Remember that the intercalated discs are not continuous with the extracellular fluid.

The studies by Huxley (1964), Simpson (1965), Franzini-Armstrong and Porter (1964), Endo (1964), and Nelson and Benson (1963) clearly show that the T-system is continuous with the plasma membrane. The experiment by Huxley is particularly interesting. He treated the frog sartorious muscle with 35% ferritin in Ringer's solution for 1 hour, after which the tissue was fixed in glutaraldehyde and stained lightly with lead citrate. Electron micrographs showed that the ferritin penetrated the cell only along the channel which formed the central element of triads. No ferritin was found in the LSR or elsewhere. Since the ferritin molecule has a diameter of about 110 Å units, is inert, and possesses a molecular weight of about 750,000, it is unlikely that a simple diffusion process was operating. Similar experiments were performed using a dye and newer methods of fixation and staining, and these studies confirmed the suggestion that specific elements of the T-system do indeed communicate with the extracellular space.

7. THE CONTRACTILE ELEMENTS

The cross striations seen in both cardiac and skeletal muscle are due to the structure of the muscle fibers as pictured in Figs. 9.2 and 9.3. Note that the muscle fiber exhibits alternate light and dark bands. The light bands are known as the "I bands" or isotropic area while the dark areas are called "A bands" or anisotropic area. The alternate band patterns are particularly evident when a polarizing microscope is used (see Chapter I). The dark portions are *birefringent*.[1]

The basic unit of contraction is the myofibril, which consists of the sarcomere, the length of which is bounded by two Z bands (Z = *Zwischinscheibe* or intervening disc) (Fig. 9.7). With the appropriate optics one

[1]Muscle has two descriptive indices, a maximal, in the direction parallel with the fiber axis, and a minimal, perpendicular to the axis. This double refraction is called birefringence.

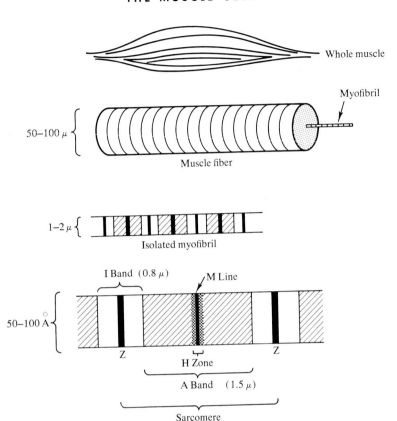

Fig. 9.7. The contractile elements of muscle.

can see the H band within the A band; the H band (Hensen's disc or Heller, which means a "brighter area") is seen only when the fiber is stretched to some degree. The center of the H band usually contains the so-called M band (*Mittelscheiber* or intermediate disc), which bisects the A band. The use of interference and polarizing microscopy combined with well-defined salt extractive procedures has shown a number of proteins to be specifically associated with the contractile apparatus (Table 9.2). Myosin is the protein associated with the "thick filaments," the diameters of which are about 110 Å units each; the "thin filaments" consist largely of actin and probably to a lesser extent, tropomyosin; they are approximately 50 Å units in diameter. The total length of the sarcomere is approximately 2.4 μ, but this does vary with the state of the muscle. The length of the A band is approximately 1.5 μ; the length of the I band is approximately 0.95μ . In living vertebrate muscles the thick filaments lie approximately 450 Å units apart in a regular hexagonal arrangement

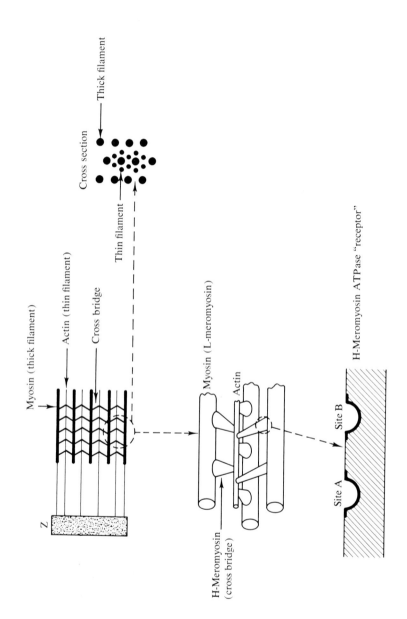

TABLE 9.2
Proteins of the Sarcomere

Protein	Band location	% of dry mass of sarcomere
Myosin	A	54
Actin	I	21
Tropomyosin	I	15
Unknown protein	Z	10

(Chapter III, Fig. 3.1A). Each thick filament is surrounded by six thin filaments, and each thin filament is shared by three thick filaments (see cross-sectional view in Fig. 9.8). Each of the thick filaments has regularly spaced short lateral projections known as *cross bridges* (Fig. 9.8). These projections extend toward and appear to touch the thin filaments, but only where the thin filaments are in the vicinity of the thick ones. There are usually six longitudinal rows of projections, which apparently are associated with the six thin filaments; these rows are situated on each of the thick filaments. There is one projection for every 400 Å. This remarkable series of observations, concerning the molecular aspects of muscular function, was obtained by Huxley and his colleagues, who used ultrathin sectioning techniques combined with electron microscopy, interference microscopy, and polarizing microscopy. The data were obtained using skeletal muscle. Cardiac muscle is less well defined at this time.

C. The Contractile Proteins

1. MYOSIN

Myosin appears to consist of at least two distinct types of proteins, *L-meromyosin* and *H-meromyosin*, both of these obtained by enzymatic digestion procedures. The letters H and L refer, respectively, to heavy

Fig. 9.8. Molecular aspects of contractile elements.

Requirements	
Site A	Site B
Ca^{2+}	EDTA activates (chelates Mg^{2+}?)
Cysteine	Isoleucine
Aspartic acid	Cysteine
Glycine	Arginine

and light. The heavy protein (HMM) has a molecular weight of about 345,000 while the light protein(LMM) is about 141,000. There is no difference in molecular weight between skeletal and cardiac myosin, both being approximately 500,000. The length of cardiac myosin is about 1,500 to 1,700 Å units. Light meromyosin is an α-helix, two-stranded rod, approximately 800 Å long and 20 Å wide. Heavy meromyosin consists of a globular and an α-helix portion. The globular region of HMM may possess the necessary flexibility to engage in cyclic interaction with actin filaments, as will be discussed later. As indicated previously, myosin possesses at least one specific type of important enzyme activity, namely an ATPase. This enzyme is located almost exclusively in the heavy-meromyosin portion, which apparently is the cross bridge (Fig. 9.8). The ATPase is a calcium-requiring enzyme and is inhibited by certain concentrations of Mg ions. It is of interest that the only difference between cardiac myosin and skeletal myosin appears to be in the ATPase activity, with the cardiac being about one third that of the skeletal muscle. There are at least two specific and different sulfhydryl-dependent sites of enzyme activity in the myosin molecule (Fig. 9.8). One site contains the amino acids cysteine, aspartic acid, and glycine; and the other, isoleucine, cysteine, and arginine. The differences in ATPase activity between cardiac and skeletal myosin may possibly be explained on the basis of differences in amino acid sequence surrounding the active sites.

Recently, an analysis of red and white skeletal muscles, and slow and fast acting muscles, revealed specific differences in ATPase activity. The red skeletal muscles, similar to cardiac muscle, possess about one fourth to one third the ATPase activity of the white muscles. As indicated above, since white muscles possess rapid but intermittent action, it would follow that enzymic activity associated with the contractile event would be higher in this type of tissue. Interestingly enough, however, both red and white muscles contain the same *total amount* of myosin and actin.

L-meromyosin (LMM) lies lengthwise in the filament and is considered the "backbone" of the A band (Fig. 9.8). It does not directly combine with actin. Remember that the ATPase activity of myosin is associated only with HMM (the lateral projections or cross bridges, which protrude at regular intervals from the L-meromyosin). Selected references for this section include Kay and Green (1964); Lowey (1964), Seidel *et al.* (1964), and Yamashita *et al.* (1964).

2. ACTIN

The molecular weight of actin is approximately 70,000; each filament is about 2 μ long and consists of about 600 molecules. Associated with actin are small amounts of *tropomyosin* (see below). There are about 1.7 molecules of actin for each molecule of tropomyosin, the molecular weight of the latter being about 53,000. The functional group of actin is ATP bound possibly to sulfhydryl groups and Ca ions. Actin itself has no ATPase activity although when actin polymerizes, there appears to be a conversion of ATP to ADP:

Actin (globular or unpolymerized)-ATP → actin (fibrous or polymerized)-ADP + inorganic phosphate.

There are 12 molecules of ATP attached to each globular actin molecule. Upon dephosphorylation, there is a net loss of a negative charge at a specific point. The polymerization of globular actin is an autocatalytic one, and it represents a very regular type of aggregation occurring perhaps, every 50 Å units along the length of the individual actin filament.

Mg and Ca ions appear to be associated with actin in a rather tight type of bonding. There are approximately six sulfyhydryl groups for each 60,000 gm of actin; two of the six are fast reacting and apparently not essential for polymerization. Selected references for this section include Barany *et al.* (1964), Hanson and Lowy (1963, 1964), and Ohnishi *et al.* (1964).

3. ACTOMYOSIN

Under the conditions of relatively low ionic strength and with the use of specific types of salt extractives, an associated combination of actin and myosin occurs in the form of actomyosin. This combined protein can be prepared in the form of threads or suspensions (usually prepared glycerinated as fibers or suspensions) and exhibits *in vitro* many of the characteristics of the whole muscle. For example, under the appropriate conditions, actomyosin suspensions can be made to superprecipitate or synerese. This process of syneresis is similar to contraction and can be inhibited by substances or processes which effect relaxation *in vivo*. The process of syneresis may be studied in two general ways: measuring the superprecipitation or "clotting" by use of a small capillary tube; recording superprecipitation by following the increase in optical density at 660 mμ. The two procedures are illustrated in Fig. 9.9. Addition of ATP plus Mg^{2+} effects a superprecipitation, just as one would predict from *in vivo* models. Addition of various inhibitors like organic mercurials or histones specifically inhibits the syneresis process; a "relaxing system" (consisting of small membrane-lined granules, derived from muscle homogenates; this will be discussed in detail below) reverses or prevents the process of superprecipitation; and the addition of Ca ions reverses the relaxing factor's activity. This is characteristic of an active actomyosin preparation and is considered to be analogous to whole muscle activity.

Another *in vitro* cardiac muscle model which has been employed by

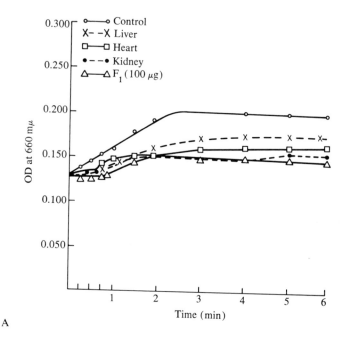

A

The "Relaxing" Effect of Histones on Superprecipitation Actomyosin and Myofibrils from Heart Muscle

Preparation	Addition	Histone	Height of precipitate (% of control)	Capillary tubes (after "centrifugation")
Actomyosin	Mg	–	100	
	Mg + ATP	–	2	
	Mg + ATP	+	68	
Myofibrils	Mg	–	100	
	Mg + ATP	–	54	
	Mg + ATP	+	80	

B

Fig. 9.9. Measurement of muscle "contraction" *in vitro*. A. Superprecipitation: optical density procedure. Additions are basic protein material from indicated tissues; F_1 is a lysine-rich histone. B. Superprecipitation: capillary tube centrifigation procedure. From A. Schwartz and C. L. Johnson, *Life Sci.* **4**, 145 (1965).

many investigators is the glycerinated muscle fiber, introduced first by Szent-Györgyi. Small muscle fibers, usually derived from papillary muscle, are soaked in a 50% solution of glycerol at $-20^{\circ}C$ for several weeks. The muscle is then washed free of glycerol and is suspended in a suitable bath, and attached to an appropriate recording device. Changes in isometric or isotonic tension under varying conditions may be studied. Actomyosin from both cardiac and skeletal muscle has been prepared by numerous laboratories and investigated in detail; most of the results indicate no significant difference between the two muscles, with the exception that, again, the ATPase activity of skeletal muscle actomyosin appears to be higher than that of cardiac muscle.

Actomyosin possesses an ATPase activity which appears to be different in requirements than that exhibited by isolated myosin. It requires Mg ions and low concentrations of Ca ions for optimal activity. It is of importance that during the process of superprecipitation (syneresis) the ATPase activity of actomyosin is markedly increased; when syneresis is either reversed or inhibited, ATPase is also inhibited concomitantly. Thus, the "contractile" and "relaxation" activity of this *in vitro* preparation is often measured by simply following changes in ATPase activity (Fig. 9.10).

4. TROPOMYOSIN AND OTHER PROTEINS

Tropomyosin, which is always found associated, *in vivo,* with actin, may possibly extend from the Z line to the thin filaments, where it may provide a template upon which the double helix of F-actin is laid with specific polarity. Tropomyosin, *in vitro,* inhibits actomyosin-ATPase activity and delays superprecipitation, in the presence of Mg ions (Katz, 1964). This may be due to the formation of a specific F-actin-tropomyosin complex. It appears, furthermore, that the well-known control of the contraction-relaxation cycle in muscle by Ca^{2+} is specifically dependent upon tropomyosin (Ebashi *et al.,* 1964). *Native tropomyosin* consists of at least two proteins, tropomyosin and troponin. The latter promotes aggregation of tropomyosin and is bound to actin via tropomyosin, distributed along the axis with a 400 Å periodicity. The sensitivity of the contractile system to Ca^{2+} apparently may be attributed to troponin and it is therefore conceivable that the interaction of myosin with actin involves this new protein (Ebashi and Nonomura, 1967). Changes in levels of bound Ca^{2+} would induce alterations in conformation of troponin which would in turn modify and therefore control myosin-actin interaction.

Two other recently identified proteins of muscle may also modify the structure of F-actin. These are the 6-S component of α-actinin and β-actinin (Ebashi and Nonomura, 1967).

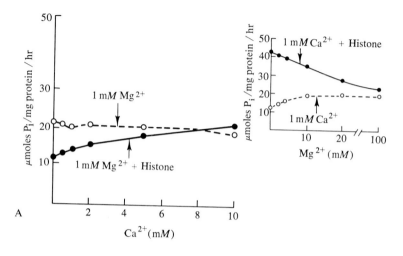

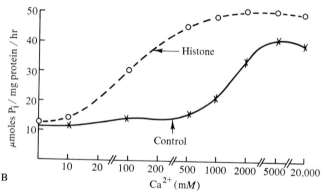

Fig. 9.10. ATPase activity of actomyosin as a measure of "contractility." A. The "reversibility" of histone-induced inhibition and stimulation of actomyosin ATPase from cardiac muscle. The histone was F_1. Ionic strength was less than 0.02. B. The effect of histone on actomyosin ATPase in the presence of varying concentrations of Ca^{2+}

Thus while the exact functions of these interesting new proteins are unknown, the possibility of modulator activities in the contractile-relaxation process is attractive.

D. FUNCTIONAL ASPECTS OF THE CARDIAC MUSCLE CELL

Cardiac energetics are directed towards fulfilling the primary function of contraction. The energetics are divided into two main aspects.

a. ENERGY PRODUCTION. This occurs, in the adult normal heart, primarily in mitochondria and is involved in the generation of ATP, which is the primary energy-rich compound involved in the contractile event (Chapter III).

b. ENERGY UTILIZATION. The ATP is utilized in muscle contraction and in the associated processes of active transport and electrical conduction.

1. STRUCTURAL CHANGES DEVELOPED DURING CONTRACTION AND RELAXATION (SKELETAL MUSCLE)

The important morphological events associated with the contractile event are:

1. The A bands do not change when a muscle contracts.
2. The I bands are observed to shorten when the fiber shortens. When the fiber is stretched, these filaments lengthen. The actin, itself, however remains at constant length.
3. The width of the H band is proportional to the stretch of the fiber.
4. When the fibril shortens to 65% (skeletal muscle) of its resting length (the maximum length of muscle in the body), the I band completely disappears, and each Z line touches the ends of two adjacent A bands.
5. As shortening continues, a dense, contraction band develops, known as the M line (Fig. 9.7).

Over a wide range of lengths, a muscle can contract without any detectable change occurring in the *overall length* in any of its filaments.

These observations, and others, have led to the development by H. E. Huxley and A. F. Huxley, of the *sliding-filament theory of* contraction, which is illustrated in Fig. 9.11. It may be seen that the actin filaments (I band) literally slide past the stationary myosin rods during the process of contraction, by means of the cross bridges, which, it should be recalled,

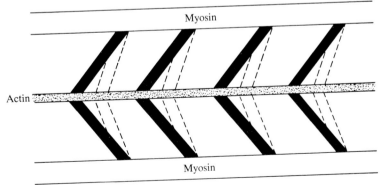

Fig. 9.11. The sliding-filament theory of muscle contraction.

consist of H-meromyosin. Muscle contraction, therefore, may be likened to a canoe propelled through the water by a large number of oarsmen. While this analogy might explain the mechanism of muscle contraction on a microscopic level, it still remains to describe the actual molecular events involved. A variety of schemes have been proposed through the years by several investigators. In almost all cases, however, the involvement of ATP is of primary significance. It is now known that Ca ions are of critical importance in both contraction and relaxation. Of course, this knowledge does not exclude the significance of both Na and K ions, which are involved during the depolarization phenomenon preceding the actual contractile event.

Since most of the chemical aspects of contraction have been described above, it is worthwhile to discuss one of the current, popular schemes which attempts to collate all of the observations into a composite picture of muscle contraction. The first event preceding contraction is the depolarization of the muscle membrane. While the mechanism of this process is not completely known, it is quite clear that the resting membrane potential is due predominately to a potassium diffusion current. In other words, the unequal distribution of potassium between the extracellular and intracellular space *effects* the resting membrane potential (Fig. 9.12). A rapid inrushing of Na ions (due to change in permeability) down their chemical potential gradient facilitated by some type of sodium carrier system not requiring energy, is responsible for the development of the "spike" or rapidly rising phase of the action potential. The sodium carrier system (which may, in fact, be some type of phosphorylated compound) is then inactivated by an unknown process, "shutting off" the sodium current, leading to the plateau phase. Potassium then diffuses down its potential gradient, i.e., moving from the inside of the cell to the outside, an event which is associated with the repolarization phase. The recovery phase, as diagrammed in Fig. 9.12, involves the *active* pumping

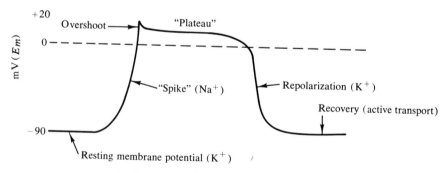

Fig. 9.12. A cardiac muscle action potential.

of sodium and potassium back into their respective compartments. This latter process involves the utilization of energy and is discussed fully in Chapter VII. The wave of depolarization spreads from the plasma membrane of the sarcolemma down through the transverse elements of the sarcoplasmic reticulum (Fig. 9.1) where it releases, by some mechanism, a specific amount of *calcium ions* from sites in the SR. The distance from the SR to the actual contractile elements is small enough so that calcium can very rapidly diffuse to a site between the crossbridge (HMM) of myosin and the ADP-actin (i.e., polymerized actin or actin-F), forming a link or chelate between the myosin and actin. This results in the development of tension. The calcium chelate effectively diminishes the negativity of the HMM which causes a contraction of the extended HMM. The contraction is due to the electrostatic *pulling* by the fixed, negative charge (Fig. 9.13). The contraction may come about by the formation of an α-helix. As soon as the crossbridge comes in contact or in close apposition to the ATPase site of the LMM, a dephosphorylation of the ATP on the HMM takes place and the link between actin and myosin is broken. The resulting ADP, associated with myosin, is then rephosphorylated by cytoplasmic reactions, possibly involving creatine phosphate. This reaction effectively recharges the HMM with a negative site, resulting in an electrostatic repulsion. The crossbridge is once again extended. The contractile cycle is primed once again. Repeated link formations occur, resulting in a movement of the actin in the direction indicated in Fig. 9.13.

As will be discussed below, the relaxation process is associated with the active pumping of Ca ions back to the sacroplasmic reticulum, resulting in a termination of the active contractile state.

In summary, the primary molecular event in contraction is the formation of chelate links between the bound ATP of the extended crossbridge (HMM) and the bound ADP of the F-actin, resulting in electrostatic bonds between actin and myosin fibrils; the neutralization of the terminal phosphate of ATP causes an abolition of the coulombic repulsion within the heavy meromyosin. The extended polypeptide in the crossbridge contracts and forms an α-helix by twisting and forming hydrogen and hydrophobic bonds. The ATPase activity of HMM is inside the crossbridge. The contraction of the polypeptide chain brings the bound ATP into the vicinity of the ATPase.

This is a particularly useful molecular model for contraction, and while it may not be entirely correct, at least it effectively brings together most of the known biochemical and physiological facts about muscle.

Selected references for this section include Daries (1963), Bing (1965), Fanburg and Gergely (1965), Mommaerts (1911), Starling (1918), and Sonnenblick *et al.* (1965).

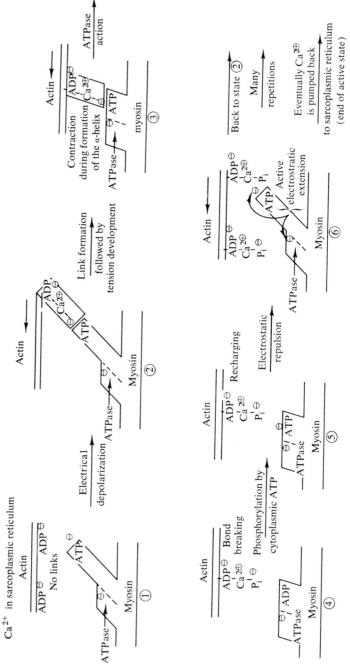

A

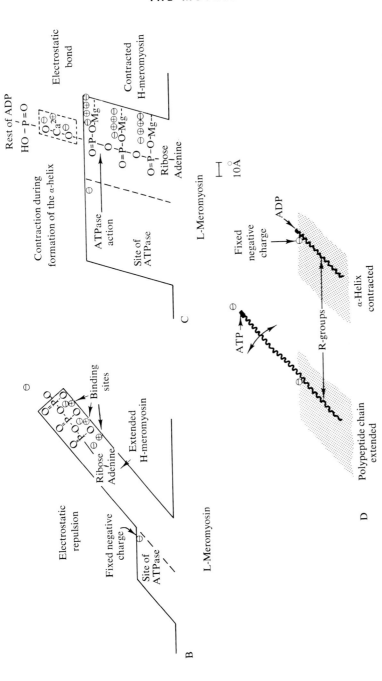

Fig. 9.13. A molecular mechanism of muscle contraction. A. Ca²⁺-dependent contraction during α-helix formation plus ATP-dependent extension of the cross bridges. B. Resting extended state. ATP bound to H-meromyosin. C. Contracted H-meromyosin maintaining tension by the bound ADP of actin. D. Diagram of a flexible polypeptide chain of an H-meromyosin in an extended and contracted state. From R. E. Davies, *Nature* **199**, 1608 (1963).

2. THE RELAXING SYSTEM OF MUSCLE

In 1951 Marsh reported the presence of a substance or substances in muscle supernatants, which modified the effect of ATP upon a muscle fiber. At that time Bendall also reported a similar observation. Two effects were noticed. One, a relaxation of the muscle fiber upon addition of the soluble substance; and two, a significant inhibition of ATPase activity associated with the muscle fiber. Both of these events were reversed or prevented by the addition of calcium ions. Thus, the concept of a *relaxing* or *Marsh-Bendall factor* developed. Since that time, countless investigations have implicated a variety of compounds as possible factors, including myokinase, pyrophosphate, α-glycerol phosphate, creatine phosphokinase, carnosine, pyruvate kinase, cylic $3',5'$-AMP, etc. None of these, however, have been confirmed or accepted. The relaxing phenomenon in skeletal and cardiac muscle is attributed to sarcoplasmic reticulum elements [first reported in skeletal muscle by Ebashi and Lipmann (1962)]. It has been shown that these elements not only effect a strong binding of calcium ions but also possess a specific type of ATPase activity which is associated with the energy requirements of the system. There is a specific correlation between the removal of calcium by the SR and "relaxing factor" activity. It has also been shown that actomyosin muscle models require small amounts of calcium in order to effect superprecipitation. The amount of bound calcium to actomyosin is approximately 1.5 to 2.3 mmoles/gm protein. If this amount is reduced to 1 mmole or less per gram, both actomyosin ATPase and superprecipitation (syneresis) are completely inhibited (Fig. 9.14). The SR vesicles are capable, in fact, of reducing the concentration of ionized calcium in the medium to about 0.02 mM or less, thereby effectively fulfilling the requirement of a relaxing "factor" or system. Most of this information was derived using skeletal muscle models. Recently, a similar system has been found in the heart which appears to be much less in activity than its skeletal counterpart. This has led to the concept of a *soluble system* as well as a particulate system in heart muscle. The soluble relaxing system of cardiac muscle is apparently not influenced by calcium ions and may be associated with the regulation of contractile strength rather than directly with inhibition of contraction. In addition, as discussed in Chapter III, heart mitochondria are particularly active in sequestering calcium ions, and hence may function in the relaxing system. This aspect would not be of importance in skeletal muscle since there is a paucity of mitochondria as compared to heart.

The relaxing system of skeletal muscle, at least, can now be explained on the basis of a *calcium sink* or calcium sequestering mechanism, associated with the sarcoplasmic reticulum. The ATPase, which is calcium dependent and presumably also located in the SR, is a particularly active one and is associated with the calcium uptake (Fig. 9.14) as well as with an ATP-ADP exchange reaction. Both the splitting of ATP and the accumulation of calcium apparently cease when the free calcium concentration in the medium is reduced to about 10^{-8} or 10^{-9} M. The rate of calcium binding by the SR may be in excess of 12 millimoles of calcium per milligram of nitrogen per minute; this rate is entirely consistent with the rapid processes involved in the relaxation event.

In confirmation of the particulate relaxing system as defined above, recent studies have provided actual evidence that a calcium "sink" does, in fact, operate in living muscle and is associated with the SR. Podolsky, utilizing a sarcolemmal-removed muscle fiber, applied calcium ions to local discrete areas with a microelectrode and observed a triggering of a localized contraction wave. Three to five seconds later relaxation was complete despite the very large concentration of calcium presumably present in local areas. Therefore, contraction in this system was restricted to a local area of application and did not spread across the diameter of the fiber as would be expected if relaxation were simply due to calcium dilution through a diffusion process. The extremely rapid rate of relaxation also vitiates the diffusion argument. This preparation also shows that the plasma membrane or sarcolemma is not needed for the release of calcium ions. Therefore, the terminal step in the activation process is a transient increase in the concentration of free calcium ions in the "Z" region of the contractile elements (Fig. 9.1).

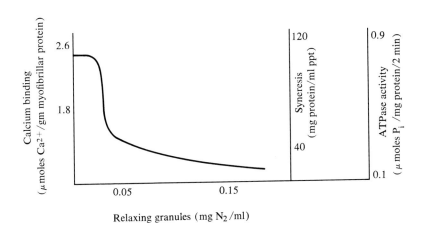

Fig. 9.14. Effect of SR (relaxing system) on calcium binding, syneresis, and ATPase of myofibrils. Incubation time: 130 seconds; medium: $^{45}CaCl_2$, 0.02 mM, ATP-Mg, 4 mM, oxalate, 1 mM, myofibrils, protein = 1.98 mg/ml. From A. Weber, R. Herz, and I. Reiss, *J. Gen. Physiol.* **46**, 679 (1963); *Proc. Roy. Soc.* **B160**, 489 (1964).

Those elements of the sarcoplasmic reticulum of muscle which are associated with the relaxing system appear to be morphologically the same as the active transport membrane ATPase which is discussed in Chapter VII. However, the relationship between the pump ATPase (which is ouabain-sensitive) and the relaxing system (which is ouabain-insensitive) is not clear. Nevertheless, it is attractive to postulate a link between the enzymic process of active transport and the relaxing system. It is *possible,* for example, that the depolarization wave (which initiates in the plasma membrane) releases sodium ions from binding sites in the plasma membrane T-system as well as in specific regions of the SR. The sodium ions diffuse inward (assisted by a so-called sodium carrier system) and interact with the calcium sites or in the SR membrane. The calcium ions, then, may easily diffuse to the required sites adjoining the contractile elements. The membrane ATPase (ouabain-sensitive) becomes activated during the process of recovery wherein sodium must be pumped out of the cell and/or perhaps back into the original sites. This pump enzyme complex might also play a role in the reinstatement of calcium into the elements of the sarcoplasmic reticulum. Since sodium and calcium *may* compete for similar or the same sites, changes in the activity of the Na^+,K^+-ATPase may result in corresponding increases or decreases in intracellular Na^+. This would effect decreases or increases, respectively, in SR Ca^{2+} binding. The problem is that most investigators find that the ATPase which is specifically associated with the relaxation event, *in vitro,* is a calcium-sensitive enzyme and one which is presumably *not* responsive to ouabain. However, during the process of isolation of these various enzyme systems, it is entirely possible that the ouabain-sensitive membrane ATPase is dissociated from the SR-relaxing elements. The resultant "calcium-sensitive ATPase" *in vitro* may be a part of the ouabain-sensitive enzyme complex. This is, of course, highly speculative and must await a resynthesis, as it were, of the various enzyme moieties that have been studied extensively in the isolated systems.

The molecular approach to muscle contraction and relaxation has provided logical explanations for basic events and may extend to other systems. These may include mitochondrial, bacterial and viral movement, and even the coiling and uncoiling of DNA. Selected references for this section include Weber *et al.* (1963, 1964), and Ebashi and Lipmann (1962).

3. METABOLISM OF THE CARDIAC MUSCLE CELL

The normal adult cardiac muscle cell derives its energy from substrates, including proteins, carbohydrates, and fats, depending upon the nutritional state of the organism. For example, in a fasting animal, the primary substrate is fatty acid, derived mostly from plasma triglycerides or a nonesterified fatty acid fraction (NEFA). In a well-fed animal, however, the primary substrate is glucose. Consequently, if available, carbohydrate is the preferred substrate when the animal is in the *fed* state.

The arterial level of carbohydrate is an active one, i.e., the level determines the amount of substrate taken up. Therefore, one speaks of a glucose threshold concentration *in vivo*.

If needed, the heart can effectively utilize acetate, ketone bodies as well as amino acids; the extent depends upon the arterial blood concentration, the state of nutrition, and the endocrine balance.

The myocardial extraction of fatty acids (triglycerides and NEFA) appears to vary with the activity of myocardial lipoprotein lipase located at the cell membrane. The enzyme activity depends upon nutritional state, arterial concentration of the fatty material, the carbon chain length, and the number of double bonds in the fatty acids. Oleic acid is an active substrate; this fact may be related to its ability to bind to serum albumin in the transport of the fatty acid. The nonesterified fatty acid fraction is bound to serum albumin in plasma. It is of interest that in the presence of pyruvate or acetoacetate, the uptake of free fatty acids is depressed about 50%. These fatty acids are then incorporated into neutral lipids for storage and subsequent use.

Fatty acids also depress glucose uptake in the heart and divert glucose to other pathways such as synthesis of glycogen and lactic acid.

The phosphofructokinase (PFK) step in the glycolytic scheme is responsible for the conversion of fructose 6-phosphate into fructose 1, 6-diphosphate (see Chapter III, Appendix) and is a critical control step in cellular metabolism. The oxidation of fatty acids may be responsible for the inhibition of phosphofructokinase; hence, fatty acid levels may function as part of a cellular energetic control system. Citric acid, e.g., derived from Krebs' cycle oxidations, inhibits this sensitive step in glycolosis, while cyclic 3′,5′-AMP stimulates the PFK step (see Chapter III for discussion of metabolic control mechanisms). Selected references for this section include Coffey *et al.* (1961), Gold and Spitzer (1964).

Both substrate and oxygen uptake in the intact heart depend upon coronary blood flow. The latter is particularly sensitive to variations in tissue oxygen tension, so that tissue hypoxia produces an almost immediate dilatation of the coronary arteries, resulting in an increased blood flow. The primary determining factor of myocardial energy requirements, measured in terms of its oxygen utilization, appears to be the *tension* produced by the contracting heart muscle, usually in the isometric or isovolumic phase of contraction in the left ventricle (phase of contractile cycle where the volume of blood in the ventricles is constant).

The energy liberated from various substrates enters the Krebs' cycle present in mitochondria (Chapter III) and is converted to the stored, energy rich compound, ATP. The general metabolic state of heart muscle may be evaluated according to the content of energy rich phosphate compounds, although this evaluation gives an impression of steady-state levels only. When rapid freezing techniques are used, approximate concentrations (μmoles/gm wet weight) in mammalian heart are:

ATP	Creatine phosphate	Inorganic phosphate	ADP	AMP
6	13	< 1	< 1	< 1

Creatine phosphate is the most sensitive phosphate compound. Diminution of oxygen tension in the cardiac muscle cell is reflected in an almost immediate drop in creatine phosphate content, with an attendant rise in inorganic phosphate. Selected references for this section include Bing (1965), Coffey *et al.* (1964), Gold and Spitzer (1964), and Wollenberger (1947).

4. ELECTROPHYSIOLOGY

The electrical aspects of cardiac muscle are basically no different from skeletal muscle or nerve (see Chapter VII).

In this section a brief discussion of the function of the conducting system in heart muscle is appropriate.

The electrical impulse arises in the SA node (sinuatrial node), located in the atrium, and spreads like a pebble dropped into water, to the AV (atrioventricular) node where it is delayed for a short period, after which the impulse traverses to the bundle of His and then to the Purkinje conducting system. Thus, the rapid spread of electrical impulses throughout the heart effects a more or less uniform ventricular contraction. The shape of the action potentials in each of the areas of the muscle is presented in Fig. 9.15.

The EKG (electrocardiogram) is an expression of the electrical activity of the whole heart muscle and is pictured below. The P wave is due to a depolarization of the atrium; the QRS complex is due to a ventricular depolarization, and the T wave is an expression of the repolarization phase of the muscle. Analyses of EKG tracings frequently yield information concerning disease processes in heart muscle. The reader is referred to the paper by Hoffman *et al.* (1963) for further information.

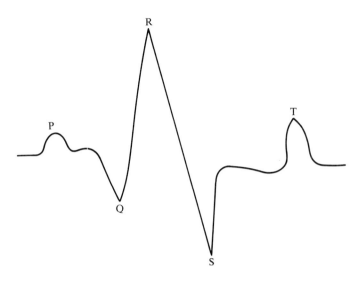

5. Protein Synthesis in Heart Muscle

The heart, like other organs, exhibits the usual characteristics of protein synthesis, described in detail in Chapter V. The muscular nature and somewhat heterogeneous quality of the tissue, however, does not lend itself to easy manipulation; therefore, most of the studies on protein synthesis have employed tissues other than heart muscle. In addition, the normal adult heart is not a growing organ. Most of its activity is directed toward its primary function of contraction; consequently, its protein synthetic activity is somewhat lower than most other organs. However, it is highly important that an intimate knowledge of its protein synthetic *potentiality* is known, since in certain disease states the heart can increase in mass up to six times normal.

Another difficulty involved in studies of protein synthesis in heart muscle is the fact that conventional fractionation procedures for preparation of ribosomes and polysomes yield particles with very low RNA-to-protein ratios and rather poor ability to catalyze protein synthesis *in vitro*. Recently, ribonucleoprotein particles with high activity and good yield were isolated from heart muscle. The procedure involves a pretreatment of the muscle homogenate with deoxycholic acid in a medium of high ionic strength. These particles catylyze the transfer of large amounts of radioactivity from sRNA-^{14}C-phenylalanine to protein, in an energy-dependent reaction which is inhibited by RNase and by puromycin (Chapter V). The reaction is stimulated by the addition of polyuridylic acid. Hence, the reaction in heart muscle is qualitatively similar

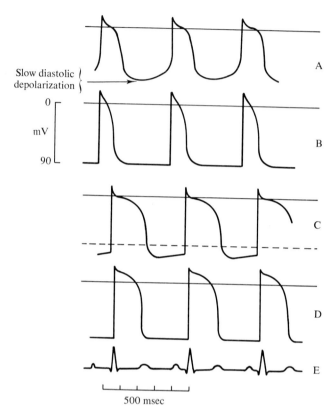

Slow diastolic depolarization

Fig. 9.15. Cardiac action potentials. The membrane potential in the course of two cardiac cycles of a fiber of the sinoatrial node (A); of the atrium (B); of the Purkinje system (C); and of the ventricular myocardium of a dog heart (D); drawn on the same time axis as the electrocardiogram (E). From W. Trautwein, *Pharmacol. Rev.* **15**, 277 (1963).

to the protein synthetic mechanism in liver and other tissues (Chapter V). Quantitative differences are quite probably related to a peculiarity of protein regulating processes (Chapter V) in heart muscle. It is of interest in this regard that fetal and newborn heart are rapidly growing organs and exhibit a very high protein synthetic capacity, both *in vivo* and *in vitro*. In addition, the metabolism of these organs appears to be predominantly a glycolytic one rather than oxidative. This is understandable when it is recalled that an essential requirement of protein synthesis is the production of RNA, an important constituent of which is ribose (Chapter V). The latter is synthesized via the hexose monophosphate shunt pathway of glycolysis (see Chapter III, Appendix). In addition, since most of the activities of the heart muscle at this stage are directed toward growth, production of reduced nicotinamide-adenine dinucleotide (NADH and

NADPH) is required. These are also acquired primarily through glycolytic pathways. At some stage in the development of the muscle it is likely therefore that some type of regulatory mechanism is activated which effectively shuts off synthesis. However, the capacity for reactivation is undoubtedly present through the normal adult life span because when a suitable stress is placed upon the heart, hypertrophy occurs. Current research activities in this regard are directed toward elucidation of the nature of the regulatory mechanism(s) in heart muscle and particular species of RNA involved in the process of derepression (Chapter V). Figure 9.16 exhibits an experimental approach to this problem. Concentrations of RNA and protein synthesis are high in the newborn; then the concentrations progressively descend to a low control level.

According to current biological dogma, the nucleus of the cell is the "heart" of control of protein synthetic activities (Chapter V). Hence, it is essential that an intimate knowledge of heart muscle nuclei be available, particularly in states in which the heart changes in mass. Recent isolation technique modifications have made available relatively intact heart muscle nuclei for studies of this type (Figs. 9.5 and 9.6). Isolation and characterization of DNA and RNA are of prime importance in understanding normal and dysfunctioning heart muscle. During the hypertrophic growth process of the adult heart, DNA presumably does not

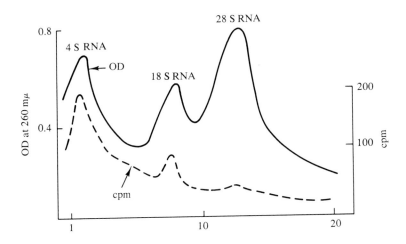

Fig. 9.16. Analysis of RNA from hearts of 2-day-old rats. Method: Pulse-labeled with uridine-^{14}C for 15 to 25 minutes; killed by decapitation. RNA extracted by phenol-SDS method; dissolved in water and applied to a linear sucrose gradient (5−40%) and centrifuged at 25,000 rpm for 15 hours in model L-2 Spinco. UV absorption at 260 mμ monitored by an ISCO device, collecting 1 ml aliquots which are subsequently counted in a Tri-Carb Spectrometer. 18 S + 28 S RNA, ribosomal RNA components.

change in synthetic capacity or content, indicating a lack of hyperplasia, while changes in RNA are expected.

Heart muscle mitochondria are particularly active both in the incorporation of amino acids into protein and in RNA synthesis (see Chapter III). There is also a relatively high content of specific DNA in heart mitochondria, the function of which is, at present, unknown. It is possible that the mitochondria in heart muscle may direct their activity toward *reductive synthesis* in the young, growing heart, while most of their activities would be directed toward *energy production* in the normal adult heart. When something goes wrong with the heart and growth becomes necessary, the mitochondria may be of prime significance. Selected references for this section include Gluck *et al.* (1964a,b).

II. THE CEREBRAL CORTICAL CELL

A. Anatomical Features

Nervous tissue consists of nerve cells, otherwise known as neurons, each of which has a cell body containing a nucleus and possessing several cytoplasmic extensions. One of the cytoplasmic extensions, known as the *axon,* is rather long; the rest are called *dendrites,* most of which are short-branched filaments. The mammalian cerebral cortex is complicated by the presence of numberous cell types. In 1840 Baillarger, a French psychiatrist, found that when the human cerebral cortex is sliced in thin sections and placed between plates of glass, six discrete layers are seen and the relative width of each layer varies considerably. Utilizing modern microscopic techniques and staining procedures. Ramon y Cajal, Campbell, and others have subdivided the cortex into areas of specific structure. The overall purpose of such anatomical subdivision lies in the hope that morphological differences can define functional specificities. Within each subdivision there are two primary divisions: the *external lamina,* having four layers, and the *internal lamina* with two layers.

Each part of the cortex receives incoming impulses and gives rise to outgoing impulses. The impulses are carried by the *dendrites,* and the outgoing "messages" are transmitted by the *axons.* Each part of the cortex, therefore, is both a termination of some afferent pathway and the start of some efferent path.

It is of interest that while there are specific differences in each part of the cortex, the basic structures of all the areas are quite similar. Lorente de No has stated that, "The cortical neuronal chains are in no way different from chains of internuncial neurons in any part of the central nervous system."

All neurons are surrounded by *glial* cells with their neuroglial processes. The exact function of the glial cells is unknown, although it has been suggested that they function in supportive aspects, in transmission of materials, and in memory processes.

Adult neurons are very large; the cell body, for example, may be as much as 30 μ across, possessing a volume of about 10^{-4} μ^3 and a mass of 10^{-8} gm. The cell body has an attached axon which is approximately 7 μ in diameter and may be centimeters or decimeters long. Its length, therefore, is approximately 10^4 or 10^5 times as great as its diameter, and the total volume may be 1000 times that of the cell body. Interestingly enough, the nucleus of the cell body or neuron is very much larger in the cerebral cortical cell than in most other cells.

The cell bodies within the cerebral cortex occupy only about 3% of the total volume, the remainder consisting of extracellular fluid, glial cells, and various fibers and processes from neurons. Nevertheless, these large, elongated structures (adult neurons) remain as single cells. The neurons of the cerebral cortex are somewhat unusual and include various specific biochemical characteristics. For example, the supply of material from the nucleus involves transportation over distances very large in relation to most ordinary cell types. Yet such transportation does, in fact, occur at a relatively rapid rate. Material is presumably synthesized in the neuronal cell body and then is transported to various portions of the cell along axons. There is much evidence, therefore, in favor of *axonal flow*. The proteins of the axoplasm originate or are synthesized in the cell body itself and flow down or are transported through the axonal material at the average rate of $1-2$ mm a day. This value is consistent with estimated nerve regeneration time (4 mm/day) from the hypothalamus to the posterior lobe of the pituitary as well as with the neurosecretory movement concept.

The neurons contain a greater abundance of cell constituents, particularly nuclear components, than most other adult cells in the body. This fact may be related to the rather large amount of cytoplasm usually associated with a given nucleus.

In general, the components of the cell consist of the following: nucleus, nucleolus, chromatin, mitochondria, endoplasmic reticulum. Nissl bodies (possibly parts of rough endoplasmic reticulum), Golgi formation, myelin, and soluble cytoplasm.

The *nucleus* is the most prominent component of the cell body and is typically very large and spherical. The diameter is about $15-20$ μ. The high content of DNA is characteristic. Usually, the nucleus is located in the center of the cell, but it may become displaced during various changes in activity. The nucleolus is a well-defined central body appearing in the nucleus of the neuron and may be from 0.5 to 2 μ in diameter, a rather large size compared to nucleoli of other cell types. Interestingly, the nucleus in neuronal cells is always present whereas in most other cells it is seen only during a particular stage of development or division. The nucleolus is rich in polynucleotides and basic proteins (histones). It appears that the nucleolus is more active in liberating phosphate from thiamine pyrophosphate, ATP, and glycerophosphate than is the rest of the nucleus. The nucleolar-associated chromatin is quite distinct. During periods of active growth the nucleolus increases tremendously in size. During periods when the cell is either not growing or exhibiting diminished growth rate, nucleolar-associated material appears to be absent or minimal.

Mitochondria of the neuronal cell are distributed uniformly in the cytoplasm, usually around the nucleus. Although they are generally spherical in shape, sometimes they appear as rods. It is somewhat difficult to prepare relatively pure mitochondrial fractions from cerebral tissues, although more recently, modifications of existing procedures have yielded fairly good preparations. The main difficulty is that the mitochondria appear to be firmly associated with glycolytic enzymes. In fact, some investigators have postulated that the mitochondria actually contain glycolytic enzymes. However, such phenomena probably are due to contamination, since purification procedures involving density gradient centrifugation have yielded intact mitochondria without the presence of glycolytic intermediates. In addition to oxidative phosphorylation, the organelles, like other mitochondria, contain some DNA as well as RNA. It is of interest that mitochondria of the cerebral cortical cell are most active in energy liberation. Oxygen consumption proceeds at a rate of anywhere from 5 to 15 μmoles per milligram nitrogen per hour when standard substrates are used (see Chapter III). This is a rate which is somewhat similar to heart mitochondria. Most cerebral mitochondrial preparations are capable of utilizing glucose as a substrate, due to the presence of contaminating glycolytic enzymes. The mitochondria are about 0.2 μ in diameter but may vary considerably.

The Golgi network is usually observed around the nucleus of the cell body. It apparently contains material of reducing properties, perhaps similar to ascorbic acid; it also contains lipids and a lecithin-like substance as well as some protein. It is possible that the Golgi bodies are associated with secretory processes. It has been suggested that mitochondria and other membrane-bound cellular organelles originate from this region.

The Nissl bodies are stained rather intensely by basic dyes, such as methylene blue and toluidine blue. These granules are normally found in the cell body as well as in some of the larger dendrites but are not found in the axon. They usually appear after a period of rapid cell division and possibly indicate the transition stage of the embryonic neuroblast to the adult neuron. These bodies contain nucleoprotein, possibly of the RNA type, and this property would explain the affinity for the basic dyes. They may in fact be fragmented rough endoplasmic reticulum.

Myelin fragments are very easily separated from other subcellular particles of brain homogenates because of their low density. They are easily recognized by their laminated structure, which is apparent on microscopic examination.

Lysosomes are somewhat similar to the lysosomes of other tissues already described (see Chapter VI). The enzymes, acid phosphatase and β-glucuronidase, present in low concentration in brain tissue, are found in the lysosomes.

Glial cell processes send out long processes described previously, similar in length to the dendrites or the axons.

Selected references for this section include Curtis (1963), McIlwain (1959, 1963), and Whittaker (1965).

B. Physiological Aspects

It is particularly characteristic that the brain maintains an almost constant blood flow regardless of the peripheral "situation." Unlike most of the other organ systems, a prolonged severe hypertension or hypotension is required to alter the flow of blood through the brain. The brain depends almost completely upon glucose as substrate and, hence, requires an adequate blood flow as well as adequate oxygenation. The existence of a *blood-brain barrier* serves to protect the brain from external influences. Thus, ingested foreign substances often do not find their way into the cerebral blood flow system.

The blood volume flowing through the brain represents over 15% in normal adults even though the brain is only 3% of body weight. The tremendous requirement for oxygen is apparent when one examines the arterial-venous differences (A-V). Over $20-25\%$ of blood oxygen is removed by the cerebral tissues. The average arterial oxygen concentration in ml/100 ml of blood is 19.6; the venous content is 12.9. This is a very large proportion for an organ which does no obvious external work, and it clearly indicates the very high metabolic activity involved. The cerebral blood vessels, particularly the *pial* vessels, are exquisitely sensitive to carbon dioxide, responding instantaneously with an increase in diameter, thus effecting an increased blood flow. The only other vessels in the body responding so quickly to a chemical substance are the coronary vessels in the heart. In the case of the coronaries however, the vessels are sensitive mostly to oxygen lack rather than carbon dioxide. An examination of the

A-V CO_2 difference in the brain reveals almost exactly the same figure as that found with oxygen. In this case, of course, the amount in venous blood is higher than that in arterial blood; the average figure is 54.8 ml of carbon dioxide per 100 ml of blood in the venous system and 48.2 in the arterial system. These figures give further evidence that the main substrate for the brain is glucose (RQ = 1).[2] Approximately 9.8 mg of glucose are taken up by the brain for every 100 ml of blood flow and about 85% of the glucose removed by the brain is oxidized completely, with the resulting formation of energy. Normal cerebral blood flow is approximately 50 to 60 ml/100 gm of tissue/minute. The normal cerebral respiratory rate is 4 ml of oxygen/100 gm of tissue/minute. The reader is referred to McIlwain (1959, 1963) and McIlwain and Rodnight (1962) for additional information.

C. Biochemical Aspects

During the past 25 years, efforts have been directed by many investigators toward elucidating various biochemical characteristics of the cerebral cortex. Modern biochemical techniques such as differential centrifugation, density gradient centrifugation, chromatography, and histochemistry have yielded information concerning most of the subcellular organelles in the neuron. The relative content of a number of important compounds and enzymes of cerebral cortical tissues from guinea pig, rat, and rabbit is listed in Table 9.3.

Some of the more important constituents are discussed below. Selected references for these sections include Barondes (1964), DeRobertis et al. (1963), Gray and Whittaker (1962), McIlwain (1959, 1963), Schwartz et al. (1962), Tobin and McIlwain (1965), Waelsch and Lajtha (1961), Wallgren (1963); and Whittaker (1965).

1. LIPID MATERIAL

Cholesterol is the predominant lipid present in the brain and is found in both the microsomal and the mitochondrial fractions. Phospholipids, as a class, are very high in brain, located in membranes as part of the structure. Some investigators have implicated specific types of phospholipids as part of an active transport system (see Chapter VII). The gangliosides are another group of lipids associated with membranes, particularly those of the endoplasmic reticulum and mitochondria. The active moiety of the ganglioside molecule is probably N-acetylneuraminic acid (sialic acid). This is an electronegative compound which has been implicated in a molecular theory of active transport (see Chapter VII).

[2]Respiratory quotient = $CO_2 \div O_2$.

TABLE 9.3

Some Chemical Constituents of Isolated Cerebral Tissues[a]

Fraction	RNA (μg/mg protein)	Gangliosides (μg N-acetylneuraminic acid per gm fresh tissue)	Acetylcholine (μmmoles/gm fresh tissue)	Choline acetylase (μmoles/gm/hour)	Cholinesterase (μmoles/mg protein/hour)
Nuclei	—	75	21	0.15	2.3
"Heavy" mitochondria	—	72	59	0.15	4.1
"Light" mitochondria	24	116	12	2.20	9.7
"Microsomal"	81	373		0.35	5.8
Supernatant	—	39	23	1.18	—

[a] From H. McIlwain, "Chemical Exploration of the Brain." Elsevier, Amsterdam, 1963.

Other lipid substances include *cerebroside sulfate* (sulfatide-A) which may be as high as 50% in brain mitochondria.

2. ACETYLCHOLINE (ACH)

The presence of this substance in membranous regions of the cerebral cell is of particular interest since this hormone is confined mostly to neural tissues. Acetylcholine is not only a synaptic transmitter at the peripheral synapses; it is quite possibly involved in central transmission at synapses (Fig. 9.17). Acetylcholine is associated with *synaptic vesicles*, which apparently have a density close to that of mitochondria. The crude mitochondrial fraction (originally developed by Brodie and Bain in 1952) can be subjected to various types of osmotic treatments and density gradient centrifugations. The acetylcholine rich fraction is found not in

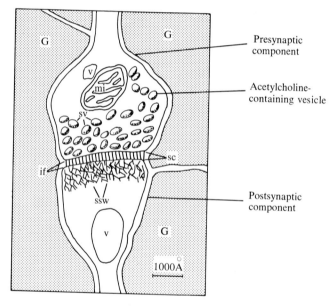

Fig. 9.17 A synapse of the brain cortex. The presynaptic component shows one mitochondrion (mi), a vacuole (v), and numerous synaptic vesicles, some of which are adjacent to the presynaptic membrane. The synaptic cleft (sc) is crossed by parallel intersynaptic filaments (if) of about 50 Å and separated by 100 Å intervals. These filaments are fixed to both the pre- and subsynaptic membranes, which are slightly thickened and denser than the other surface membranes. Within the postsynaptic component there is a web of filaments (or canaliculi) of about 80 Å, that is implanted on the subsynaptic membrane on one side and extends at a varying distance into the postsynaptic cytoplasm. This is the so-called subsynaptic web (ssw). G, glial processes that surround the synapse.) From E. De Robertis, G. R. De Lores Arnaiz, L. Salganicoff, A. Pellegrino de Iraldi, and L. M. Zieher, *J. Neurochem.* **10**, 225 (1963).

mitochondria but in discrete granules, identified as synaptic vesicles or "synaptosomes" $(0.02-1 \ \mu$ in size). Acetylcholine and the enzyme responsible for its synthesis (*choline acetylase*) are concentrated in this fraction. The synaptic vesicles are regarded as morphological units, probably compartmentalized. Acetylcholinesterase, an enzyme which hydrolyzes acetylcholine, is located at a site other than the synaptic vesicle proper. Therefore, the synthesis and degradation of acetylcholine takes place at spatially separate sites (see equations below).

At the nerve terminal, small amounts or *quanta* of acetylcholine are continually released from synaptic vesicles and represent the "carrier" of the electrical impulse, designated as end plate potential (EPP), which is associated with the process of depolarization. The immediate destruction of acetylcholine by acetylcholinesterase prevents prolonged depolarization. Drugs which inhibit this enzyme prolong the action of acetylcholine and find usefulness in diseases involving acetylcholine deficiency or muscular weakness, such as muscular dystrophy, multiple sclerosis and myasthenia gravis. The importance of the membrane in the depolarization phenomenon and active transport has already been stressed (see Chapter VII).

Enzymes Involved in Acetylcholine Metabolism

(1) $CH_3COOH + CoA - SH + ATP \xrightarrow[\text{synthetase}]{\text{acetylcoenzyme A}} CH_3CO \cdot S \cdot CoA + AMP + PP$

(2) $CH_3CO \cdot S \cdot CoA + HO-CH_2-N \overset{\diagup CH_3}{\underset{\diagdown CH_3}{-CH_3}} \xrightarrow[\text{acetylase}]{\text{choline}} CH_3COO \ CH_2CH_2N \overset{\diagup CH_3}{\underset{\diagdown CH_3}{-CH_3}} + CoA - SH$

(3) $CH_3COO \ CH_2CH_2N \overset{\diagup CH_3}{\underset{\diagdown CH_3}{-CH_3}} + H_2O \xrightarrow[\text{esterase}]{\text{acetycholine}} CH_3COOH + HOCH_2CH_2N \overset{\diagup CH_3}{\underset{\diagdown CH_3}{-CH_3}}$
     (Ach)

3. OTHER HUMORAL AGENTS

a. SEROTONIN (*5-hydroxytryptamine*). This substance also has received attention with regard to transmitter substance in the central nervous system, although its function is still in doubt. The exact subcellular localization is still not completely known, although it appears to reside in a fraction similar in density to the acetylcholine-containing particles (synaptosomes).

b. NOREPINEPHRINE. This hormone has been implicated in central nervous system transmission, and the distribution is also somewhat similar to that of acetylcholine in the brain.

c. HISTAMINE. This is another chemical possibly involved in central nervous system transmission. Its localization appears to be different from that of norepinephrine and acetylcholine.

d. DOPAMINE. This substance is a precursor of norepinephrine and is present in high concentration in the caudate nucleus. Its localization is not definitively known, although it appears to be different from that of acetylcholine and norepinephrine.

e. AMINO ACIDS. The most important of these is glutamate which not only may serve as a substrate for cerebral metabolism but may be involved in the transmission phenomenon with GABA (γ-aminobutyrate), a substance which also has been implicated in electrical activity in the central nervous system. Glutamate and GABA have a distribution which is rather similar to that of lactic dehydrogenase, suggesting a soluble cytoplasmic localization. Other abundant amino acids in the brain are aspartic, glycine, serine, alanine, and threonine; all appear to be located in the soluble fraction of the cell. Their function in transmission is unknown.

f. SUBSTANCE P. This is a polypeptide which may be involved in central nervous system transmission. Its subcellular distribution is somewhat similar to acetylcholine, being maximally located in the synaptosome fraction.

g. ATP. It is possible that ATP may be involved in the binding of various amines, involved in transmission, to various subcellular particles in the brain cell. Therefore, the localization of this nucleotide is of importance. The greatest amount of ATP is in fact in the synaptosome fraction.

4. OTHER ENZYMES INVOLVED IN CELLULAR ACTIVITY

a. MONOAMINE OXIDASE. This enzyme appears to be involved in the metabolism of the sympathetic amines, and is located in mitochondria (outer membrane).

b. AMINO ACID DECARBOXYLASES. These enzymes are involved in the decarboxylation reactions which lead to the formation of serotonin, norepinephrine, and histamine, compounds which have been implicated in transmission. The location of these enzymes appears to be in the soluble cytoplasmic portion of the cell.

c. GLUTAMIC ACID DECARBOXYLASE. This is a specific enzyme which is unique to nervous tissue and which is involved in the formation of GABA from glutamic acid; possibly, it is of significance in synaptic transmission in the central nervous system. It could function either presynaptically and/or postsynaptically (Fig. 9.17). The enzyme might also participate in a metabolic shunt around the α-ketoglutarate oxidase system of the mitochondria (Chapter III), and the shunt could account for up to 40% of the oxidative metabolism of the brain. The significance of the enzyme, therefore, could be primarily metabolic. The enzyme has been recovered mainly in a fraction of the cell containing mitochondria and some myelin, along with some incompletely disrupted synaptosomes.

d. ATPASES. The ATPases are of particular importance and are localized in the membranous regions of various subcellular fractions. Of these, probably the most important is the ouabain-sensitive, membrane ATPase, which is found almost completely in the microsomal fraction of the brain cell (see Chapter VII). The activity of this enzyme is particularly high in the cerebral cortex and, in fact, is probably among the highest in all the tissues examined. The function and characteristics are described completely in Chapter VII.

e. NONSPECIFIC ESTERASES. A nonspecific esterase, of unknown function, appears to be specifically localized in the membranes associated with the "microsomal" fraction. It is frequently used as a "biochemical marker" for the microsomal pellet.

5. PROTEINS IN THE BRAIN

A knowledge of protein metabolism in the brain cell is essential for an understanding of transmission phenomenon, memory and other functions, as well as disease processes.

What are the aspects of protein metabolism which differentiate this cell from others? The most important difference between the brain cell and other cells of the body (with regard to free amino acids) is the preponderance of glutamic acid, which together with glutamine and glutathione accounts for more than 50% of the α-amino nitrogen. The carbon skeleton of glutamic acid and of other nonessential amino acids is derived from glucose. The most important aspect of measurement of protein metabolism is the existence of the blood-brain barrier, which impedes the flux of amino acids and other nutriments into the brain. This barrier represents a homeostatic mechanism which isolates brain metabolism from external influences. The existence of this barrier, however, does not preclude the potentiality for the utilization of all substrates by the brain.

The brain of newborn mice incorporates, *in vitro* as well as *in vivo*, glucose into essential and nonessential amino acids; however, the adult brain incorporates glucose only into the amino acids whose carbon skeletons are derived through the citric acid cycle. In other words, in the adult the carbon of glucose is incorporated into amino acids mainly through the operation of the Krebs citric acid cycle.

It is known that the adult central nervous system does not have the capacity to renew neurons by cell division. However, since the neuron is considered a secretory cell, similar to the liver or pancreas, the cell body must be called upon to produce proteins required to maintain its metabolism and, in addition, probably is responsible for the proteins of the axon. Ribosomes, isolated from brain homogenates, are similar in composition to ribosomes from liver. Incorporation of labeled precursors into ribosomes proceeds at a rate quite comparable to or even higher than incorporation into liver ribosomes. Therefore, the brain tissue undoubtedly possesses a high potential for protein synthesis. Like other tissues, the microsomal fraction shows a higher rate of incorporation of amino acids than other subcellular portions of the cell. The turnover of cerebral proteins is comparable to that found in other tissues, and the somewhat low or even insignificant incorporation of labeled amino acids administered in vivo is probably due to slow penetration across the blood-brain barrier. If the amino acids are administered directly into the brain (intracisternally or intraventricularly), the incorporation rate is quite high. The half-lives for total brain proteins are longer than that of liver proteins but shorter than muscle protein. The proteins of the cortical areas of the brain have a somewhat higher metabolic activity than those of the underlying areas of the brain. Autoradiography of brain parts after the administration of labeled methionine shows incorporation in the following descending order: cerebellum, cortex, pons, spinal cord, peripheral nerve. Gray matter is twice as active as white matter after 24 hours. In terms of protein synthesis, the most active cells in the brain and, in fact, in the body itself are the ganglion cells. The high rate of turnover of the proteins of the cerebral cortical cell is indicative of the very active metabolic capacity.

In addition to the synthesis of proteins, the degradation of proteins in the brain is of particular significance, particularly with regard to function. The brain contains more peptidases than muscle. The cell bodies of the neurons and glia are the principal sites of peptidase activity in the cerebral cortex.

D. The Pharmacology and Biology of Isolated Cerebral Slices

Characteristics of brain function may be studied in vitro, using morphologically intact tissues. These may be obtained easily by slicing the tissues so that the diameter is small enough to permit access of nutriments and oxygen. Such preparations not only respire at a rate quite comparable to that found in vivo, but exhibit interesting electrical phenomena, which might be compared to that found in the intact organ. For example, placing electrodes on the tissue slice (Fig. 9.18) and stimulating them under appropriate conditions produces changes in oxygen consumption, energy-rich phosphate compounds, proteins, and intermediary substances (Tables 9.4 and 9.5). These changes may be completely reversed upon the

TABLE 9.4

Metabolic Response of Cerebral Tissues
to Electrical Stimulation[a]

Property measured	Electrical stimulation produces[b]	Depressant drugs[c] In vitro (correlates)	In vivo (correlates)
Respiration	+	−	−
Lactic acid	+	−	−
Creatine phosphate	−	+	+
Inorganic phosphate	+	−	−
$^{32}P_i$ incorporation into phosphoprotein	+	−	

[a]From P. J. Heald, Nature 193, 151 (1962); H. McIlwain, "Biochemistry and the Central Nervous System," 2nd ed. Little, Brown, Boston, Massachusetts, 1959.
[b]+ = increase, − = decrease.
[c]The depressant agents at low concentrations have no effect on unstimulated, respiring cerebral tissues.

TABLE 9.5

Effect of Specific Agents on Oxygen Consumption of
Electrically Stimulated Cerebral Slices[a]

Agent(s)	Effect on metabolic response[b]
Trimethadione	−
Reserpine	0
Chlorpromazine	−
Oligomycin	−
Phenobarbital	−
Ethanol	−

[a]From R. B. Tobin and H. McIlwain, Biochim. Biophys. Acta 105, 191 (1965); H. McIlwain, "Biochemistry and the Central Nervous System," 2nd ed. Little, Brown, Boston, Massachusetts, 1959; H. Wallgren, J. Neurochem. 10, 349 (1963).
[b]See Table 9.4 for the metabolic response pattern. These agents have no significant effect on unstimulated, respiring cerebral slices.

cessation of the electrical impulse. The incorporation of isotopically labeled precursors in the medium can yield information concerning the nature of intermediary substances involved in the stimulation phenomenon. Upon electrical stimulation, a certain labile, phosphate compound incorporates inorganic ^{32}P to an extent greater than any other phosphate compound. This compound is a phosphoprotein, which has been implicated in active cation transport (see Chapter VII).

A   B   C   D

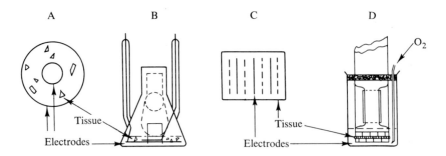

Fig. 9.18. Device used for electrically stimulating tissues. Electrode arrangements: (A), electrodes dip to test tube or beaker; (B), electrodes are fused to manometric vessel; (C), electrodes are wound to small plastic frame, removable from manometric vessel; (D), electrodes hold tissue in small beaker.

Incubation of tissue slices in the cold causes a loss of response to electrical stimulation. Investigation of this phenomenon led to the discovery that a basic protein (histone) migrates from the nucleus of the cerebral cortical cell to acidic sites in the endoplasmic reticulum and is responsible for the loss of response to stimulation. Pretreatment with or addition of gangliosides, highly acidic compounds, reverses this effect. These results have led to a plausible molecular concept of active transport (see Chapter VII).

The normal, respiring tissue slice is unresponsive to a great many pharmacological agents which have a rather pronounced effect on the central nervous system. These compounds include chlorpromazine, phenobarbital, amphetamine, ergot alkaloids, ethanol, various anticonvulsants, reserpine, and chlorpromazine. These agents might act *in vivo* by converting a higher level of activity to a lower level of activity; therefore, actions on isolated cerebral tissue might not be seen. However, electrical stimulation of these tissues *in vitro* converts their metabolic activity to higher levels; hence, many of these agents then have significant effects. Table 9.5 exhibits a number of these activities.

GENERAL REFERENCES

HEART—BIOCHEMICAL ASPECTS

Bing, R. J. (1965), *Physiol. Rev.* **45**, 171.
Coffey, R. G., Cheldelin, V. H., and Newburgh, R. W. (1964). *J. Gen. Physiol.* **48**, 105.
Davies, R. E. (1963), *Nature* **199**, 1068.
Ebashi, S., and Lipmann, F. (1962). *J. Cell Biol.* **14**, 389.
Fanburg, B., and Gergely, J. (1965). *J. Biol. Chem.* **240**, 2721.
Gold, M., and Spitzer, J. J. (1964). *Am. J. Physiol.* **206**, 153.
Gluck, L., Talner, N. S., Gardner, T. H., and Kulovich, M. V. (1964a). *Nature* **202**, 770.

Gluck, L., Talner, N. S., Stern, H., Gardner, T., and Kulovich, M. V. (1964b). *Science* **144**, 1244.

Weber, A., Herz, R., and Reiss, I. (1963). *J. Gen. Physiol.* **46**, 679.

Weber, A., Herz, R., and Reiss, I. (1964). *Proc. Roy. Soc.* **B160**, 489.

Wollenberger, A. (1947). *J. Pharmacol. Exptl. Therap.* **91**, 39.

HEART— PHYSIOLOGICAL ASPECTS

Beznak, M. (1964). *Circulation Res.* **15**, Suppl. II, 141.

Endo, M. (1964). *Nature* **202**, 1115.

Franzini-Armstrong, C., and Porter, K. (1964). *Nature* **202**, 355.

Hoffman, B. F., Moore, E. N., Stuckey, J. H., and Cranefield, P. F. (1963). *Circulation Res.* **13**, 308.

Huxley, H. E. (1964). *Nature* **202**, 1067.

Mommaerts, W. F. H. M. (1961). *Ann. Rev. Physiol.* **23**, 529.

Nelson, D. A., and Benson, E. S. (1963). *J. Cell Biol.* **16**, 297.

Simpson, F. O. (1965). *Am. J. Anat.* **117**, 1.

Sonnenblick, E. H., Braunwald, E., and Morrow, A. G. (1965). *J. Clin. Invest.* **44**, 966.

Starling, E. H. (1918). "The Linacre Lecture on the Law of the Heart." Longmans, Green, New York.

"The Sarcoplasmic Reticulum." (1961). *J. Biophys. Biochem. Cytol.* **10**, No. 4, Part 2.

HEART—ACTOMYOSIN, MYOSIN, ACTIN, AND NEWER CONTRACTILE PROTEINS

Bárány, M., Koshland, D. F., Springhorn, S. S., Finkelman, F., and Therattil-Antony, T. (1964). *J. Biol. Chem.* **239**, 1917.

Ebashi, S., and Nonomura, Y. (1967). *Proc. 7th Intern. Congr. Biochem., Tokyo, Abstr. II*, p. 329.

Ebashi, S., Ebashi, F., and Maruyama, K. (1964). *Nature* **203**, 645.

Hanson, J., and Lowy, J. (1963). *J. Mol. Biol.* **6**, 46.

Hanson, J., and Lowy, J. (1964). *Proc. Roy. Soc.* **B160**, 449.

Katz, A. M. (1964). *J. Biol. Chem.* **239**, 3304.

Kay, C. M., and Green, W. A. (1964). *Circulation Res.* **15**, Suppl. II, 38.

Lowey, S. (1964). *Science* **145**, 597.

Ohnishi, P. T., Kawamura, H., Takeo, K., and Watanabe, S. (1964). *J. Biochem. (Tokyo)* **56**, 273.

Seidel, J. C., Sreter, F. A., Thompson, M. M., and Gergely, J. (1964). *Biochem. Biophys. Res. Commun.* **17**, 662.

Yamashita, T., Soma, Y., Kobayashi, S., Sekine, T., Titani, K., and Narita, K. (1964). *J. Biochem. (Tokyo)* **55**, 576.

BRAIN

Barondes, S. H. (1964). *J. Neurochem.* **11**, 663.

Curtis, A. S. G. (1963). *Endeavour* **22**, 134.

De Robertis, E., De Lores Arnaiz, G. R., Salganicoff, L., Pellegrino de Iraldi, A., and Zieher, L. M. (1963). *J. Neurochem.* **10**, 225.

Gray, E. G., and Whittaker, V. P. (1962). *J. Anat.* **96**, 79.

McIlwain, H. (1959). "Biochemistry and the Central Nervous System," 2nd ed. Little, Brown, Boston, Massachusetts.

McIlwain, H. (1963). "Chemical Exploration of the Brain." Elsevier, Amsterdam.

McIlwain, H., and Rodnight, R. (1962). "Practical Neurochemistry." Little, Brown, Boston, Massachusetts.

Schwartz, A., Bachelard, H. S., and McIlwain, H. (1962). *Biochem. J.* 84, 626.
Tobin, R. B., and McIlwain, H. (1965). *Biochim. Biophys. Acta* 105, 191.
Waelsch, H., and Lajtha, A. (1961). *Physiol. Rev.* 41, 709.
Wallgren, H. (1963). *J. Neurochem.* 10, 349.
Whittaker, V. P. (1962). *Proc. 1st Intern. Pharmacol. Meeting, Stockholm, 1961* Vol. 5, p. 61. Pergamon Press, Oxford.
Whittaker, V. P. (1965). *Progr. Biophys. Mol. Biol.* 15, 39.

AUTHOR INDEX

SUBJECT INDEX

A

A-bands, 427
Absorption theory of active transport, see Sorption
Acetabularia, life cycle, 172, 173
Acetylcholine, 454–455
Acetyl coenzyme A, see Coenzyme A
Acetylglucosaminidase in lysosomes, 287
Acetylneuraminic acid
  formation of, 290
  of gangliosides, 288–289
N-Acetylneuraminic acid, see Gangliosides
Acid deoxyribonuclease in lysosomes, 287
Acid phosphatase
  Gomori method and, 33, 34, 36
  in lysosomes, 281, 287
Acid ribonuclease in lysosomes, 287
Acridine orange, 31, 32, 174
Actin, 387, 430
Actinomycin D, 205–207
  binding to DNA, 206
  effect on nucleoli, 206
  inhibition of RNA synthesis, 206
  as inhibitor of ribosomal RNA
    synthesis, 227
  structure, 205
Active transport, 88, see also Chapter VII
  acyl phosphate in, 341
  aldosterone on, see Aldosterone
  ATPase in, see Adenosinetriphosphatase
  chemi-osmosis in, 350–351
  criteria of, 306
  effect of drugs on, 352–357
  gangliosides in, 343
  histones in, 343
  methods of measurement, 306–309
  phosphoproteins in, 339
Actomyosin, 431
Acyl phosphate, see Active transport

Adenosine 5′-monophosphate, 198, 200
Adenosinetriphosphate (ATP), formula of, 113
Adenosinetriphosphatase (ATPase), see also Mitochondria
  allosteric effects, 345–347
  importance in active transport, 319–347
  localization in tissues, 359
  method of preparation, 326
  morphology of, 338–339
  partial reactions of, 329, 342
Adenosinetriphosphatase A in nucleoli, 197
S-Adenosylmethionine, utilization in t-RNA synthesis, 240–241
Alanyl-t-RNA structure, 236, 238
Aldosterone, effects on transport, 356–357
Alkaline phosphatase, 31
  stains for, 33–35
Alkylating agent, 192
Alleles, definition, 398
Allosteric inhibition on enzyme regulation, 266–267
Amino acid transport, 351–352
Amino acyl synthetase, 233–234
Amino acyl t-RNA, binding to ribosome, 254, 256, 257
Aminopeptidase, 35, 38
Amoeba proteus, 171
Amytal, see Inhibitors
Anaphase, 379, 380, 381, 388–390
  chromatids in, 388
  chromosomal aberrations and, 402
  endoplasmic reticulum in, 389
  meiotic, 391, 395, 396
  mitotic apparatus in, 385, 386
  RNA and, 388–389
Anesthetics, action of on membranes, 85, 93

471